# SATELLITE SAR DETECTION OF SUB-MESOSCALE OCEAN DYNAMIC PROCESSES

## ADVANCED SERIES ON OCEAN ENGINEERING

Series Editor-in-Chief
**Philip L-F Liu** (*Cornell University*)

---

*Published*

Vol. 33 Random Seas and Design of Maritime Structures (Third Edition)
*by Yoshimi Goda* (Yokohama National University, Japan)

Vol. 34 Coastal Dynamics
*by Willem T. Bakker* (Delft Hydraulics, Netherlands)

Vol. 35 Dynamics of Water Waves: Selected Papers of Michael Longuet-Higgins Volumes 1–3
*edited by S. G. Sajjadi* (Embry-Riddle Aeronautical University, USA)

Vol. 36 Ocean Surface Waves: Their Physics and Prediction (Second Edition)
*by Stanislaw R. Massel* (Institute of Oceanology of the Polish Academy of Sciences, Sopot, Poland)

Vol. 37 Computational Wave Dynamics
*by Hitoshi Gotoh* (Kyoto University, Japan),
*Akio Okayasu* (Tokyo University of Marine Science and Technology, Japan)
*and Yasunori Watanabe* (Hokkaido University, Japan)

Vol. 38 An Introduction to Hydraulics of Fine Sediment Transport
*by Ashish J. Mehta* (University of Florida, USA)

Vol. 39 Liquefaction Around Marine Structures
(With CD-ROM)
*by Mutiu Sumer* (Technical University of Denmark, Denmark)

Vol. 40 Design and Construction of Berm Breakwaters
*by Van der Meer, Jentsje* (Van der Meer Consulting BV, The Netherlands & UNESCO-IHE, The Netherlands) *and*
*Sigurdur Sigurdarson* (IceBreak Consulting Engineers ehf, Iceland)

Vol. 41 Dynamics of Coastal Systems (Second Edition)
*by Job Dronkers* (Netherlands Centre of Coastal Research, The Netherlands)

Vol. 42 Theory and Applications of Ocean Surface Waves (Third Edition)
(In 2 Volumes)
*by Chiang C. Mei* (Massachusetts Institute of Technology, USA),
*Michael Stiassnie* (Technion-Israel Institute of Technology, Israel) *and*
*Dick K.-P. Yue* (Massachusetts Institute of Technology, USA)

Vol. 43 Japan's Beach Erosion: Reality and Future Measures (Second Edition)
*by Takaaki Uda* (Public Works Research Center, Japan)

Vol. 44 Satellite SAR Detection of Sub-Mesoscale Ocean Dynamic Processes
by *Quanan Zheng* (University of Maryland, College Park, USA)

Advanced Series on Ocean Engineering — Volume 44

# SATELLITE SAR DETECTION OF SUB-MESOSCALE OCEAN DYNAMIC PROCESSES

**Quanan Zheng**
University of Maryland, College Park, USA

World Scientific

JERSEY • LONDON • SINGAPORE • BEIJING • SHANGHAI • HONG KONG • TAIPEI • CHENNAI • TOKYO

*Published by*

World Scientific Publishing Co. Pte. Ltd.
5 Toh Tuck Link, Singapore 596224
*USA office:* 27 Warren Street, Suite 401-402, Hackensack, NJ 07601
*UK office:* 57 Shelton Street, Covent Garden, London WC2H 9HE

**British Library Cataloguing-in-Publication Data**
A catalogue record for this book is available from the British Library.

**Advanced Series on Ocean Engineering — Vol. 44**
**SATELLITE SAR DETECTION OF SUB-MESOSCALE OCEAN DYNAMIC PROCESSES**



ISBN 978-981-4749-00-8

Typeset by Stallion Press
Email: enquiries@stallionpress.com

Printed in Singapore

To the memory of my parents, and to Zhenfang, Lily, friends and colleagues for their encouragement and inspiration.

# Preface

As an oceanographer, I always would like to emphasize significances of the ocean and ocean research to human beings. The ocean covers nearly three-quarters of the globe. The ocean has served as a fishery ground providing seafood and a convenient navigation waterway for humans since ancient times. The ocean is an original water source of land. The ocean contains vast mineral and energy resources that may meet the human demands for thousands of years. The coastal waters and beaches are the best places for development of tourist industries, enjoyed by millions of people for swimming, boating and other recreational activities. These facts indicate a close tie between the human society and the ocean.

The ocean is an engine that drives the global climate through storing huge amount of solar irradiation and redistributing its significant portion poleward. Small shifts in ocean circulation can cause significant changes in regional climate, which can lead to changes in the marine ecosystem. The ocean also serves as an important sink of greenhouse gases that helps to modify human impacts on global climate through absorbing and storing about 30% of industrial carbon dioxide emissions from the atmosphere. Thus, understanding the ocean, gradually clarifying its changing laws and providing accurate, believable forecasting for ocean states and long-term variability trend would be grand challenges facing oceanographers.

To understand the ocean as a huge, restless moving ware body, the first step is collecting the data, which are required to have the spatial coverage as broad as possible, the resolution as high as possible and the repeat observation period as short as possible. This demand promotes the development of satellite remote sensing technologies. The SEASAT satellite launched by the USA on June 28, 1978 was a milestone for ocean observation. Its payload included a group of microwave sensors, which were specially designed for ocean detection: scanning multichannel microwave radiometer, microwave scatterometer, radar altimeter and synthetic aperture radar (SAR). Following research results indicate that the SAR is a powerful tool for detecting ocean dynamical processes, not only because of all weather and all day/night detection as well as high-resolution imaging abilities, but also its high sensitivity to sea surface micro-scale variability.

On the other hand, seawater is a dielectric medium of the high conductivity for all bands of the radar waves. Based on the principles of electromagnetic wave propagation, microwave pulses, which the radars emit to detect the targets on the earth surface, are capable of penetrating only into a thin layer of seawater with a maximum depth on the order of centimeters. Thus, the radar echoes are formed by backscattered energy only from the ocean surface. Moreover, the SAR image is a single parameter (radar backscattering coefficient) product. Different atmospheric and oceanic phenomena and ocean bottom topography, all may form imprints or patterns on SAR images. To identify these phenomena physically is a fundamental effort for SAR image interpretation and application. Due to the technical limitation of the swath of images (<500 km), state-of-the-art SARs are only suitable to detect the ocean processes with length scales smaller than 200 km.

Writing this book is aimed at sharing the research results and experience in the field of SAR detection of mesoscale ocean dynamical processes with readers, who are interested in this topic. The book consists of 11 chapters. For the convenience of the readers with various backgrounds, Chapter 1 provides radar and SAR principles including fundamental concepts and definitions, brief development history of space-based SAR used for ocean detection, physics of ocean SAR imaging including ocean wave-radar modulation transfer function. In Chapter 2, the scales of ocean dynamical processes are defined first followed by SAR image interpretations of two cases of sub-mesoscale ocean vortex trains. Chapter 3 contains the analysis results of spirals on the sea observed by Space Shuttle photography and satellite SAR images. In addition, a three-dimensional structure of an ocean spiral in the East China Sea showing up on the SAR image is derived from near simultaneous cruise observations. Chapter 4 presents the analysis results of ocean near-inertial oscillations in the Gulf of Mexico observed by the mooring project and SAR images. Chapter 5 includes SAR image interpretation and dynamical analysis of estuarine outflow plumes off the Delaware Bay mouth. Chapters 6 and 7 are devoted to the ocean internal waves, which are one of the ocean phenomena earliest identified on the satellite SAR images. Chapters 8–10 are devoted to SAR detection of ocean bottom topography. Three cases, longitudinal flow passing over parallel topographic corrugations, traverse flow passing over periodical topography and stratified flow passing over isolated topography, are analyzed, respectively. Finally, Chapter 11 is an introduction to selected marine atmospheric boundary layer processes, which may generate footprints on satellite images of the ocean surface. In this special case, SAR imaging principles of atmospheric processes are first discussed. Island-induced atmospheric solitary wave packets, atmospheric frontal waves including linear coastal lee waves and solitary wave packets, as well as ocean surface rainfall are examined. Chapters 2–11 include theories for SAR imaging and methods for dynamical parameter extraction.

The book is written mainly on the basis of papers published by the author and collaborators, whose names appear in the co-authored papers listed in the Bibliography. I am indebted to their contributions. Besides, the book cites a large number of literatures. I am particularly grateful to the following authors for their excellent works being cited: Munk *et al.* [2000] in Chapter 3, Li *et al.* [2004, 2013] in Chapters 4 and 11, Zhao [2005, 2014] and Wang *et al.* [2012] in

Chapters 6 and 7. Many thanks to Lingling Xie, Xiaomin Ye, Ying Feng and Junyi Li for that help in data processing.

Quanan Zheng
*University of Maryland*
*College Park, Maryland*
*United States*
*December 2015*

# List of Tables

# List of Figures

# Contents

# Chapter 1

# Principles of Radar Imaging of Ocean Processes

## 1.1. Radar Principles

The term *radar* is an acronym for an electronic system that is used for Radio Detection and Ranging of remote targets [Skolnik, 1970]. As shown in Figure 1.1, a radar system comprises two elementary subsystems: a transmitter and a receiver with a correlator and display. The transmitter serves as a generator of radar waves. The receiver serves as a collector of echo signals returned from targets. The antennas perform transducers. Transmitting antenna converts the electrical currents fed by the generator into the electromagnetic (EM) fields and emits the EM waves to the free space with required radiation beam patterns. Vice versa, receiving antenna captures echo signals in the form of the EM fields from the space and converts the EM fields into the electrical currents and feeds the signals to the receiver. The receiver also has functions for data storage, processing and display.

In terms of remote sensing technology, the radar is a kind of active remote sensors. It illuminates targets with the EM energy transmitted by itself rather than by sunlight or infrared (IR) radiation of targets. Thus, the radar system has the ability to work day and night. The band of EM waves used by radars has strong penetration ability through cloud, fog, aerosol, smoke, rainfall and snowfall in the atmosphere. In other words, the radar system is able

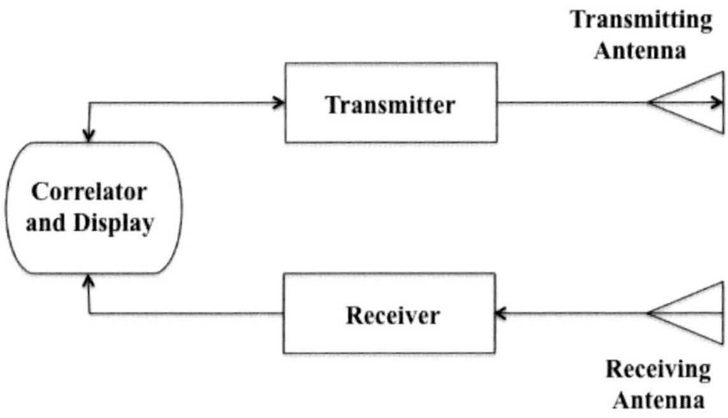


Fig. 1.1 Elements of a radar system

to work at all weather conditions. In addition, the radar waves also have good penetration ability through vegetation cover, dry soil, dessert, snow and ice cover and possibility being resonant with the target surface roughness. Thus, radars, including space-based radar, air-borne radar and ground-based radar, have become advanced remote sensors for detecting of the earth resources and environment. In particular, space-based radars are the most powerful tools for the ocean detection, owing to their sensitivities to the ocean dynamical processes to be discussed later.

The radar system shown in Figure 1.1 has two separated antennas. In fact, for many types of operational radars, the transmitter and the receiver share the same antenna as shown in Figure 1.2. In this case, the relationship between EM power density $P_r$ received by the antenna and transmitted EM power density $P_t$ is formulated by

$$P_r = \frac{P_t G \sigma}{(4\pi)^2 R^4} A_e, \tag{1.1.1}$$

where $G$ is the antenna gain, $\sigma$ is the radar cross section (RCS), or scattering coefficient, of the target, $A_{\mathrm{e}}$ is the effective aperture (area) of the antenna, and $R$ is the slant range from the antenna to the target. This is the so-called radar equation, which constitutes a theoretical fundamental for radar science and technology [Skolnik, 1970].

From Eq. (1.1.1), one can see that the intensity of radar echo that is represented by the received EM power density depends linearly on the RCS of the target. This implies that the radar echo contains the

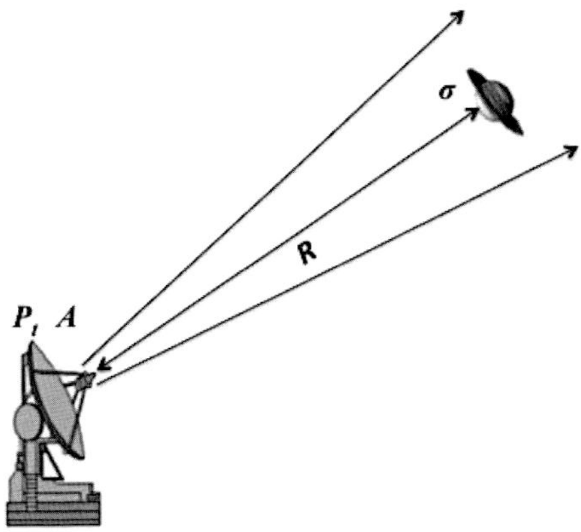


Fig. 1.2 Schema for principles of a sharing antenna radar

information on electrical properties and surface status of the target. Thus, RCS is the most important parameter for radar detection and target recognition functions. Physically, the RCS is a measure of ability of a radar target to reflect or scatter the radar waves in the direction of the radar antenna. For ocean radar remote sensing, the average RCS of per unit area of the ocean surface is a suitable parameter, which is a dimensionless, related quantity called the normalized RCS and defined as

$$\sigma^0 = \mathrm{RCS}/A_0, \tag{1.1.2}$$

where RCS is the radar cross-section of small waves on the ocean surface, and $A_0$ is an area of the ocean surface illuminated by the radar waves, so-called radar antenna footprint. $\sigma^0$ (sigma naught) usually is a very small quantity, but may vary within a quite large range. In application fields, therefore, its logarithmic form is commonly used, *i.e.*,

$$\sigma^0(\mathrm{dB}) = 10\log_{10}\sigma^0. \tag{1.1.3}$$

In addition, other two parameters in Eq. (1.1.1), effective aperture $A_e$ and gain $G$ of the antenna, determine the efficiency of EM energy transfer of the antenna. In fact, antenna gain $G$ is closely linked to $A_e$ through the relationship of

$$G = \frac{4\pi A_e}{\lambda r^2}, \tag{1.1.4}$$

where $\lambda_r$ is the wavelength of the EM waves, and

$$A_e = K_a A, \tag{1.1.5}$$

where $K_a$ is the antenna aperture efficiency, and $A$ is the aperture physical size. This implies that in the process of EM wave power transferred to the space, the effective aperture of radar antenna plays a critical role, which is a determinative parameter used for a measure of the performance of the antenna [The Ohio State University, 2014].

Equation (1.1.4) shows that for a given $A_e$, antenna gain $G$ decreases quickly with the square of the wavelength of EM waves. This implies that there is an upper limit for the wavelength of EM waves that can be efficiently radiated to the space by the radar antenna. Besides the antenna gain, other factors, such as requirements for antenna radiation pattern, target characteristics and attenuation of the EM field in the air, must be also considered when choosing a radar wavelength band. The wavelength range used by state-of-art radars is from 1 to 100 m, *i.e.*, the frequency ranges from 3 MHz to 300 GHz.

With this band of EM waves, the radar has two transmitting modes to be chosen: continuous wave transmitting mode and pulse wave transmitting mode, depending on mission requirements. Figure 1.3 shows a set-up process of a radar return signal in the case of pulse wave transmitting mode. The antenna transmitted a square pulse with amplitude $a$ and frequency $f$ of carrier waves at time $t_0$, and received a return pulse from the target at times $t_1$.

First, based on the time delay of the return pulse, the distance between the radar and the target can be calculated by

$$R = (t_1 - t_0)\, C/2, \tag{1.1.6}$$

where $C(= 2.998 \times 10^8 \text{ m s}^{-1})$ is the speed of light. Then, one can see that the return pulse no longer has the same shape as the transmitted

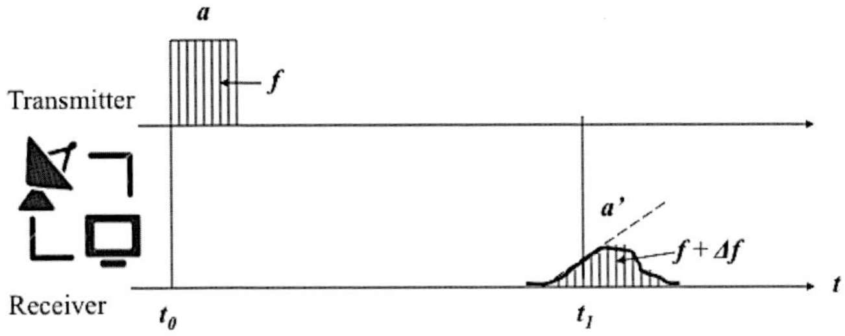


Fig. 1.3 Set-up process of a radar return signal in the case of pulse wave transmitting mode.

square pulse, but modified. From this return pulse, we can retrieve the following data and information:

*RCS of target*: Amplitude function $a'(t)$ of return pulse can be used for calculation of the EM power density $P_r$ received by radar antenna. Substituting the value of $P_r$ into radar Eq. (1.1.1), RCS $\sigma$ of the target can then be derived.

*Target surface roughness*: The widened frontal edge of return pulse contains the information of target surface roughness. In the case of the ocean, the frontal edge function is used to extract the significant wave height (SWH) of the sea surface wave field [Hayne, 1980].

*Doppler frequency shift*:Compared to the transmitted pulse, the carrier wave frequency of return pulse gains Doppler frequency shift $f_D$ owing to relative motion between the radar and the target, which is

$$\Delta f = f_D = -2f(v_r/C), \tag{1.1.7}$$

where $v_r$ is the radial velocity component and $C$ is the speed of light [Colwell, 1983]. This is the principle to measure the speed of moving target by radars, and also makes the synthetic aperture radar (SAR) possible.

Polarization is another important property of the radar system. The so-called "radar polarization" actually refers to polarizations of transmitted and received EM waves. Base on principles of EM dynamics, all EM waves comprise two components, the electric field

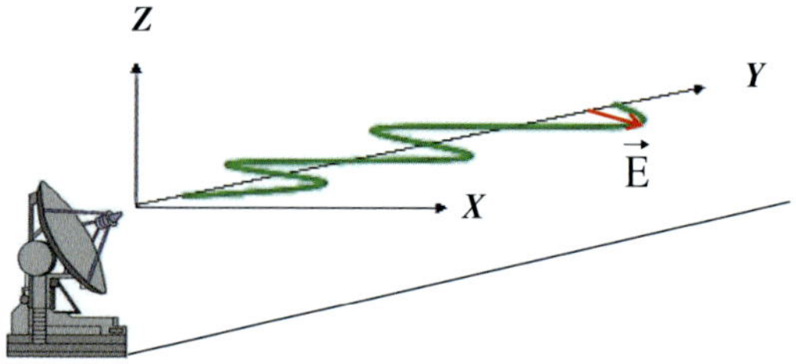


(a)

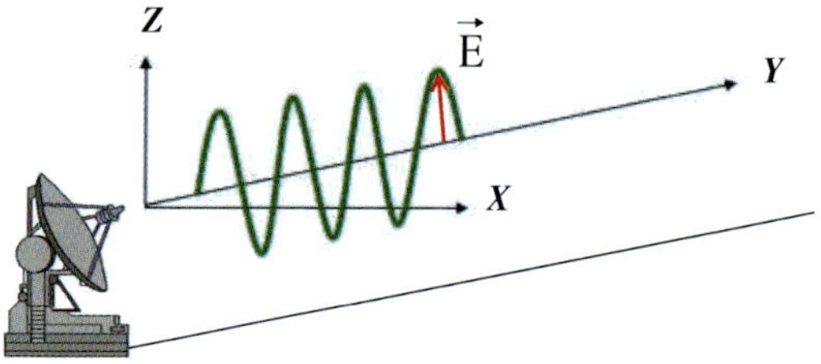


(b)

Fig. 1.4 Two polarization modes of radar antenna: (a) horizontal polarization and (b) vertical polarization. Amplitudes of electric victors are in arbitrary units

and the magnetic field, which are usually represented by vectors $\vec{E}$ and $\vec{H}$, respectively. The two vectors oscillate in phase and perpendicularly to each other and the direction of wave propagation. Among the three, the direction of the electric field is defined as the direction of polarization of the EM waves.

Polarization of transmitted EM waves is controlled by internal configuration design of the antenna system as shown in Figure 1.4. Figure 1.4(a) shows that a plane EM wave propagates along the $Y$-direction, and its electric field $\vec{E}$ oscillates in the $X-Y$ plane that is parallel to the ground reference plane. This case of the EM wave polarization is defined as the horizontal polarization (H). Figure 1.4(b) shows that another plane EM wave propagates also along the $Y$-direction, but its electric field $\vec{E}$ oscillates in the $Y-Z$ plane that is perpendicular to the ground reference plane. This case is defined as the vertical polarization (V). However, radars are incapable of controlling the polarization of received EM waves, because the backscattered EM waves from the target are unable to maintain polarization of originally transmitted radar waves, but depolarized by the target with a variety of polarizations in most of cases. The receiving antenna is capable of selecting an EM wave component with a required polarization from the backscattered EM waves with a variety of polarizations.

Both theories and measurements have revealed that RCS of all materials has direct dependence on EM wave polarizations. Thus, radars may collect distinct and complementary information on physical properties of targets using different combinations of polarizations of transmitted and received EM waves. Based on this idea, a special technology of radar remote sensing is developed, *i.e.*, radar polarimetry. For a dual polarized radar system, *i.e.*, the antenna is automatically switchable between H and V polarizations, the following four combinations of transmitting and receiving polarizations are commonly used:

HH: for horizontal transmission and horizontal reception,
VV: for vertical transmission and vertical reception,
HV: for horizontal transmission and vertical reception and
VH: for vertical transmission and horizontal reception.

The first two polarization combinations are referred to as "like-polarized", because the transmitting and receiving polarizations are the same. The last two combinations are referred to as "cross-polarized" [Natural Resources Canada, 2013].

## 1.2. Ground Resolution of Radars

Radar resolution is an important parameter used for characterizing the ability for the radar to "recognize" targets correctly. For the space-based imaging radar, its ground resolution determines the dimensions represented by an imagery cell, *i.e.*, imagery resolution, therefore, determines the scope of applications that require certain image resolutions. This section will discuss fundamental concepts and definitions of parameters that determine radar resolution and brief ideas of SAR.

### 1.2.1. *Real aperture radar*

As discussed in Section 1.1, the radar emits EM waves to the free space through its transmitting antenna. In the space, the EM

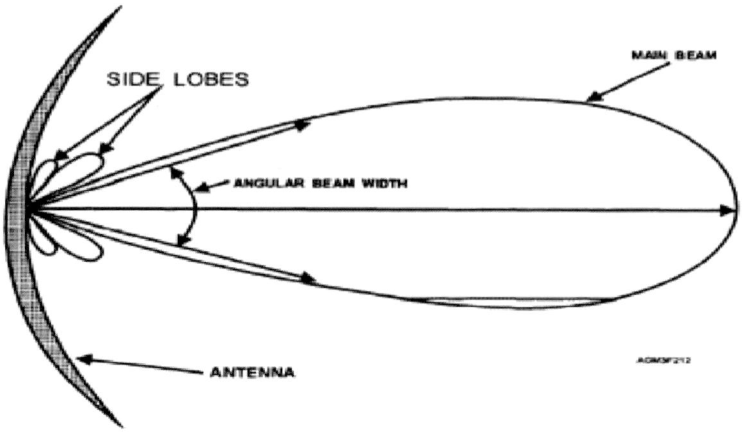


Fig. 1.5 Radar antenna radiation patterns (download from www.tpub.com)

energy propagates in the form of the EM field, which is confined to a definite radiation pattern. The three-dimensional shape of the radiation pattern depends on the design of antenna. The parabolic dish antenna is the most frequently used design in the radar systems. The radiation pattern of this sort of antenna contains a major lobe (or main beam) and several minor lobes as shown in Figure 1.5. The antenna emits most of the EM energy through the major lobe, and the axis of the major lobe is the direction of maximum radiated power. The side lobes are smaller beams radiating in undesired directions. Thus, the main beam carries all detection tasks.

The beamwidth of the major lobe is defined as an angular extent of the half power width of the radiation beam, at which the EM field strength has fallen by 3 dB compared to the maximum EM field strength. The beamwidth, in degrees, of an antenna is given by

$$\theta_B = \frac{\lambda r}{D}, \tag{1.2.1}$$

where $\lambda_r$ is the wavelength of radar waves, and $D$ is the maximum linear dimension of the antenna. For example, if the wavelength is 3 cm and the antenna dimension is 1 m, from Eq. (1.2.1) we obtain the beamwidth as 1.7°.

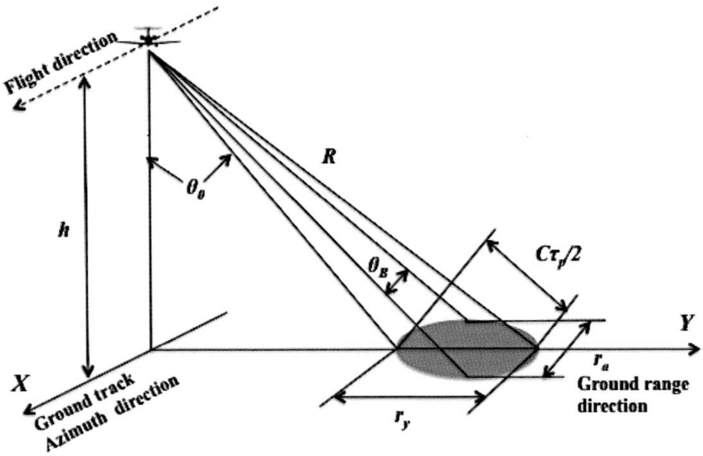


Fig. 1.6 SLAR imaging geometry

The beamwidth of the antenna is a determinative parameter for radar ground resolution, which is a parameter used for characterizing the radar ability to display targets correctly. Azimuthal resolution or directional resolution is used for characterizing the radar ability to display along-track targets correctly. While range resolution is used for characterizing the radar ability to display along-beam targets separately. As example, Figure 1.6 shows imaging geometry of real aperture side-looking airborne radar (SLAR) (here "real" means "true" for distinguishing from "synthetic"). One can see that azimuthal resolution $r_a$ is controlled by the beamwidth, *i.e.*,

$$r_a = \theta_B R = \frac{\theta_B h}{\cos\theta_0}, \tag{1.2.2}$$

where $h$ is the height of the SLAR above the local ground plane, and $\theta_0$ is the incident angle of the radar beam. Range resolution $r_y$ is controlled by the radar pulse length or duration, *i.e.*,

$$r_y = \frac{C\tau_p}{2\sin\theta_0}, \tag{1.2.3}$$

where C is the speed of light and $\tau_p$ is the radar pulse length [Ulaby *et al.*, 1982].

Following two examples may give the reader an intuitive feel about ground resolution of real aperture radar:

(1) Consider the $X$ band radar with a 100-ns pulse length and a 10-mrad beamwidth antenna (dimension of 3 m) onboard an aircraft flying at height of 10 km. For the incident angle of the radar beam is 30°, Eqs. (1.2.2)–(1.2.3) give azimuthal resolution of 120 m and range resolution of 30 m, respectively.
(2) The same radar, but onboard a satellite at an orbit height of 1000 km. Azimuthal resolution and range resolution are 12 and 3 km, respectively.

These calculations imply that if the real aperture radar were used as an imaging remote sensor, the use of its image products would be limited due to low ground resolution. To break through this technological "bottle neck", new technologies such as pulse-narrowing technology and synthetic aperture technology are developed as discussed in the next subsection.

### 1.2.2. *Basic ideas of synthetic aperture*

As shown in Figure 1.6, imagery resolution of the imaging radar is determined by the sectional area of antenna beam at the ground or called "antenna footprint". To obtain fine imagery resolution, therefore, the antenna footprint must be reduced as small as possible. The above analysis has shown that dimensions of this two-dimensional pattern are dependent on different mechanisms. The radar transmitted pulse length controls one dimension along the range (or cross-track) direction, and the antenna beamwidth controls another dimension along the azimuthal (or along-track) direction. Thus, the way to obtain fine range resolution is to adopt pulse-narrowing technology to the radar system, *i.e.*, narrow pulses yield fine range resolution. On the other hand, however, the way to obtain fine azimuthal resolution is complex. Theoretically, based on Eq. (1.2.1), to enlarge dimension of the antenna would efficiently reduce the beamwidth and then obtain finer azimuthal resolution. Practically, however, it is unfeasible for a spacecraft because

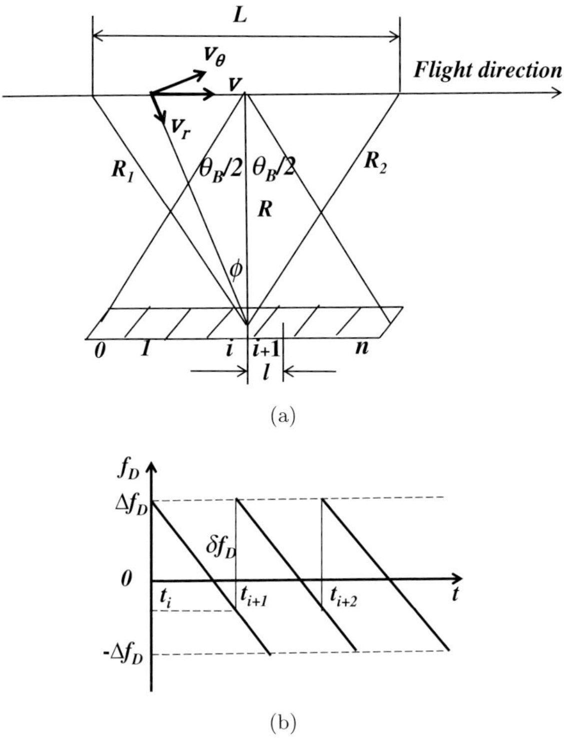


Fig. 1.7 Concept of Doppler beam-sharpening radar. (a) Geometry for Doppler frequency shift signal formation and (b) Doppler frequency shift history of three adjacent ground resolution cells

limited payload capacity requires dimensions of radar antennas as small as possible. It is synthetic aperture technology to solve this contradiction.

We use Doppler beam-sharpening approach shown in Figure 1.7 to illustrate principles of synthetic aperture technology. Figure 1.7(a) shows a slant plane, in which the radar with a constant velocity $v$ flies along a horizontal track at slant range $R$ to a linear ground target, which is parallel to the flight track. The target consists of $n$ continuously distributed ground resolution cells with equal center-to-center distance $l$. The maximum field of view within that the $i$th ground resolution cell may "see" the radar is bounded by $R_1$ and $R_2$, which represent the earliest and the latest radiant lines for the antenna beam to illuminate the cell. Thus, its angular width equals to antenna beamwidth $\theta_B$, and the corresponding length along the flight track and also the length of the linear target is $L$. On the other hand, at any point of the flight track, the velocity $v$ may always be decomposed into two components perpendicular to each other: a radial component $v_r$ and a tangential component $v_\theta$. For the radar arriving at angle of view $\phi$ of the $i$th ground resolution cell, the radial velocity component is

$$v_r = v \sin \phi. \tag{1.2.4}$$

At this moment, from Eq. (1.1.5), we calculate the Doppler frequency shift of the radar waves received by the $i$th cell as

$$f_D = -\frac{2fv \sin \phi}{C} \simeq -\frac{2fv}{C}\phi, \tag{1.2.5}$$

where $f$ is the frequency of radar waves, and $C$ is the speed of light.

Figure 1.7(b) shows the curves of Doppler frequency shift history of three sequential ground resolution cells. One can see that Doppler frequency shift $f_D$ is a linear function of time with top and lower limits of

$$\pm\frac{2fv}{C}\theta_B, \tag{1.2.6}$$

respectively. Thus, we obtain a functional form of $f_D$ as

$$f_D = \frac{2fv\theta_B}{C}\left(1 - \frac{2v}{L}t\right). \tag{1.2.7}$$

The time interval for the radar flies from the $i$th cell to the $(i+1)$th cell is $\delta t = l/v$. Thus, the difference of Doppler frequency shift of the $i$th cell from that of the $(i+1)$th cell is

$$\delta f_D = \frac{2fv\theta_B}{C}\left(1 - \frac{2l}{L}\right). \tag{1.2.8}$$

From Eq. (1.2.8), one can see that when the difference of Doppler frequency shifts of two sequential ground resolution cells $\delta f_D$ is detected, the distance between the two is determined. This implies that the Doppler beam-sharpening radar is capable of recognizing two adjacent ground resolution cells with a separate distance of $l$, *i.e.*, the radar achieves fine ground resolution of $L/n$ or $R(\theta_B/n)$. In other words, the effective beamwidth of Doppler beam-sharpening radar is

$$\theta_s = \frac{\theta_B}{n}. \tag{1.2.9}$$

From Eq. (1.2.1), we obtain the effective aperture of Doppler beam-sharpening radar becomes $n$ times of that of real aperture radar, *i.e.*,

$$D_s = nD. \tag{1.2.10}$$

This is the fundamental principle of the SAR.

### 1.2.3. *Brief development history of space-based SAR*

The engineering achievement of synthetic aperture concept in 1950s laid the foundation for space-based radars with acceptably small antennas. The SEASAT satellite launched by the USA on June 28, 1978 began the evolution of space-based SAR that continues to this day [McCandless and Jackson, 2004]. Table 1.1 provides a brief overview of development history of space-based SARs, of which the image data are suitable for ocean remote sensing applications.

A SAR system operated with an L band (1.275 GHz), HH polarization and a fixed incident angle was first launched into the space as an important payload onboard the SEASAT satellite. The SEASAT SAR was originally designed for monitoring of the global ocean surface wave field and polar sea ice conditions [Lame and Born, 1982]. During its 105 mission days, approximately 42 h of SAR data were collected. Following research indicated that not only ocean surface waves and polar sea ice, but also a variety of important atmospheric and oceanic processes were manifested on the SEASAT SAR images [Fu and Holt, 1982; Vesecky and Stewart, 1982]. Since then, the spaceborne SAR has been recognized as a powerful remote sensor for ocean applications. Following SEASAT, the USA launched the Shuttle Imaging Radar-A (SIR-A) onboard

Table 1.1 Brief overview of space-based SARs

| Name (Country) | Year | Band | Wavelength (cm) | Polarization | Resolution Range × Azimuth (m×m) |
|---|---|---|---|---|---|
| SEASAT (USA) | 1978 | L | 23.5 | HH | 7.9 × 6 |
| SIR-A (USA) | 1981 | L | 23.5 | HH | 24.9 × 6.5 12.5 × 6 |
| SIR-B (USA) | 1984 | L | 23.5 | HH | 24.9 × 6.5 12.5 × 6 |
| ERS-1/2 (Europe) | 1991/1995 | C | 5.7 | VV | 9.7 × 25 |
| ALMAZ (USSR) | 1991 | S | 10 | HH | 15 × 15 |
| JERS-1 (Japan) | 1992 | L | 23.5 | HH | 10 × 30 |
| SIR-C (USA, Germany, Italy) | 1994 | L | 24.2 | HH, HV | 15 × 7.5 |
| | | C | 5.7 | VH, VV | |
| | | X | 3.1 | VV | 7.5 × 6 |
| RADARSAT-1 (Canada) | 1995 | C | 5.7 | HH | 12.9 × 28 ∼ 5 × 100 |
| SRTM (USA, Germany, Italy) | 2000 | C | 5.7 | HH, VV | 7.5 × 15 |
| | | X | 3.1 | VV | 18.7 × 8/12 |
| ENVISAT ASAR (Europe) | 2002 | C | 5.331 | HH, HV | 16.6 × 6/150/1000 |
| | | | | VH, VV | |
| ALOS-1 PALSAR-1 (Japan) | 2006 | L | 23.6 | HH, HV, VH, VV | 10 × 10, 100 × 100 |
| RADARSAT-2 (Canada) | 2007 | C | 5.6 | HH, HV, VH, VV | 10 × 9 ∼ 100 × 100 |
| TerraSARX/ TanDEM-X (Germany) | 2007/2010 | X | 3.1 | Full Polar. Modes | 1, 3, 16 |
| Cosmo-SkyMed-SkyMed (Italy) | 2010 | X | 3.1 | HH, HV, VH, VV | 1 × 1, 30 × 30, 100 × 100 |
| ALOS-2 PALSAR-2 | 2014 | L | 22.9 | Full Polar. Modes | 3 × 10, 100 × 100 |

Space Shuttle *Columbia* and Shuttle Imaging Radar-B (SIR-B) onboard Space Shuttle *Challenger* in 1981 and 1984, respectively. Both SIR-A and SIR-B radars were dedicated for testing of variations on the SEASAT SAR specifications. For example, SIR-B had an added capability of operating at variable incident angles.

1990s were a booming age for a significant expansion of SAR missions. During the period of 10 years, five earth remote sensing SAR satellites and two of shuttle imaging radar missions were launched by Europe, USSR, Japan, USA (joint with Germany and Italy) and Canada. European Remote Sensing satellites (ERS)-1/2 launched in 1991 and 1994, respectively, operated identical C band SAR systems with a single wavelength (5.7 cm) and single polarization (VV) [ESA, 1991]. Russian ALMAZ SAR, an S band (10 cm) and single polarization (HH) system launched in 1991, had the ability to operate at variable incident angle [Li and Raney, 1991]. The Japanese Earth Resources Satellite (JERS) SAR was an L band and HH polarization system similar to the SEASAT SAR launched in 1992 [Nemoto *et al.*, 1991]. The Canadian RADARSAT-1 satellite SAR, a C band and HH polarization system launched in 1995, had more advances in technology including the ability to operate at variable incident angles and seven beam modes offering different swath widths from 50–500 km and different image resolutions from 10 to 100 m [Raney *et al.*, 1991].

The Shuttle Imaging Radar-C/X band SAR (SIR-C/XSAR), a joint US-Germany–Italy mission flown twice in April and October 1994 on Space Shuttle *Endeaver*, has been the most complex space-based SAR since the SEASAT SAR [Jordan *et al.*, 1991]. SIR-C/XSAR could simultaneously operate at three bands (L, C and X) with C band and L band having the ability to alternate transmitting and receiving polarizations, *i.e.*, full polarimetric capability, which allows the SAR to collect complete polarization characteristics of the targets [Zebker and Van Zyl, 1991]. The C and X band portions of the SIR-C SAR were flown again for the Shuttle Radar Topography Mission (SRTM) in 2000. On September 23, 2014, the US announced that the highest-resolution topographic data generated from SRTM would be released globally.

Satellite SARs launched in the 21st century are regarded as new generation SARs. This new generation of satellite SARs is characterized by the following advantages. (1) Developing polarimetric technology. European ENVISAT Advanced SAR (ASAR) launched in 2002 had the capability of dual polarizations. The Japanese ALOS-1 (Advanced Land Observing Satellite) PALSAR-1 (Phased Array type L band SAR) launched in 2006, the Canadian RADARSAT-2 SAR launched in 2007 and the Italian Cosmo-SkyMed SAR had the capability of quadrature polarimetry. The German X band TerraSARX and TanDEM-X SARs launched respectively in 2007 and 2010, and the Japanese ALOS-2 PALSAR-2 launched in 2014 had the capability of full polarimetric modes. (2) Having ScanSAR operating modes. Following the RADARSAT-1 SAR, the maximum nominal swatch width of RADARSAT-2 SAR kept 500 km (compared to 100 km of the SEASAT SAR). The maximum imaging swatch width of ALOS-1/2 PALSAR-1/2 reached 350 km. (3) Higher ground resolutions. Ground resolutions of the new generation of satellite SARs have been getting higher, from the order of 10 m before 2007 to the order of 1 m after 2007. For example, TerraSARX, TanDEM-X and Cosmo-SkyMed SARs achieved the highest ground resolution of up to 1 m. (4) Providing image products for operational and commercial uses. The SAR image data have been developed into different data products per request and requirements of users. For example, RADARSAT-2 SAR image data product package contains 17 products with spatial resolutions selectable from 1 to 100 m and corresponding nominal swatch widths from 18 to 500 km, as well as selectable polarization combinations.

Figure 1.8 shows a comparison of imagery qualities of a RADARSAT-2 ScanSAR image to NASA Aqua satellite Moderate Resolution Imaging Spectroradiometer (MODIS) IR and true color images. All images were acquired in the same ocean area, the East China Sea (ECS), and selected as high quality images from their databases, respectively. The RADARSAT-2 ScanSAR image acquired on March 6, 2008 (Figure 1.8(a)) has the nominal ground resolution of 100 m. The MODIS color-coded IR and true color images acquired on August 1, 2010 (Figures 1.8(b) and (c)) have

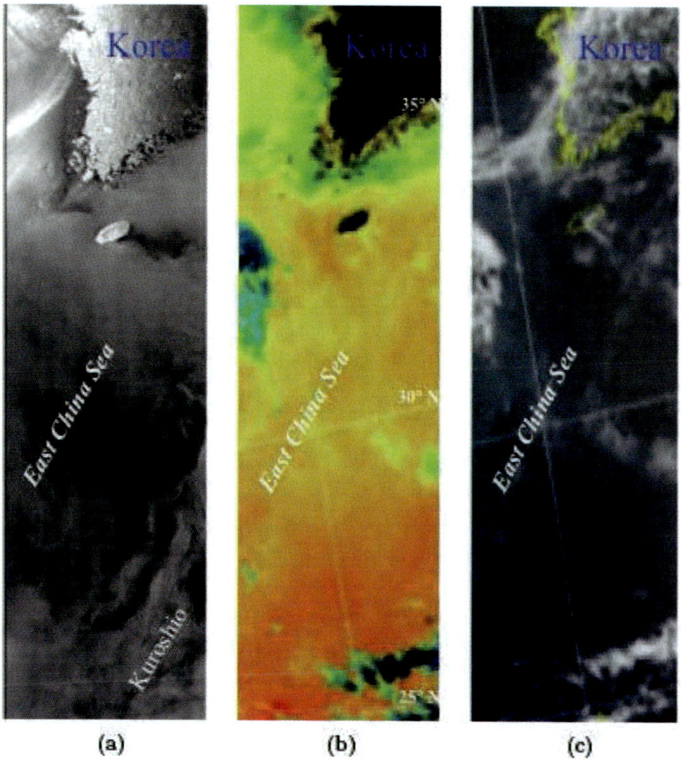


Fig. 1.8 Comparison of RADARSAT-2 ScanSAR image on March 3, 2008 (a) (Credit: NOAA NESDIS) to NASA Aqua satellite MODIS IR (b) and true color (c) images on August 1, 2010 of the ECS (Credit: Key Laboratory of Geographical Information Science, Ministry of Education, East China Normal University, Shanghai, China).

the nominal ground resolution of 1 km. One can see that the ScanSAR image looks much clearer than the other two. In particular, the ScanSAR image contains abundant information on the ocean circulation on the ESC continental slope and in the Kuroshio region. The Kuroshio stream path, fronts and fine structure clearly show up. On the color-coded IR image and true color image, the Kuroshio shows up as a high sea surface temperature region and a dark blue water area, respectively, but its fine structure is vague. These comparison results indicate that the SAR image is more suitable for the ocean circulation analysis.

## 1.3. Radar Wave Scattering on the Sea Surface

### 1.3.1. *Reflection, refraction and scattering*

Figure 1.9(a) shows the behavior of EM waves incident on a smooth boundary between two media. One can see that incident ray $B_i$ is divided into two branches at the boundary: reflected ray $B_r$ and refracted ray $B_f$. Reflected ray $B_r$ continuously travels in medium 1 but changes the direction of travel. The law of reflection states that the angle of reflected ray is always in the plane defined by the incident ray and the normal to the boundary surface. Refracted ray $B_f$ transmits into medium 2 and changes the direction of travel. The law of refraction, also known as Snell's law, describes the relationship between incident angle $\theta_i$ and refraction angle $\theta_f$.

From Figure 1.9(a), one can see that if the target surface is an ideally smooth plane, the radar cannot receive any return signal. In fact, most natural surfaces may be physically treated as a rough surface, which is defined as that average height $\overline{h}$ of radar scatterers on the ocean surface exceeds the Rayleigh criterion, *i.e.*,

$$\overline{h} > \frac{\lambda_r}{8\cos\theta}, \tag{1.3.1}$$

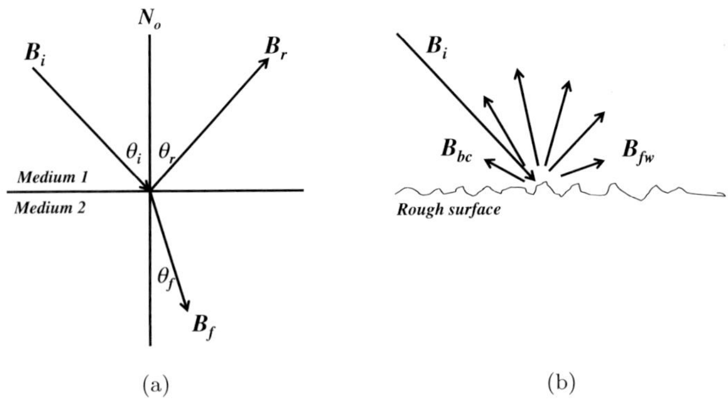


Fig. 1.9 Behavior of EM waves incident on the boundary between two media (a) reflection and refraction at a smooth boundary surface and (b) scattering at a rough surface

where $\lambda_r$ is the radar wavelength and $\theta$ the incident angle of radar waves [Ulaby *et al.*, 1982]. For example, if the radar wavelength is 3 cm and the incident angle is 30°, $\overline{h}$ is 0.4 cm, implying that most natural targets are physically rough for an X band radar. As shown in Figure 1.9(b), for a rough surface, the incident ray is reflected in many different directions. This phenomenon is defined as the scattering of EM waves. The rays that are scattered to travel forward, with respect to the direction of the incident ray, comprise the forward scattering waves, and that to travel backward comprise the backscattering waves. It is the backscattered EM radiation to form the radar echo, and carry the information of detected targets back to the radar.

### 1.3.2. *Radar wave reflectors and scatterers on the sea surface*

Seawater is a dielectric medium of the high conductivity for all bands of the radar waves. Based on the principles of EM wave propagation, microwave pulses, which the radars emit to detect the targets on the earth surface, are capable of penetrating only into a thin layer of seawater with a maximum depth on the order of centimeters. Thus, the radar echoes are formed by backscattered energy only from the ocean surface. Table 1.2 lists potential radar wave reflectors and scatterers on the sea surface.

Table 1.2 Potential radar wave reflectors and scatterers on the sea surface

| No. | Ocean Feature | Parameter/Object/Process |
|---|---|---|
| 1 | Manmade, static | Platforms, towers, buoys |
| 2 | Manmade, dynamic | Ships, boats, debris, wreckages, oils, surfactants, pollutants |
| 3 | Natural, static | Sea ice, icebergs, oil seeps, large-size algal layer, foam layer, biofilms |
| 4 | Natural, dynamic | Wind waves, swells<br>Wave-breaking products: whitecaps, bubbles, sprays<br>Rainfall products: stalks, crowns, ring waves, water drops |
| 5 | Seawater background | Temperature, salinity |

Owing to the fluid property of seawater, the ocean surface, as a boundary between the ocean and the atmosphere, is in a restlessly motional state under external actions including the astronomic tidal forcing, solar heating and the atmospheric forcing. In particular, broad distribution of wind-driven ocean surface waves makes the ocean surface a typical rough surface for a wide spectrum of EM waves from the microwave band to the high frequency (HF) band. Meanwhile, the radar return signals would have response to the sea surface wave field under certain conditions to be discussed in the next section.

### 1.3.3. *Bragg resonance scattering of sea surface waves*

Under the low wind conditions ($U_{10} < 10\,\mathrm{m\,s^{-1}}$, to here, $U_{10}$ represents the wind speed measured at 10 m above the mean sea level), the capillary waves and the short gravity waves with wavelengths ranging from 1 to 20 cm are broadly distributed on the ocean surface. If the wavelengths of radar waves are in the same range as that of the ocean surface waves, a special scattering mechanism, *i.e.*, Bragg resonance scattering or Bragg scattering for short, will occur.

Figure 1.10 shows the principle for Bragg scattering. Two parallel radar rays $R_1$ and $R_2$ illuminate peaks of two adjacent ocean waves

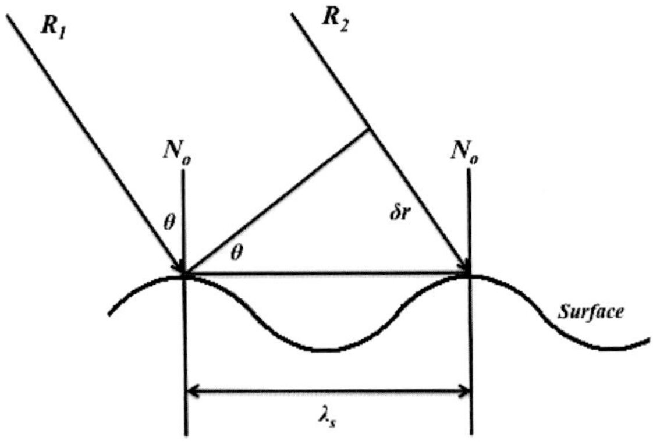


Fig. 1.10 Geometry for Bragg resonance scattering

with the same incident angle $\theta$. From the geometry, we obtain one-way path difference $\delta r$ of the two radar rays as

$$\delta r = \lambda_s \sin\theta, \tag{1.3.2}$$

where $\lambda_s$ is the wavelength of the ocean surface waves.

If $\delta r$ satisfies

$$\delta r = \lambda_r/2, \tag{1.3.3}$$

where $\lambda_r$ is the wavelength of the radar waves, the backscattered waves of $R_1$ and $R_2$ will have a round-trip phase difference of $2\pi$. Thus, the backscattered waves will in-phase superimpose each other, and generate the maximal echo signal. This is Bragg resonance scattering, which is one of the most important reasons for all SARs listed in Table 1.1 to select the same wavelength band as that of the ocean surface waves.

From (1.3.1) and (1.3.3), we obtain the occurrence condition for Bragg resonance scattering in the form of the wavelength is

$$\lambda_r = 2\lambda_s \sin\theta, \tag{1.3.4}$$

or in the form of the wavenumber is

$$k = 2k_0 \sin\theta, \tag{1.3.5}$$

where $k$ $(= 2\pi/\lambda_s)$ and $k_0 (= 2\pi/\lambda_r)$ are the wavenumbers of the ocean surface waves and the radar waves, respectively. Theoretically, if one-way path difference $\delta r$ satisfies

$$\delta r = \frac{n\lambda_r}{2}, \quad n = 1, 2, 3, \ldots. \tag{1.3.6}$$

Bragg scattering should also occur. In this case, the occurrence condition becomes

$$\lambda_r = \frac{2\lambda_s \sin\theta}{n}, \tag{1.3.7}$$

or

$$k = \frac{2k_0 \sin\theta}{n}. \tag{1.3.8}$$

The Bragg scattering mechanism lays a physical foundation for ocean radar remote sensing, which links radar physics to physical oceanography.

## 1.4. Ocean Wave-Radar Modulation Transfer Function

### 1.4.1. *Plant's model*

Generally speaking, the tasks of ocean radar remote sensing are (1) to detect or recognize the ocean processes and (2) to extract the data and information of the processes from radar echo signals or radar images. Physically, a radar image can be considered to be a two-dimensional presentation of the intensity of radar return signals. If we assume that the radar receiving system is a perfectly linear system, the intensity of radar return signal should depend linearly on the roughness of the target surface. In the case of the ocean, the two-dimensional (Cartesian) wavenumber spectral density of the ocean surface wave field, which satisfies the Bragg resonance scattering condition, is used to parameterize the ocean surface roughness. In order to set up an analytical relationship between the radar signals and the ocean wave spectral density, *i.e.*, ocean wave-radar modulation transfer function, the previous investigators have developed theoretical models, which are in the different forms, but all based on the Bragg resonance scattering mechanism [Plant *et al.*, 1983; Plant, 1990; Romeiser and Alpers, 1997; Romeiser *et al.*, 1994; Thompson, 2004; Wright *et al.*, 1980]. In this section and the following chapters of this book, we adopt ocean wave-radar modulation transfer function by Plant [1990] as a baseline.

According to Plant [1990], the intensity of radar return signals from the ocean is parameterized by the backscatter cross section per unit area in the form of

$$\sigma_0(\theta)_{ij} = 16\pi k_0^4 |g_{ij}(\theta)|^2 \psi(0, 2k_0 \sin\theta), \tag{1.4.1}$$

where $\theta$ is the incident angle, $k_0$ is the wavenumber of the radar waves, and $\psi$ is the two-dimensional (Cartesian) wavenumber spectral density of the ocean surface wave field, which satisfies the Bragg resonance scattering condition, the incident radiation is in the $x$–$z$ plane ($z$ being the vertical direction and $x, y$ are the horizontal coordinates), indices $ij$ denote polarizations of incident and backscattered radiations, respectively, and $g_{ij}(\theta)$ are the first-order scattering coefficients. For horizontal polarization

$$g_{\mathrm{HH}}(\theta) = \frac{(\varepsilon_r - 1)\cos^2\theta}{[\cos\theta + (\varepsilon_r - \sin^2\theta)^{1/2}]^2}, \tag{1.4.2}$$

for vertical polarization

$$g_{\mathrm{VV}}(\theta) = \frac{(\varepsilon_r - 1)[\varepsilon_r(1 + \sin^2\theta)]\cos^2\theta}{[\varepsilon_r\cos\theta + (\varepsilon_r - \sin^2\theta)^{1/2}]^2}, \tag{1.4.3}$$

where $\varepsilon_r$ is the relative dielectric constant of seawater [Klein and Swift, 1977; Saxton and Lane, 1952].

### 1.4.2. *Derivation of the wavenumber spectral function of the sea surface wave field*

Equation (1.4.1) shows that for given radar wavenumber $k_0$ and incident angle $\theta$, the intensity of the radar return signal depends only on two-dimensional wavenumber spectral density of the ocean surface wave field $\psi$. The spatial variability of $\psi$ results in the intensity variability on radar images, so-called image patterns or signatures. For example, the Kuroshio fronts appear as alternating bright–dark, vortex-like patterns on the SAR image as shown in Figure 1.8(a). Thus, the key to quantitatively interpret an ocean process showing up on a radar image, no matter real aperture radar or SAR, is to determine the functional form of $\psi$, which is modulated by the variable current field of the process of interest. Considering the modulation of short ocean surface waves by wave–current interaction, Yuan [1997] derived the wavenumber spectral density function of the HF ocean surface wave field as follows.

The measurements by Zhang and Cox [1994] indicated that the wave field still has a small slope for ocean waves with wavelengths within a range from millimeters to centimeters. Thus, the growth process of the wave spectrum in the wavenumber space can be described by the wavenumber spectrum balance equation in the form of

$$\frac{\partial\psi(k)}{\partial t} + (U + C_g)\nabla\psi(k) = S_{in}(k) + S_{nl}(k) + S_{ds}(k) + S_{cu}(k), \qquad (1.4.4)$$

where $U$ is the mean flow velocity, $C_g$ is the group velocity, $k = 2\pi/\lambda$ is the wavenumber of ocean surface waves and terms on the right-hand side $S_{in}, S_{nl}, S_{ds}$ and $S_{cu}$ represent the wind input source function, the wave–wave nonlinear interaction source function, the dissipation source function and the wave–current interaction source function, respectively. Suppose that the HF band of the wave spectrum is in an equilibrium state, therefore, the wave spectrum does not change with time, and is spatially uniform. Thus, the wavenumber spectrum balance Eq. (1.4.4) becomes

$$S_{in}(k) + S_{nl}(k) + S_{ds}(k) + S_{cu}(k) = 0. \qquad (1.4.5)$$

In fact, all these source functions have been derived in analytical forms by the previous investigators. Plant and Wright [1977] gave an expression for the wind input source function in the form of

$$S_{in}(k) = m\left(\frac{u_*}{c}\right)^2 \omega\psi(k), \qquad (1.4.6)$$

where $m$ is a dimensionless constant $(= 0.04)$, $u_*$ is the sea surface wind friction velocity, $\omega$ is the angular frequency and $c(= \omega/k)$ is the wave phase speed. On the basis of the resonance condition of three-wave interaction and four-wave interaction suggested by McGoldrick [1965], we have

$$S_{nl}(k) = \begin{cases} -m_3 k^{-4} B^2(k) & \text{for the gravity–capilary wave band} \\ -m_4 k^{-4} B^3(k) & \text{for the capilary wave band,} \end{cases} \qquad (1.4.7)$$

where $m_3$ and $m_4$ are dimensionless constants to be determined, and $B(k)[= k^4\psi(k)]$ is defined as the dimensionless wavenumber spectrum. According to the generally accepted concept [see, for example,

Plant, 1982], the dissipation source function can be written as

$$S_{ds}(k) = -4\gamma k^2 \psi(k), \tag{1.4.8}$$

where $\gamma$ represents the viscosity of seawater. For the wave-current interaction source function, Phillips [1977] gave

$$S_{cu}(k) = -\left(S_{\alpha\beta}\frac{\partial U_\beta}{\partial x_\alpha}\right)\psi(k), \tag{1.4.9}$$

where $U_\beta$ represents the velocity components of the large scale current field and $S_{\alpha\beta}$ represents the residual momentum flux tensor. Substituting Eqs. (1.4.6)–(1.4.9) into Eq. (1.4.5) yields

$$\psi_g(k) = m_3^{-1}\left[m\left(\frac{u_*}{c}\right)^2 - 4\gamma k^2\omega^{-1} - \left(S_{\alpha\beta}\frac{\partial U_\beta}{\partial x_\alpha}\right)\omega^{-1}\right]k^{-4}, \tag{1.4.10}$$

for the gravity–capillary wave band and

$$\psi_c(k) = m_4^{-1/2}\left[m\left(\frac{u_*}{c}\right)^2 - 4\gamma k^2\omega^{-1} - \left(S_{\alpha\beta}\frac{\partial U_\beta}{\partial x_\alpha}\right)\omega^{-1}\right]^{1/2}k^{-4}, \tag{1.4.11}$$

for the capillary wave band, in which $m, m_3$, and $m_4$ are coefficients, $u_*$ is the friction velocity, $c$ is the wave phase speed, $\gamma$ is the viscosity of seawater, $\omega$ is the angular frequency of the ocean surface waves, and $k$ is the wavenumber of ocean surface waves. For the HF band, the ocean can always be treated as the deep water. In this case, the tensor calculation yields

$$\left(S_{\alpha\beta}\frac{\partial U_\beta}{\partial x_\alpha}\right) = \left[\frac{\partial u}{\partial x}\cos^2\phi + \left(\frac{\partial u}{\partial y} + \frac{\partial v}{\partial x}\right)\right. \\ \left.\times \cos\phi\sin\phi + \frac{\partial v}{\partial y}\sin^2\phi\right]\Big/2, \tag{1.4.12}$$

where $u$ and $v$ are the velocity components in a local Cartesian (x-y) coordinate system and $\phi$ is the wave direction. Equations (1.4.1)–(1.4.3) and (1.4.10)–(1.4.12) comprise a complete analytical expression of ocean wave-radar modulation transfer function, which will serve as theoretical foundations for quantitative and dynamic interpretations of SAR images in the following chapters.

Here, it is worth noting that the above Plant's model and others are derived based on two important assumptions: (1) the small sea surface waves, including the capillary waves and the short gravity waves, are dominant radar scatterers, and (2) Bragg scattering is a dominant mechanism for formation of radar echoes. Point (1) stands only for the low sea states or under the low wind conditions ($U_{10} < 10\,\mathrm{m\,s^{-1}}$), in other words, the model would not work well beyond these conditions. For the high sea states or under the high wind conditions ($U_{10} > 10\,\mathrm{m\,s^{-1}}$), the breaking wave products as listed in Table 1.2 might become dominant radar scatterers, and the radar scattering mechanisms and analytical or numerical models are on the way to be developed. Point (2) states that the Bragg waves have dominant contribution to formation of radar echoes. This does not mean that the non-Bragg waves do not have any contribution. According to calculations by Holliday *et al.* [1986], contribution of the non-Bragg waves may reach a maximal 10% of that of the Bragg waves at low incident angle 23° for X and L band radars.

# Chapter 2

# Sub-Mesoscale Ocean Vortex Train

## 2.1. Introduction

### 2.1.1. *Scales of ocean dynamic processes*

As mentioned in Section 1.3.2, owing to the fluid property of seawater, the ocean is in a restlessly motional state driven by external forcing. As listed in Table 2.1, motions in the ocean appear in various forms with different length, time and speed scales. These scales are closely associated with properties of external forcing, boundary conditions, horizontal and vertical inhomogeneity of seawater bodies. Thus, to recognize and interpret an ocean phenomenon showing up on satellite SAR images, the first step is to examine its imagery patterns and to determine its characteristic scales. Furthermore, all textbooks of physical oceanography and geophysical fluid dynamics teach the readers the momentum equations and the vortisity equation that are used for the dynamic analysis of ocean processes or phenomena [Pedlosky, 1979]. The general forms of these equations, however, are too complex to be solved by analytical methods or numerical modeling. In this case, investigators may extract the scales of the phenomenon from the satellite SAR images and other data sources, and then use the scale analysis method to simplify the governing equations to solvable forms.

In geophysical fluid dynamics, a dimensionless Rossby number is used as a criterion to determined the motion scale, which is

Table 2.1 Statistics of length, time and speed scales as well as the Rossby numbers of ocean dynamic processes

| **Ocean Process** | **Length Scale (m)** | **Time Scale (s)** | **Speed Scale ($m\ s^{-1}$)** | **Rossby Number** |
|---|---|---|---|---|
| Capillary waves | $1 \times 10^{-3} - 2 \times 10^{-2}$ | $1 \times 10^{-3} - 1 \times 10^{-1}$ | | N/A |
| Wind waves | $2 \times 10^{-2} - 1.5 \times 10^{2}$ | $1 \times 10^{-1} - 1.2 \times 10$ | | N/A |
| Swells | $1.5 \times 10^{2} - 7 \times 10^{2}$ | $1.2 \times 10 - 2.4 \times 10$ | | N/A |
| Tsunami | 100–800 k | $3 \times 10^{2} - 3.6 \times 10^{3}$ | 200–300 | 2–10 |
| Internal waves | 500–5 k | $3 \times 10^{2} - 8.64 \times 10^{4}$ | 0.2–0.5 | 0.5–5 |
| Topography waves | 1–10 k | $1 \times 10^{2} - 1 \times 10^{3}$ | 0 | N/A |
| Sub-mesoscale vortex train | 5–20 k | $1 \times 10^{5} - 3 \times 10^{5}$ | 0.2–0.3 | 0.05–0.3 |
| Near-inertial oscillation (NIO) | 20–50 k | $1 \times 10^{5} - 7 \times 10^{5}$ | 0.1–0.5 | 0.01–0.125 |
| Mesoscale shear waves | 20–100 k | $4 \times 10^{4} - 10^{5}$ | 0.5–1 | 0.1–0.25 |
| Western boundary currents (WBCs) | 100–200 k | $3 \times 10^{6} - 10^{8}$ | 0.5–2 | 0.0125–0.1 |
| Tides and associated processes | 100–2000 k | $10^{4} - 10^{5}$ | 0.05–5 | $1.25 \times 10^{-4} - 0.25$ |
| Mesoscale eddies (ME) | 200–500 k | $2 \times 10^{5} - 10^{6}$ | 0.2–0.5 | 0.002–0.0125 |
| Continental shelf waves (CSW) | 300–3000 k | $2 \times 10^{5} - 10^{6}$ | 0.1–10 | $1.7 \times 10^{-4} - 0.17$ |
| Planetary waves (PW) | 500–2000 k | $3 \times 10^{6} - 10^{8}$ | 0.06–0.5 | $1.5 \times 10^{-4} - 5 \times 10^{-3}$ |

defined as

$$\varepsilon = \frac{U}{2\Omega L'} \tag{2.1.1}$$

where $\Omega$ is the angular velocity of earth rotation, $U$ and $L$ are the speed scale and the length scale of the motion, respectively. The motion is treated as a large-scale process if the Rossby number satisfies $\varepsilon \leq 1$. Table 2.1 lists length, time and speed scales as well as the Rossby numbers of ocean dynamic processes except the sea surface waves and the sea bottom topography waves.

Figure 2.1 shows a space-time distribution of ocean dynamic processes listed in Table 2.1. Dashed box shows a range of sub-mesoscale processes from 1 to 200 km, which is defined only by the length scales of the processes in this book. The top limit of the length scale is near the top detectable limit of the maximum imaging swatch

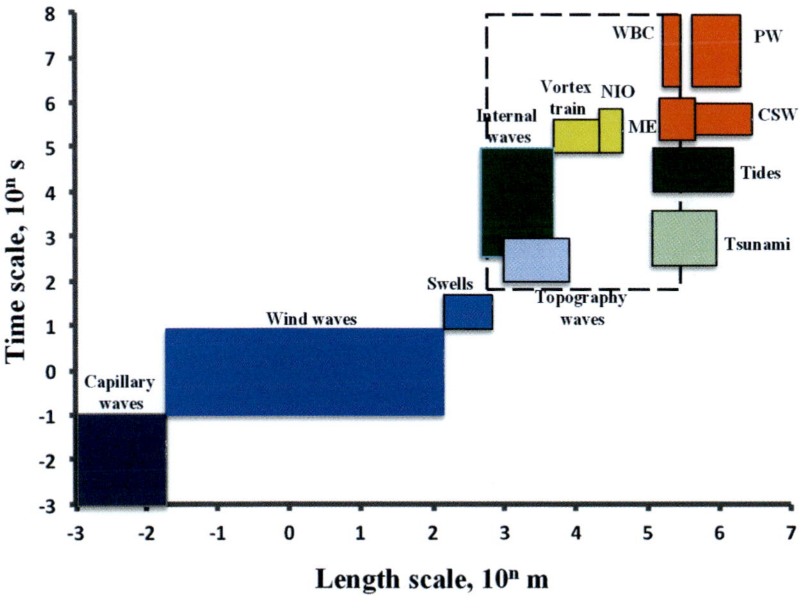


Fig. 2.1 Time scale versus length scale of ocean dynamical processes Boxes represent length and time ranges of the processes. Dashed box shows a range of sub-mesoscale processes defined in this book

widths of the state-of-art ScanSAR image products (350 km of ALOS-1/2 PALSAR-1/2 and 500 km of RADARSAT-2 SAR). On the other hand, from the Rossby numbers listed in Table 2.1, one can see that most processes within this range dynamically satisfy the large scale flow criterion except the short portion of internal waves and ocean bottom topography waves. This chapter will interpret sub-mesoscale ocean vortex trains (OVTs) on satellite SAR images, which have the length (wavelength) scales from 5 to 20 km, the time scales from 1 to 3 days, the speed scales from 0.2 to 0.3 $\mathrm{m\,s^{-1}}$ and the Rossby number from 0.05 to 0.3.

### 2.1.2. *OVT*

The vortex train is a phenomenon well documented in classical textbooks of fluid dynamics [*e.g.* Kundu, 1990]. Field observations and laboratory simulations indicate that when an object is placed in a steadily flowing fluid, the object will form a source of disturbance to the fluid flow. The downstream flow behind this obstacle will become unstable. Once the flow speed $U$ and the object width $D$ satisfy a certain relation (Reynolds number $UD/\nu >> 10$, in which $\nu$ is the viscosity of fluid), two parallel rows of alternating vortices with opposite rotations will be generated. This kind of vortex train is commonly called a Kármán vortex street (KVS) [Chopra and Hubert, 1964; Kundu, 1990; Taneda, 1965; von Kármán, 1954].

In recent years, the KVS in the lower atmosphere, occurring on the lee side of isolated mountain peaks or islands, has been studied using the satellite images. On visible band images, the atmospheric KVS shows up through modulated cloud patterns [Thomson *et al.*, 1977; Tsuchiya, 1969; Turner and Warren, 1988]. On the SAR images, the atmospheric KVS shows up through modulation to the spatial distribution of ocean surface small waves [Li *et al.*, 2000]. These satellite images can provide a two-dimensional, plane-view of KVS. By interpreting the satellite images, the fine structures and dynamic parameters of KVS can be obtained [Li *et al.*, 2000].

Meanwhile, the KVS may also occur in the ocean, which is termed the OVT in this book. This chapter will examine two cases of sub-mesoscale OVTs: one was observed on ENVISAT ASAR images of the Luzon Strait, which is sandwiched by the Pacific and the South China Sea (SCS) [Zheng *et al.*, 2008], and the second one was on ERS-1 SAR images of the Kuroshio region southeast of Taiwan Island [Zheng and Zheng, 2014]. The significance of this chapter is not only to report the observed phenomena, but also to develop methods for derivation of the dynamical parameters of OVT and mean flow. For a close-up view of the OVTs, Figure 2.2 shows a visible band image of two OVTs in the lee of Taiwan Island on June 30, 2001 [Ho *et al.*, 2001].

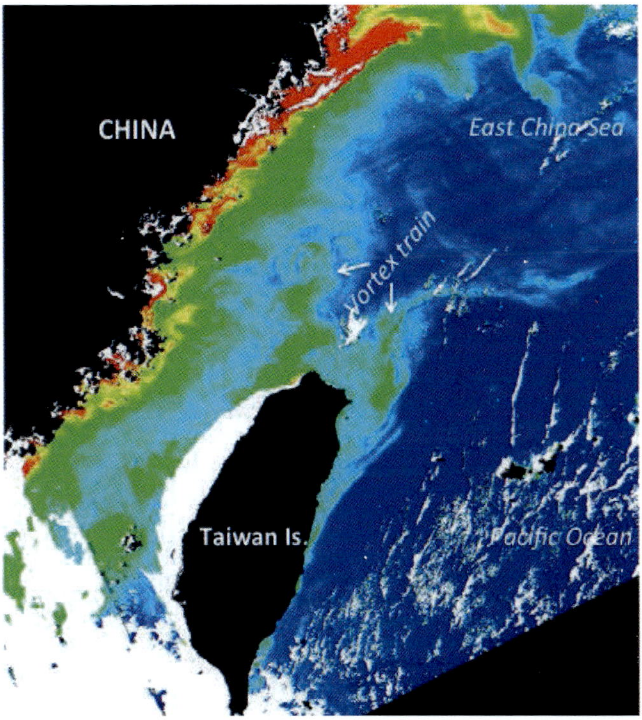


Fig. 2.2 Image of seawater chlorophyll distribution in the East China Sea on June 30, 2001 (color codes are in arbitrary unit) Original visible-band ocean color data were taken by ROCSAT-1 satellite Ocean Color Imager (OCI), which was launched by Taiwan on January 27, 1999. The spatial resolution of the image is 800 m. Two OVTs in the lee of Taiwan Island showed up clearly (Courtesy: Chung-Ru Ho, National Taiwan Ocean University)

## 2.2. OVTs in the Luzon Strait

### 2.2.1. *Study area*

The Luzon Strait lies between Taiwan Island and Luzon Island as shown in Figure 2.3. The strait is approximately 386 km wide from north to south. The average depth is 1400 m [Guo *et al.*, 2004]. The Luzon Strait connects the SCS on the west and the Philippine Sea and the Pacific on the east. On the east side of the strait, there are two groups of islands: the Batan Islands in the north and Babuyan Islands in the south. These two groups of islands divide the strait into three smaller channels: the Bashi Channel is the northernmost one, which separates Taiwan Island from the Batan Islands, the Balintang Channel is in the middle, which separates the two islands, and the Babuyan Channel is the southernmost one, which separates Luzon Island from the Babuyan Islands.

The Luzon Strait serves as a major channel for water exchange between the SCS and the Pacific. The Kuroshio flows along the east side of the Strait. Without topographic blockage, part of the Kuroshio water branches into the Strait forming the Kuroshio intrusion and the loop current [Centurioni *et al.*, 2004; Ho *et al.*, 2004]. Recent investigations reveal that the Luzon Strait is a source region for large amplitude ocean internal waves propagating westward into the SCS [Liu *et al.*, 2004; Yuan *et al.*, 2006; Zheng *et al.*, 2008].

The OVTs to be studied in this section were observed on the northwest side of Babuyan Island, the northernmost island of the Babuyan Islands. The Babuyan Claro Volcano is a major geological feature on the islands. The vent of volcano is 1180 m high above the sea level. The location of San Dionisio, the largest town on Babuyan Island, 19.5°N 121.9°E, may serve as reference coordinates for the islands.

### 2.2.2. *Satellite-tracked drifter trajectory data*

During the period from 1979 to 2003, a great number of Argos satellite-tracked surface drifters were released in the Kuroshio and adjacent waters by the Tropical Ocean Global Atmosphere (TOGA)

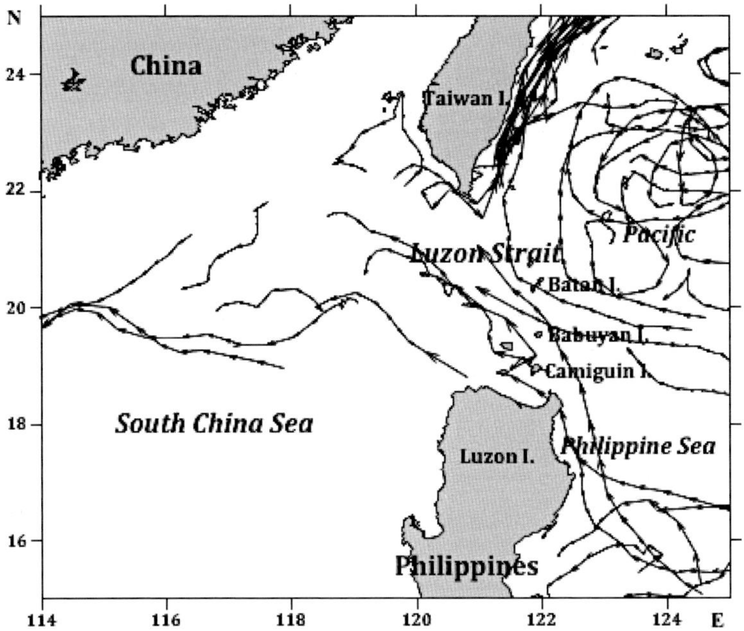


Fig. 2.3 Map of study area Lines with arrows are Argos satellite-tracked surface drifter trajectories in November. The trajectories show a train of three cyclonic vortices northwest of Babuyan Island and another one just behind Camiguin Island. The drifters were released during the period from 1979 to 2003 [Zheng *et al.*, 2008]

program, the World Ocean Circulation Experiment-Surface Velocity Program (WOCE-SVP), and Chinese oceanographic research programs. Figure 2.3 shows trajectories in and near the study area in November only. The drifter drogue was placed at a water depth of 15 m, so that the direct effects of atmospheric processes through sea surface winds on the drifter movement and current velocity calculation were ignorable [Niiler *et al.*, 1995]. The temporal resolution of drifter positioning for the data used in this study was one day [Hansen and Poulain, 1996].

From Figure 2.3, one can see that there are six trajectories passing through the southern Luzon Strait. In particular, the two trajectories passing through Babuyan Islands show a train of three cyclonic vortices distributed from 20° to 20.5°N and from 120° to

121°E, and another one just behind Camiguin Island. The average distance between two centers of consecutive cyclonic vortices in the train is 50 km. The vortices occurred in the lee of Babuyan Islands, which serves as blunt obstacles in the ocean flow, and formed a vortex train in a line behind the islands. These striking features evidence their nature of OVTs. The trajectories show that all the drifters moved northwestward (~315°TN), implying that a northwestward ocean current flows from the Pacific into the SCS. The daily average current velocities derived from the trajectories are $1.3\,\mathrm{m\,s^{-1}}$ at the east border of the strait (19° to 20.5° N, 121.5°E) and $0.6\,\mathrm{m\,s^{-1}}$ at the locations of vortex train (20° to 20.5°N, 120° to 121°E), respectively. The same features once again show up on the ENVISAT ASAR images of November 19, 2006 as shown in Figures 2.4(a) and (b) and will be analyzed in details

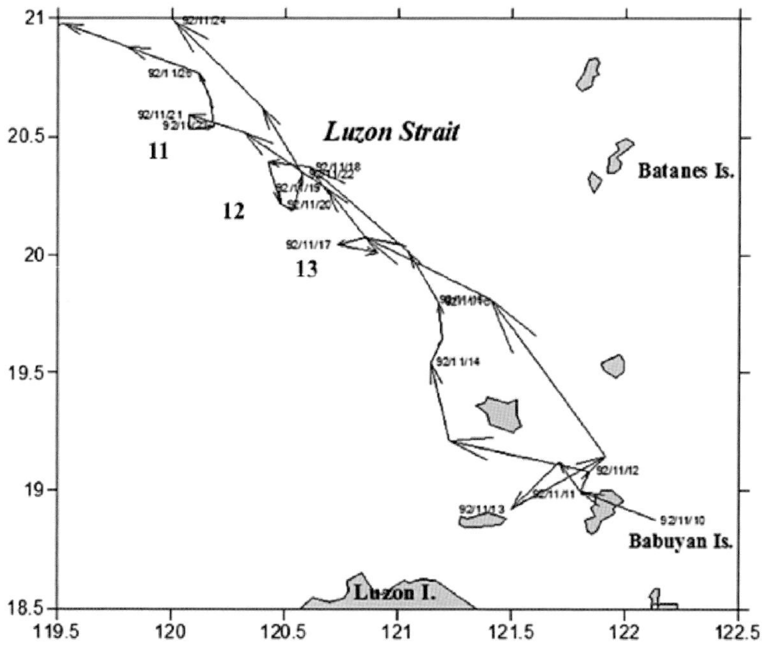


Fig. 2.4 Zoomed map showing two Argos drifter trajectories with the vortex train signal. Dates marked on the trajectories are in YY/MM/DD [Zheng *et al.*, 2008]

### 2.2.3. *Estimeted parameters of the OVT*

A zoomed map showing two Argos drifter trajectories with the vortex train signal is shown in Figure 2.4. The dynamic parameters of the vortices in the vortex train are measured for further analysis. A period of drifter rotating around a vortex $T_d$ can be determined directly from the trajectories. Thus, the rotational angular velocity of the vortex $\omega_d$ can be estimated by $2\pi/T_d$. A vortex radius at that the drifter rotated $R_d$ is determined by half of the maximum distance of vortex trajectory outedge. The radius determined by this method is an average distance of drifter to the vortex center, which does not necessarily coincide with the radius of the maximum rotational velocity. Then the drifter-determined daily mean tangent velocity can be estimated by $\omega_d R_d$. The results are listed in Table 2.2, in which the vortices are coded from number 11 in a southeastward order. One can see that the mean angular velocity of the vortices reaches $(2.66 \pm 0.56) \times 10^{-5}\,\text{s}^{-1}$, 4–5 times faster than the Gulf Stream MEs [Zheng *et al.*, 1984, 1989].

### 2.2.4. *Satellite observations and air–sea boundary conditions*

#### 2.2.4.1. *ENVISAT ASAR images*

The satellite SAR images used in this section were taken by ENVISAT satellite, which was launched into a sun synchronous orbit by European Space Agency on March 1, 2002. Its orbit

Table 2.2 Dynamic parameters of vortices estimated from drifter trajectory data

| **Vortex** | **Time (yy/mm/dd)** | **Period (d)** | **Angular Velocity ($10^{-5}\text{s}^{-1}$)** | **Radius (km)** | **Tangent Velocity ($\text{m}\,\text{s}^{-1}$)** |
|---|---|---|---|---|---|
| 11 | 92/11/21–24 | 2.2 | 3.24 | 8.97 | 0.29 |
| 12 | 92/11/18–22 | 3.8 | 1.90 | 12.56 | 0.24 |
| 13 | 92/11/16–19 | 2.6 | 2.83 | 6.28 | 0.18 |
| Mean | | 2.8±0.7 | 2.66±0.56 | 9.27±2.6 | 0.24±0.04 |

height is 786 km and inclination 98°. One of the most important payloads onboard ENVISAT is a C band (5.331 GHz) Advanced SAR (ASAR), which uses an active phased-array antenna with incidence angles between 15° and 45° to provide dual polarization, and wide swath coverage imaging data. In Alternating Polarization Mode (AP Mode), the antenna provides three choices for different transmitting and receiving polarization combinations: VV and HH, or HH and HV, or VV and VH (see Table 1.1).

The ASAR images were acquired and processed at the Hong Kong Satellite Remote Sensing Ground Receiving Station, which was established at the Chinese University of Hong Kong in 2006. The nominal spatial resolution of images is 30 m, and the swath coverage is 100 km. The study area is 600–800 km from Hong Kong Station. Thus, high quality images of low noise and low distortion may be ensured. Figure 2.5a shows a mosaic of ASAR images acquired at 13:53:45 UTC on November 19, 2006. The image coverage is shown in the inset.

From Figure 2.5(a), one can see that the southern tip of Taiwan Island and the northern tip of Luzon Island are located at the top-left corner and lower-left corner, respectively. A bright area east of Taiwan on the central top of the image shows a rainfall area, which will be analyzed in Chapter 11. The major portion of the image covers the Luzon Strait and the adjacent water of the Pacific. The landmarks of the strait, the Batan Islands and the Babuyan Islands, are as marked in the image. The major feature of interest, as enclosed by a dashed line box in the image, is a vortex train distributed on the northwest of Babuyan Island. A zoomed image of this sub-region is shown in Figure 2.5(b). One can see that the vortex train imagery consists of five pairs of dark–bright, near-circular, vortex-like patterns coded from 21 in the northeastward order and the Babuyan Island looks like a train head.

### 2.2.4.2. *Weather conditions for ASAR imaging*

The surface analysis weather charts before (12:00 UTC) and after (18:00 UTC) ASAR imaging are shown in Figures 2.6(a) and (b). One can see that over adjacent East China, a weak high pressure

(1024 mb) was moving eastward, meanwhile a cold front east of Taiwan was moving southeastward. These two features had not yet influenced the study area at the imaging time. Over the study area were a clear sky and low wind conditions (wind speed $<5\,\mathrm{m\,s^{-1}}$). Weather conditions observed at the nearest weather station, Itbayat Island located at 20°47′N 121°51′E, 80 km north of the vortex train, are listed in Table 2.3.

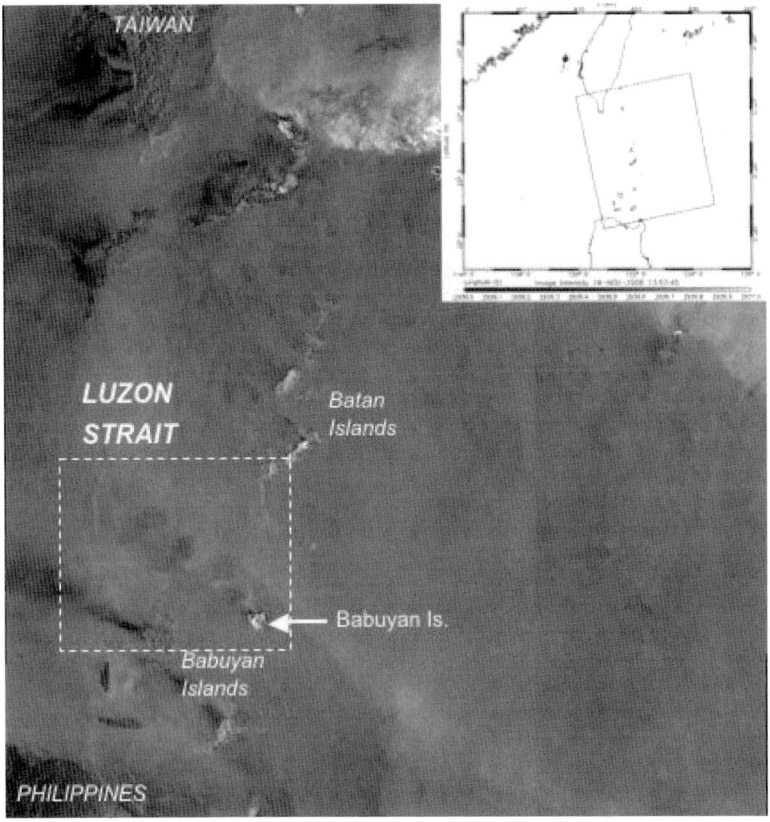


(a)

Fig. 2.5 (a) A mosaic of ASAR images acquired at 13:53:45 UTC on November 19, 2006. Image coverage is shown in an inset on top right corner. A vortex train shows up in the dashed line box (Credit: the Hong Kong Satellite Remote Sensing Ground Receiving Station at the Chinese University of Hong Kong). (b) Zoomed image of the vortex train from Figure 2.5(a)

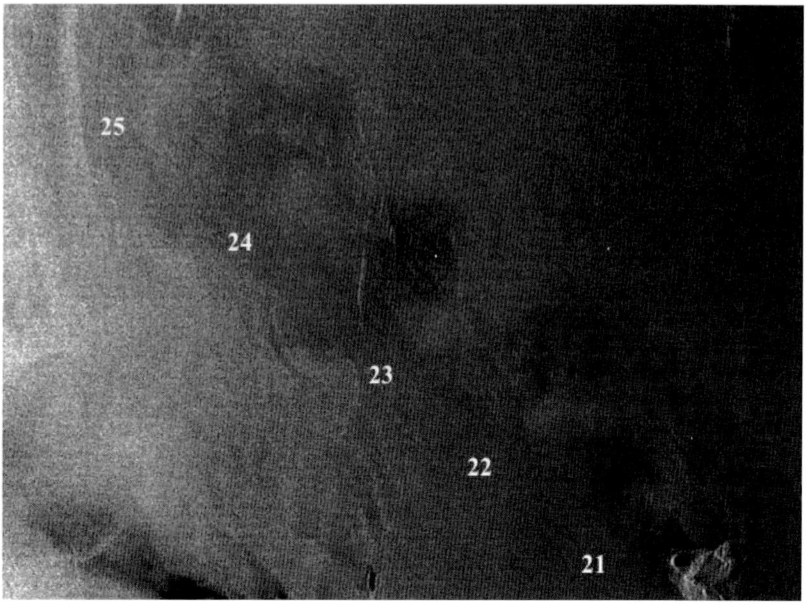


(b)

Fig. 2.5 (*Continued*)

Table 2.3 Weather conditions observed within 2 h of ASAR imaging at Itbayat Island

| Time (Nov. 19, 2006) | Cloud Coverage (%) | Air Temperature (°C) | Wind Speed ($m\,s^{-1}$) | Wind Direction |
|---|---|---|---|---|
| 12:00Z | 25 | 26 | 5 | SE |
| 15:00Z | 25 | 28 | 1 | SE |
| 13:54Z* | 25* | 27* | 2* | SE* |

*Imaging time and interpolated values.

A Moderate Resolution Imaging Spectroradiometer (MODIS) thermal infrared composite image that was near-simultaneously (51 min later) taken with ASAR image is shown in Figure 2.7. One can see again the clear sky over the study area and no strong weather systems showing up. The weather maps and satellite images all show a clear and calm weather conditions at the time of ASAR imaging. The sea surface wind speed obtained by linear interpolation was $2\,\mathrm{m\,s^{-1}}$, implying a favorable condition for imaging of the sea surface processes. Meanwhile, these boundary conditions further exclude the possibility of involvement of atmospheric processes in the vortex train images.

### 2.2.5. *ASAR image interpretation*

#### 2.2.5.1. *Recognition of OVT*

In Sections 2.2.2 and 2.2.3, the OVT in the Luzon Strait has been recognized from the Argos satellite-tracked drifter trajectories. In Figure 2.8, these trajectories are overlaid on the ASAR image. One can see that the cyclonic vortices on the trajectories are just

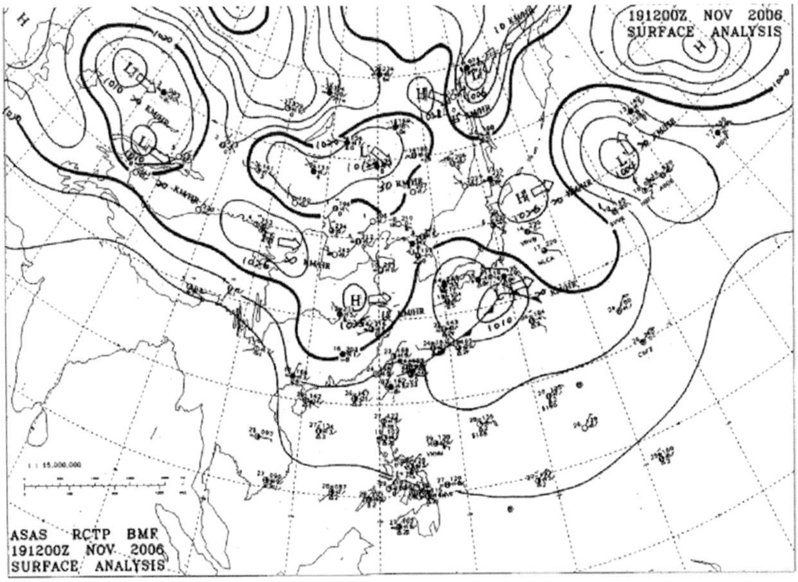


(a)

Fig. 2.6 (a) Surface analysis weather charts of East Asia 2-h before ASAR imaging at 12:00 UTC on November 19, 2006 (Courtesy: Shih-Jen Huang, National Taiwan Ocean University). (b) Surface analysis weather charts of East Asia 4-h after ASAR imaging at 18:00 UTC on November 19, 2006 (Courtesy: Shih-Jen Huang, National Taiwan Ocean University)

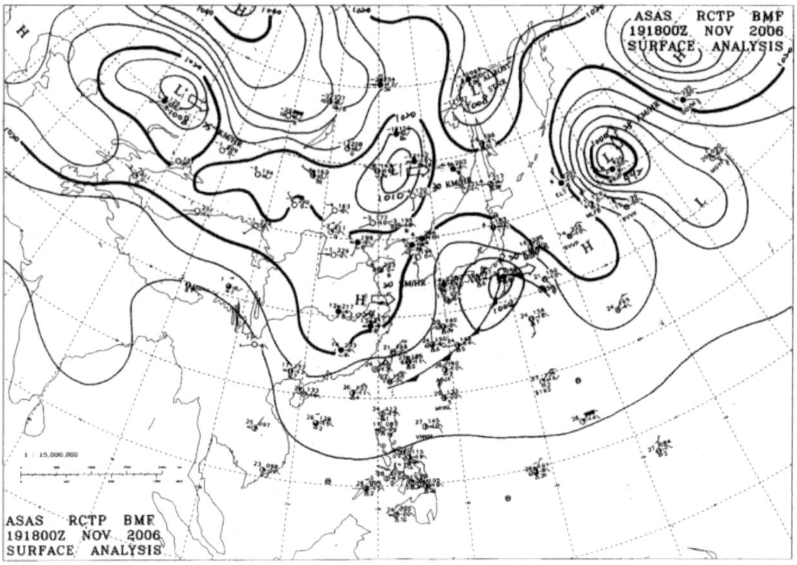


(b)

Fig. 2.6 (*Continued*)

distributed on the downstream side of vortex train on the ASAR image. The two vortex train axes constitute a continuous straight line starting from the Babuyan Island. This evidences the common nature of two vortex trains, although they were not simultaneously observed.

From the ASAR images shown in Figures 2.4(a) and (b), the following information is immediately derived. (1) The observation time was 13:53:45 UTC on November 19, 2006. (2) The OVT occurred within a range of 50 by 100 km from 19.5° to 20.0°N and from 121.0° to 122.0°E in the Luzon Strait. (3) The OVT consists of five pairs of cyclonic–anti-cyclonic vortices that are distributed in the wake of Babuyan Island. The axis of the OVT extends northwestward (∼315°) starting from the Babuyan Island, implying existence of a northwestward mean flow passing through the Luzon Strait. Later, it will be verified that the center of the dark or bright area is the center of vortex, so that the average distance between the two centers

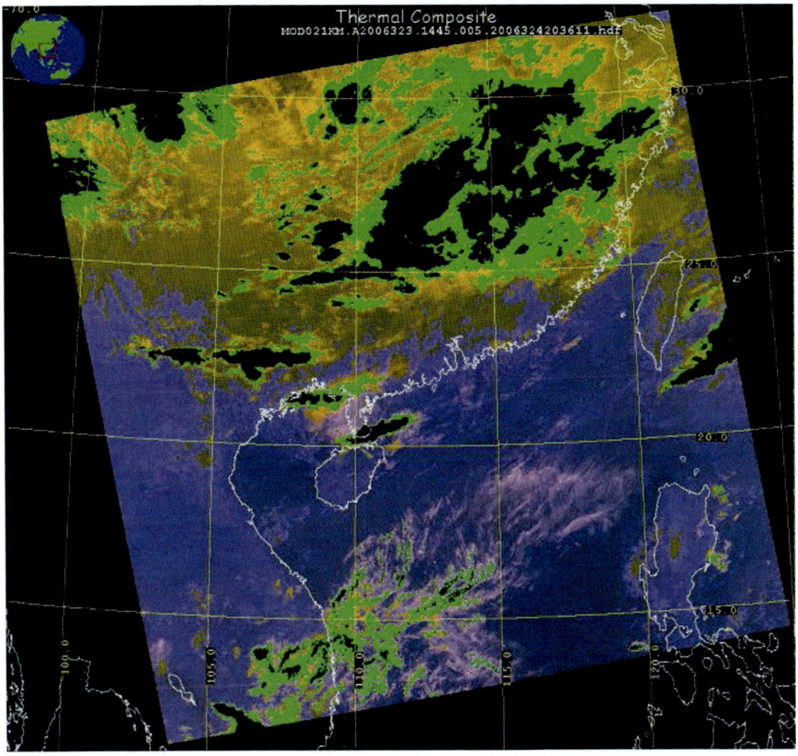


Fig. 2.7 MODIS thermal infrared composite image near simultaneously taken with ASAR image at 14:45 UTC on November 19, 2006 (Credit: US NASA)

of consecutive vortices and that between the two centers of adjacent cyclonic–anti-cyclonic vortices are measured as listed in Table 2.4.

### 2.2.5.2. *Theoretical models of ocean vortex radar image*

In order to extract dynamical parameters of the ocean vortex from the ASAR image, theoretical models of ocean vortex radar image must be derived from general radar imaging theories for ocean processes. In other words, the transformational relations between the radar imagery and the vortex truth must first be set up.

According to the theories given in Section 1.4, *i.e.*, Eqs. (1.4.1)–(1.4.3) and (1.4.10)–(1.4.11), if the wind field and seawater viscosity

Table 2.4 Distances between two consecutive vortices and two rows of vortices measured from the ASAR image

| **Vortex Pair** | **Distance Between Two vortices (km) Anti-cyclonic (dark)** | **cyclonic (bright)** | **Distance Between Two Vortex Row (km)** |
|---|---|---|---|
| 22 | | | 6.30 |
| 23 | 22.92 | 18.91 | 8.60 |
| 24 | 22.92 | 25.21 | 8.02 |
| 25 | 22.92 | 22.92 | 9.74 |
| Mean | $22.6 \pm 1.9$ | | $8.2 \pm 1.2$ |

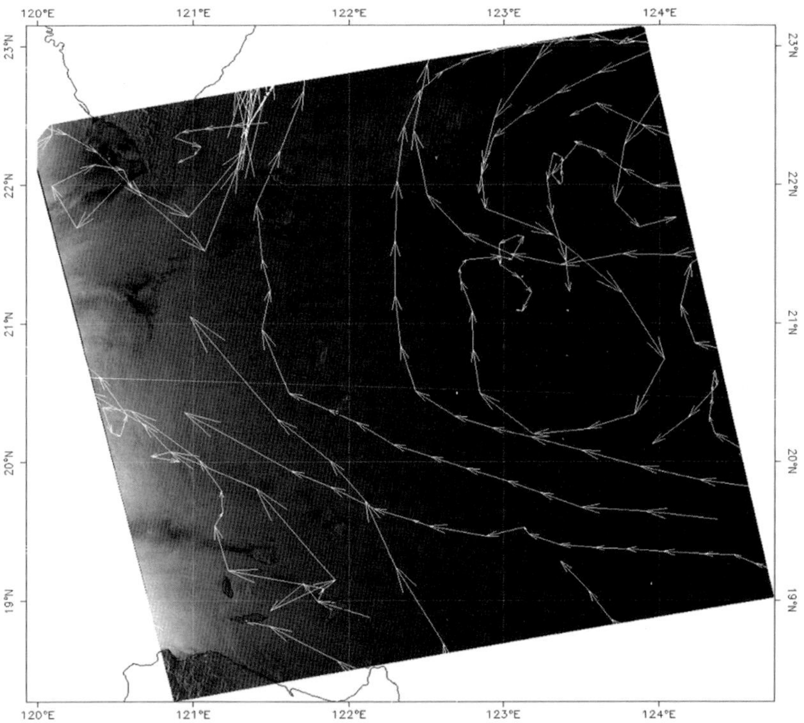


Fig. 2.8 Composite image of ASAR image with Argos satellite-tracked drifter trajectories of November [Zheng *et al.*, 2008]

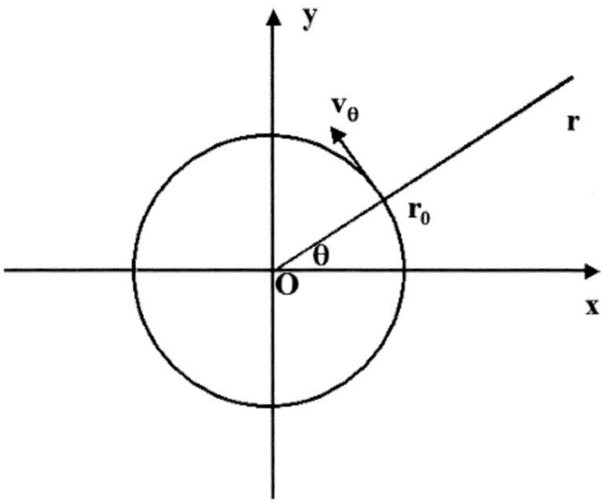


Fig. 2.9 Schema of the cyclonic vortex and the coordinate system [Zheng *et al.*, 2008]

are homogeneous, the current velocity field is a determining factor to generate the radar image of an ocean process. In the case to be studied, the ocean vortex and the coordinate system are shown as a schema in Figure 2.9.

The origin of the two-dimensional horizontal coordinate system is chosen as the vortex center. The $x$-axis is chosen as the ocean wave propagation direction, i.e., the sea surface wind direction. The $y$-direction is perpendicular to the $x$-axis. Assume that the vortex is a cyclonic eddy with a counterclockwise rotating angle $\theta$ and a radial axis $r$. The rotational velocity structure of the vortex can be written as an empirical model [Zheng *et al.*, 1984]

$$V_\theta = \frac{\omega_E r^2}{r_0} \exp\left[\frac{1}{2} - \frac{1}{2}\left(\frac{r}{r_0}\right)^4\right], \tag{2.2.1}$$

where $\omega_E$ is the rotational angular velocity, and $r_0$ is a radius, at which the velocity takes a maximum value. Figure 2.10 shows a graph of (2.2.1). The data points are taken from the drifter trajectories shown in Figure 2.3 as listed in Table 2.2 and normalized by the maximum velocity. One can see that the empirical model fits the data quite well.

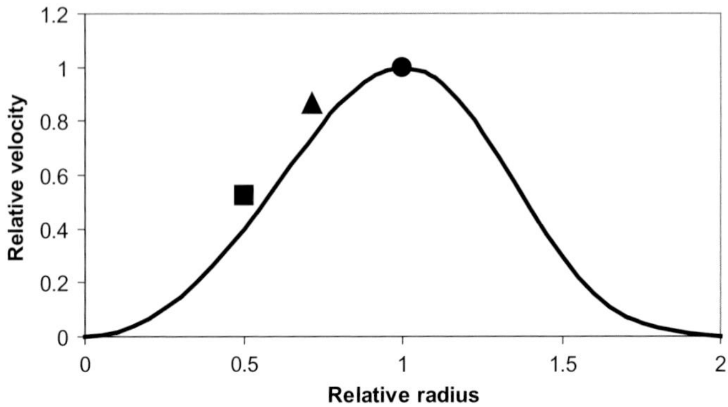


Fig. 2.10 Empirical model of ocean vortex tangent velocity distribution. Data points are taken from the Argos satellite-tracked drifter trajectories shown in Figure 2.3 and normalized by the maximum velocity [Zheng *et al.*, 2008]

On the other hand, the coordinate system setup implies that the small wave direction $\phi$ in Eq. (1.4.12) is zero. Thus, (1.4.12) is simplified as

$$S_{\alpha\beta}\frac{\partial U_\beta}{\partial x_\alpha} = \left[\frac{\partial u}{\partial x}\right]\Big/2, \tag{2.2.2}$$

in which $u = -V_\theta \sin\theta + V_r \cos\theta$ and $V_r \ll V_\theta$ so that the second term is negligible. Thus,

$$\frac{\partial u}{\partial x} = \left[\frac{V_\theta}{r} - \frac{\partial V_\theta}{\partial r}\right]\sin 2\theta. \tag{2.2.3}$$

Substituting (2.2.1) into (2.2.3) yields

$$\frac{\partial u}{\partial x} = \omega_E \sin 2\theta\left[2\left(\frac{r}{r_0}\right)^5 - \frac{r}{r_0}\right]\exp\left[\frac{1}{2} - \frac{1}{2}\left(\frac{r}{r_0}\right)^4\right]\Big/2 \tag{2.2.4}$$

Assuming that both wind field and seawater viscosity are spatially homogeneous, Eq. (1.4.10) for gravity–capillary wave band can be re-written as

$$\psi = \psi_0 + \psi_e, \tag{2.2.5}$$

where $\psi_0$ represents a mean, homogeneous wave field, and $\psi_e$ represents the variability of wave field induced by an ocean vortex, which is in the form of

$$\psi_e = m_3^{-1}\left[-S_{\alpha\beta}\frac{\partial U_\beta}{\partial x_\alpha}\omega^{-1}\right]k^{-4}. \tag{2.2.6}$$

Substituting (2.2.2) and (2.2.4) into (2.2.6) yields

$$\psi_e = m_3^{-1}k^{-4}\omega^{-1}\omega_E \sin 2\theta\left[\frac{r}{r_0} - 2\left(\frac{r}{r_0}\right)^5\right]\exp\left[\frac{1}{2} - \frac{1}{2}\left(\frac{r}{r_0}\right)^4\right]\Big/4. \tag{2.2.7}$$

Substituting (2.2.7) into (1.4.1) yields the variability of radar backscatter cross section per unit area

$$\sigma_{0e} = 4\pi k_0^4\left|g_{ij}\left(\theta_0\right)\right|^2 m_3^{-1}k^{-4}\omega^{-1}\omega_E \cdot \sin 2\theta\left[\frac{r}{r_0} - 2\left(\frac{r}{r_0}\right)^5\right]\exp\left[\frac{1}{2} - \frac{1}{2}\left(\frac{r}{r_0}\right)^4\right]. \tag{2.2.8}$$

Equation (2.2.8) is a theoretical model of an ocean cyclonic vortex radar image being looked for. For the rotational angular velocity $\omega_E$ takes a negative value, the model represents an anti-cyclonic vortex.

Figure 2.11 shows comparison of the theoretical model with a $\sigma_{0e}$ profile taken from transect across cyclonic vortex 23 behind Babuyan Island in Figure 2.5(b). One can see that as a first-order approximation, the model fits a general trend in radar imagery, in particular, the increasing edge. Meanwhile, the imagery also contains other high frequency signals or noises, which make the imagery deviate from the vortex model.

### 2.2.6. *Inferences of theoretical models*

From the theoretical model of ocean vortex radar image Eq. (2.2.8), the following inferences can directly be derived.

#### 2.2.6.1. *Center of true vortex*

Equation (2.2.8) indicates that the radar backscatter cross section $\sigma_{0e}$ increases with $r$, for $r < r_0$. Let $r = 0$, (2.2.8) gives $\sigma_{0e} = 0$. This means that the geometric center of the vortex imagery is the center of

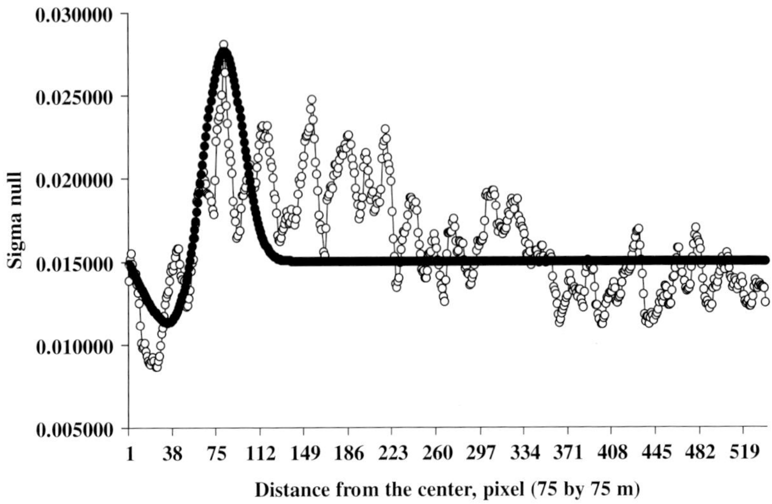


Fig. 2.11 Comparison of the theoretical model (2.2.8) (solid curve) with a $\sigma_{0e}$ profile (data points) taken from transect across cyclonic vortex 23 behind Babuyan Island in Figure 2.5(b) Pixel size is 75×75 m [Zheng *et al.*, 2008]

the true vortex. Thus, the distances between two consecutive vortices and two rows of vortices can be measured from the ASAR image. These data are important parameters for the dynamic analysis of OVT. The measured results are listed in Table 2.4.

### 2.2.6.2. *Maximum rotational velocity radius*

Equations (2.2.1) and (2.2.8) indicate that the maximum rotational velocity radius $r_0$ is a key parameter in determining the velocity structure of vortex. However, this parameter cannot directly be measured from the vortex radar image. A related parameter that can be measured from the radar image is a maximum radar backscatter radius defined as $r_m$. According to the definition, $r_m$ satisfies

$$\left.\frac{\partial \sigma_{0e}}{\partial r}\right|_{r=r_m} = 0. \tag{2.2.9}$$

Solving Eq. (2.2.9) yields the physically sound solutions

$$r_{m1} = 0.542 r_0 \tag{2.2.10}$$

with

$$\left.\frac{\partial^2 \sigma_{0e}}{\partial r^2}\right|_{r=r_{m1}} > 0 \tag{2.2.11}$$

and

$$r_{m2} = 1.307 r_0 \tag{2.2.12}$$

with

$$\left.\frac{\partial^2 \sigma_{0e}}{\partial r^2}\right|_{r=r_{m2}} < 0. \tag{2.2.13}$$

Comparing with the theoretical curve in Figure 2.10, one can see that the two solutions correspond to the minimum value at pixel 34 and the maximum at pixel 82, respectively. Thus, after $r_m$ is measured from the radar image, the maximum rotational velocity vortex radius $r_0$ can be calculated by Eq. (2.2.10) or (2.2.12). In the case shown in Figure 2.10, the maximum rotational velocity radius $r_0$ is determined to be 4.70 km.

### 2.2.6.3. *Vortex rotational angular velocity and spectral constant $m_3$*

If the radar backscatter cross section $\sigma_{0e}$ at a given $r$ is obtained from ocean vortex imagery, the rotational angular velocity $\omega_E$ and spectral constant $m_3$ can be calculated by (2.2.8), if one of them is known. Taking the case of Figure 2.10 as an example, from radar imagery transect data we obtain the radar backscatter cross section at vortex center $\sigma_{0e}(0) = 0.0149$, which serves as a background backscatter cross section, and the maximum radar backscatter section at $r = 6.15$ km $\sigma_{0e}$ (6.15 km) = 0.0276. Thus, vortex-induced increment in radar backscatter cross section at $r = 6.15$ km $\Delta\sigma_{0e}$ (6.15 km) should be 0.0127. Other parameters are as follows: $\lambda_0 = 5.7$ cm $\theta_0 = 30°$ $|g_{ij}(\theta_0)|^2 = 0.929$ (salinity 35 psu, water temperature 22°C,

HH polarization), $\omega = 35.26\,s^{-1}$, $\theta = -45°$ $r_0 = 4.70\,\text{km}$ and

$$\left[\frac{r}{r_0} - 2\left(\frac{r}{r_0}\right)^5\right] \exp\left[\frac{1}{2} - \frac{1}{2}\left(\frac{r}{r_0}\right)^4\right]_{r=6.15\,\text{km}} = -2.42.$$

Substituting these parameters into (2.2.8) yields

$$\omega_E m_3^{-1} = 1.54 \times 10^{-2} \tag{2.2.14}$$

On the other hand, the data listed in Table 2.2 give that a reasonable estimate of the vortex angular velocities should be $\omega_E = 3 \times 10^{-5}\,s^{-1}$. Thus, from (2.2.14) we obtain an estimate of spectral constant $m_3$ as $1.95 \times 10^{-3}$.

### 2.2.7. *Dynamic analysis of the OVT*

A physical model of the OVT is schematically shown in Figure 2.12. The parameters are defined as follows: $U_0$, the current velocity of mean or undisturbed flow; $U_e$, the propagation velocity of vortex; $a$, the distance between two consecutive vortices; and $h$, the distance between two rows of vortices.

Based on this model, the vortex propagation problem can be solved by solving the following two equations. The first equation describes the relation between $U_0$ and $U_e$ [Chopra and Hubert, 1964]

$$U_0 - U_e = \frac{K}{2a} \tanh\left(\frac{\pi h}{a}\right), \tag{2.2.15}$$

where $K$ is defined as the vortex circulation strength

$$K = 2\pi r_0 V_\theta. \tag{2.2.16}$$

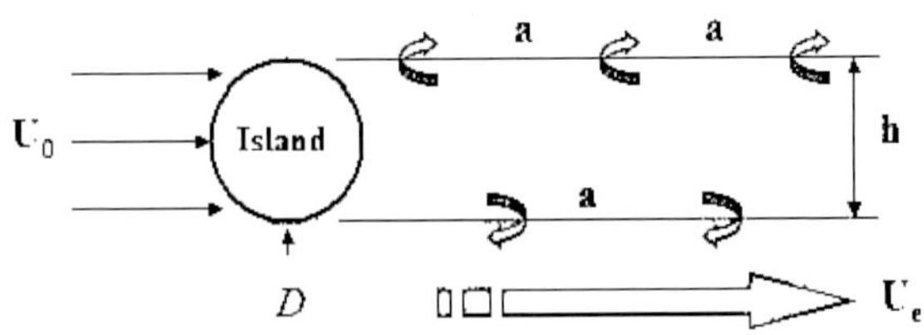


Fig. 2.12 Schema of physical model of OVT

The second equation is derived by Tsuchiya [1969]

$$(2B - A)\chi^2 + (2A - 3B)\chi + (B - A + B/4A) = 0, \qquad (2.2.17)$$

where $\chi = U_e/U_0$, $A = \coth(\pi h/a)$ and $B = \pi h/a$. In the case of interest, $h/a = 0.363$, so that $A = 1.229$, and $B = 1.139$. Substituting these parameters into (2.2.17) and taking a larger solution yields [Li *et al.*, 2000]

$$\chi = \frac{U_e}{U_0} = 0.885. \qquad (2.2.18)$$

The reasonable estimate of angular velocity should be $3 \times 10^{-5}\,\mathrm{s}^{-1}$. The maximum velocity radius has been determined as $r_0 = 4.70\,\mathrm{km}$. Substituting these parameters into (2.2.16) yields the vortex circulation strength $K = 4161\,\mathrm{m}^2\mathrm{s}^{-1}$. Substituting the value of $K$ into (2.2.15) yields

$$U_0 - U_e = 0.0749\,\mathrm{m}\,s^{-1}. \qquad (2.2.19)$$

Solving (2.2.18) and (2.2.19) yields $U_0 = 0.65\,\mathrm{m\,s}^{-1}$ and $U_e = 0.58\,\mathrm{m\,s}^{-1}$. On the other hand, the drifter trajectories in November shown in Figure 2.3 give an average mean flow velocity $(0.70 \pm 0.25)\,\mathrm{m\,s}^{-1}$, which coincides with the above estimate. Meanwhile, an estimate of the vortex shedding rate $F(= U_e/a)$ is derived as $2.57 \times 10^{-5}\,\mathrm{s}^{-1}$.

The Reynolds number is defined as [Kundu, 1990]

$$R_e = \frac{UD}{A_H}, \qquad (2.2.20)$$

where $U[=O(50\,\mathrm{cm\,s}^{-1})]$ and $D[= O(1 \times 10^6\,\mathrm{cm})]$ are the characteristic velocity of mean flow and length scale of island diameter, respectively, and $A_H$ is the horizontal turbulent viscosity. Thus, in the case of interest a reasonable estimate of the Reynolds number is $R_e = O(50\text{–}500)$, if $A_H$ is generally chosen as $1 \times 10^5\mathrm{cm}^2\,\mathrm{s}^{-1}$ to $1 \times 10^6\,\mathrm{cm}^2\,\mathrm{s}^{-1}$ for the mesoscale or sub-mesoscale ocean processes [Zheng and Yuan, 1989].

The Rossby number defined as

$$\varepsilon = \frac{U}{fL}, \tag{2.2.21}$$

where $f(=2\Omega \sin \Theta)$ is the local Coriolis parameter and $L$ is the length scale of phenomenon, is used to estimate the dynamical scale of a geophysical fluid phenomenon. From the Argos satellite-tracked drifter trajectory data, the mean angular velocity of vortices is determined as $(2.66 \pm 0.56) \times 10^{-5}\,\mathrm{s}^{-1}$. In the above subsection, the vortex shedding rate is estimated as $2.57 \times 10^{-5}\,\mathrm{s}^{-1}$ from ASAR images. The local Coriolis parameter $f(\Theta = 20^\circ)$ is $4.97 \times 10^{-5}\,\mathrm{s}^{-1}$. Thus, for an individual vortex the Rossby number $\varepsilon$ is $O(0.4)$, and $O(0.5)$ for the OVT.

### 2.2.8. *Discussion*

There are similarities between the atmospheric KVS and oceanic vortex train. However, their spatial scales are distinct. For example, the distance between two consecutive atmospheric vortices usually is $O(100\,\mathrm{km})$ [Thomson *et al.*, 1977; Tsuchiya, 1969; Turner and Warren, 1988], five times larger than that in the case of interest. The vortex life spans, estimated by the length of KVS divided by the vortex propagation velocity, are also distinct. In the atmosphere, it is on the order of a couple of hours [Li, *et al.*, 2000b]. In the studied ocean case it may last much longer, *e.g.* longer than five days. On the satellite images, they have distinct imagery signatures. The atmospheric KVS are usually visualized by zigzag low-level cloud patterns, but not for the OVT.

Satellite-tracked drifter trajectories and OVT signatures on ASAR images used here reveal that a strong current flows directly from the Pacific (Philippine Sea) to the SCS passing through the south Luzon Strait in November. The average surface current velocity is around $0.7\,\mathrm{m\,s}^{-1}$, and the direction is around 315°TN. Using the methods developed in this section, the angular velocity of ocean vortex can be calculated by the radar backscatter cross section profile taken along a vortex radial transect, if the spectral constant $m_3$

is independently determined. Then, the mean flow velocity can be calculated using the OVT dynamical models.

### 2.2.9. *Summary*

The Argos satellite-tracked surface drifter trajectory data and ENVISAT ASAR images are used to observe the OVTs in the Luzon Strait. Two cases of vortex trains that occurred in November, local fall, were observed in the lee of Babuyan Island. The first train of three cyclonic vortices showing up on drifter trajectories was distributed in the northwest of Babuyan Islands from 20° to 20.5°N and from 120° to 121°E. The second train consisting of two parallel rows, five pairs of alternating cyclonic–anti-cyclonic vortices were observed on the ASAR images of November 19, 2006. The second train occurred on the upstream side of the first one from 19.5° to 20.0°N and from 121.0° to 122.0°E. The total length of the vortex train along the axis reaches about 250 km. All vortices propagate northwestwards (~315°TN). The mean angular velocity is $(2.07 \pm 0.18) \times 10^{-5}\,\mathrm{s}^{-1}$.

Theoretical models of the ocean vortex radar image are derived from general radar imaging theories for ocean processes. These functional relations are used to extract dynamic parameters of the ocean vortex and the vortex train from ASAR imagery signatures. The results are as follows: The distances between two consecutive vortices and two rows of vortices are (22.6±1.9) and (8.2±1.2) km, respectively. The maximum rotational velocity radius is determined to be 4.70 km. The vortex rotational angular velocity $\omega_E$ is $O(3 \times 10^{-5}\, s^{-1})$, if the spectral constant $m_3$ is taken as $1.95 \times 10^{-3}$.

Dynamic analyses give reasonable estimates of OVT dynamic parameters. The current velocity of mean or undisturbed flow as $U_0 = 0.65\,\mathrm{m\,s^{-1}}$, which coincides with the average mean flow velocity $(0.70 \pm 0.25)\,\mathrm{m\,s^{-1}}$ measured by drifters, and the propagation velocity of vortex $U_e = 0.58\,\mathrm{m\,s^{-1}}$. The estimated vortex shedding rate $F(= U_e/a)$ is $2.57 \times 10^{-5}\,\mathrm{s}^{-1}$. The Reynolds number is $R_e = O(50\text{–}500)$, if $A_H$ is chosen as $1 \times 10^5\,\mathrm{cm^2\,s^{-1}}$ to $1 \times 10^6\,\mathrm{cm^2\,s^{-1}}$. For an individual vortex, the length scale is $O(5\,\mathrm{km})$, and the Rossby number is $O(0.4)$. For the OVT, the length scale is $O(20\text{–}50\,\text{km})$, and the Rossby number is $O(0.5)$, implying that vortex train observed in the Luzon Strait is sub-mesoscale in nature.

## 2.3. OVTs along the Kuroshio East of Taiwan

The Kuroshio Current is the WBC of the Pacific, which is originating from the North Equatorial Current, flows northward to the east of Taiwan year around. The pathway of Kuroshio is generally parallel to the shoreline of Taiwan Isalnd. Its width is about 100–150 km, with speeds of 1–1.5 m s$^{-1}$ [Liang *et al.*, 2013]. As shown in Figure 2.13, Taiwan Island faces eastward to the Pacific. Its east coast is a fault coast, off which isobaths of 2000 m are within only a couple of kilometers. To the southeast of the coast, there are two volcanic

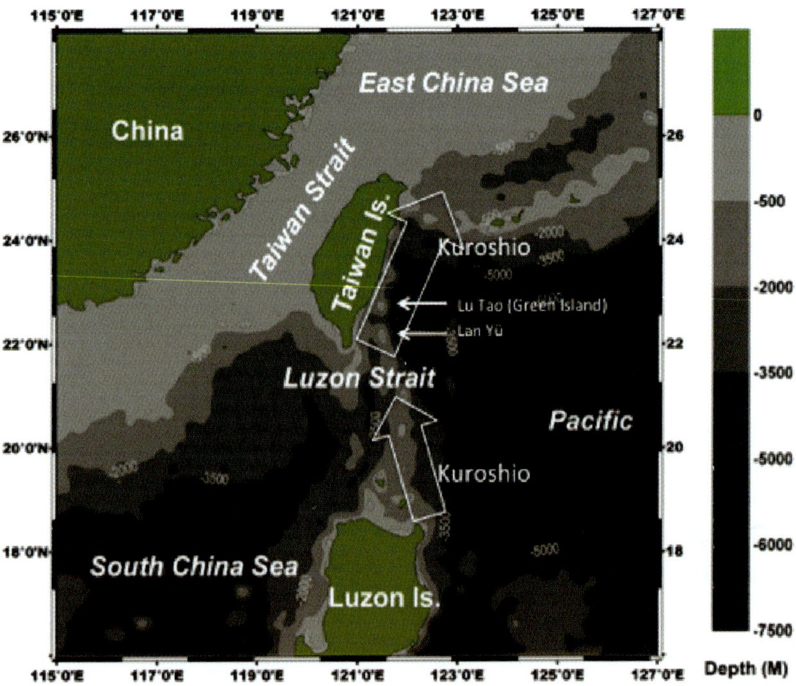


Fig. 2.13 Map of Kuroshio region east of Taiwan Island and the Luzon Strait White hollow arrows denote the Kuroshio pathway [Zheng and Zheng, 2014]

islands, Orchid Island (Lan Yü) in south and Green Island (Lü Tao) in north. Climatology of the Kuroshio velocity from *in situ* measurements reveals that the two islands are approximately located in the mainstream of the energetic Kuroshio. The two volcanic islands with maximum horizontal scales of 8 and 5 km look like two cylinders standing vertically within the Kuroshio mainstream and act as obstacles in the stream. These geographic and hydrodynamic backgrounds provide favorable conditions for generation of the OVT.

Figure 2.14 shows a composite ERS-1 SAR image of the OVTs of Orchid Island and Green Island taken with one week interval sequentially at 17:01UTC September 25 (lower) and 19:02UTC

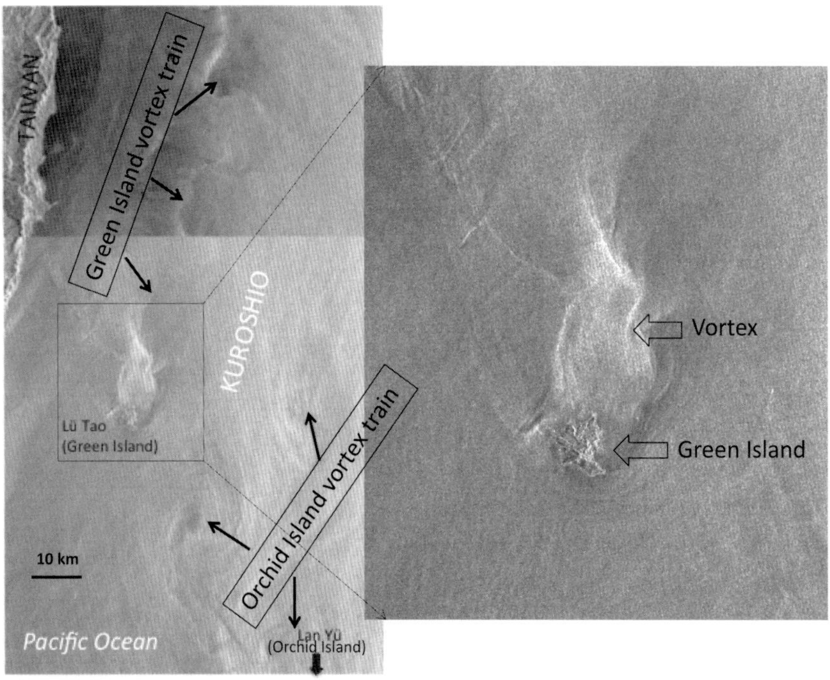


(a) (b)

Fig. 2.14 (a) Composite ERS-1 SAR image of OVTs of Orchid Island and Green Island taken at 17:01UTC September 25 (lower) and 19:02UTC on October 2 (top), 1996. (b) Zoomed Green Island and head of vortex train [Zheng and Zheng, 2014]

October 2 (top), 1996. One can see two independent OVTs. The one near the east coast of Taiwan Island, called Green Island Vortex Train, occurred downstream of Green Island, and consisted of three pairs of cyclonic vortexes appearing as bright imagery and anti-cyclonic vortexes appearing as dark imagery. From the image, we measured the data of two-dimensional structure of the OVTs. The horizontal scale of vortexes is about 10 km and the mean separate distance (wavelength) is 15 km. Another one a little far from the coast, called Orchid Island Vortex Train, occurred downstream of Orchid Island (a little south beyond the image), and consisted of three vortexes, two cyclonic (bright center) and one anti-cyclonic (dark center) vortexes of horizontal scale of about 15 km. The mean wavelength is 25 km.

Liang *et al.* [2013] studied vortex shedding due to the Kuroshio passing Green Island using the numerical model, which is based on the shallow-water equations with space-time least-squares finite element method. Chang *et al.* [2013] conducted 5-day cruise mission to survey the OVT in the lee of Green Island, and found a head vortex of the OVT. Zheng and Zheng [2014] investigated dynamic criteria to determine whether an OVT would occur in this region. A current velocity of $0.5\,\mathrm{m\,s^{-1}}$ is determined as a threshold for OVT generation. Analysis results also show that the OVTs within this region are featured by current-induced vortexes rather than wind-induced ones. In addition, the variability of OVTs shows a distinct seasonality, which is closely associated with the shoreward shift of Kuroshio mainstream driven by winter northeasterly monsoon. This scenario is verified by the vector-EOF (empirical orthogonal function) analysis during the weak OVT period in 2012. In addition to meandering of the Kuroshio, westward-propagating MEs and cross-passing typhoons also show contributions to the variability of OVTs with an intraseasonal timescale.

# Chapter 3

# Spirals on the Sea

## 3.1. Introduction

Spirals or spiral eddies are a type of sub-mesoscale ocean processes, 10–25 km in size and broadly distributed in the world's oceans [Munk *et al.*, 2000]. The distinction of spirals from sub-mesoscale vortex train examined in Chapter 2 is that spirals appear in the form of vortexes with single polarity, cyclonic overwhelmingly and vortex trains may appear in the form of vortex pairs with both polarities, cyclonic and anticyclonic [Zheng *et al.*, 2008].

The first photographs of spiral eddies were taken by astronauts on Apollo-Saturn mission 7 launched on October 11, 1968. Ten years later, in 1978, SEASAT SAR generally confirmed the discoveries from crewed space fights. The earliest planned observations of spiral eddies took place in 1981 on the first space shuttle flight of the Columbia [Stevenson, 1989]. An eddy in the Gulf of Oman had been observed on NOAA infrared imagery, and the crew of STS-1 was briefed to look for it. On the first day of the mission, the commander of the Columbia observed a series of eddies in the Gulf of Oman [Stevenson, 1989] and these compared very closely with a NOAA infrared image obtained a few hours after the Shuttle transit. On the very next flight SIR-A, SAR imaged a well-formed cyclonic spiral in the Caribbean. Photographs on the STS-8 mission revealed a field of cyclonic spiral eddies in the South Indian Ocean, each turning in a direction opposite to the anticlockwise rotation in the Northern Hemisphere [Stevenson, 1998, 1999; Munk *et al.*, 2000].

Of formation mechanisms, the spiral features shown on satellite images reveal that their formation is closely associated with instability of a shear flow as observed by laboratory experiments [Brown and Roshko, 1974; Winant and Browand, 1974] and described by fully nonlinear theories [Stuart, 1967]. Starting from parallel shear flow with an inflection point, Corcos and Sherman [1976, 1984] developed a numerical model to simulate the formation process. Qiao *et al.* [2005] carried out a cruise observation project for investigating the fine structure of circulation in the southern ECS. During the cruise, they found three interconnected spirals and gained their three-dimensional structure from cruise observations.

## 3.2. Ocean Spirals on Satellite Images

### 3.2.1. *Space Shuttle photographs*

The US space shuttles were equipped with a NASA modified Hasselblad 500 EL/M 70-mm camera. On some flights, an Aero Linhof Technika 45 camera was carried. With these cameras, the shuttle astronauts took photographs of interesting geophysical events around the world, including ocean phenomena. The shuttles flew at a typical altitude of 300 km ranging from 200 to 400 km. The resolutions of photographs are about 20–50 m depending on 250 or 100 mm focal lengths of lenses used [Zheng *et al.*, 1993]. The US NASA Johnson Space Center in Houston, Texas, archives nearly one million such still photographs. From the archives, Munk *et al.* [2000] collected 400 photographs containing spiral eddies, which are broadly distributed in the world's oceans. Most of the photographs were taken exploiting the sunglitter on the sea surface. Figure 3.1 shows a photograph coded STS41G-35-94 taken by astronauts on shuttle mission STS41G on October 7, 1984, a typical example of Munk's collection of spiral photographs. Figure 3.1(a) shows that the photograph is centered at 26.0°E 32.0°N in the eastern Mediterranean Sea off Egypt, which was under the low wind condition. Figure 3.1(b) shows that this sunglitter scene is full of bright streaks that comprise complex streamline

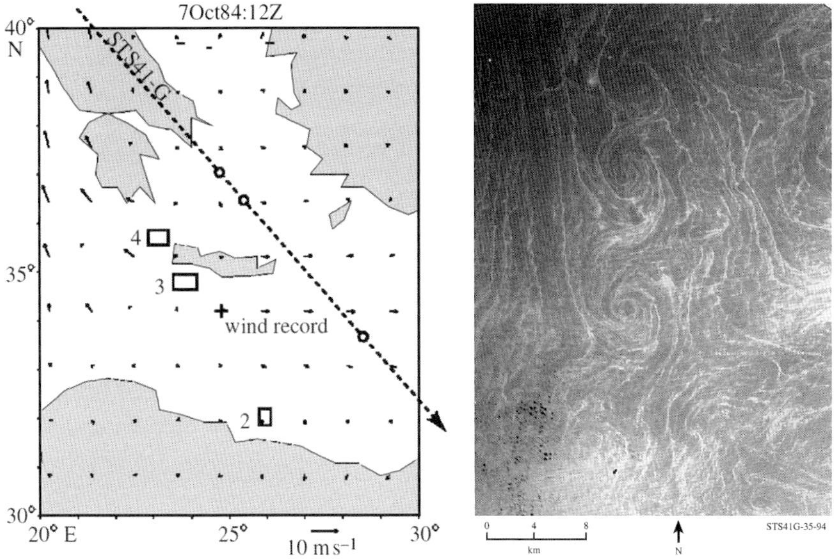


(a) (b)

Fig. 3.1 (a) Ground track of the US space shuttle mission STS41G crossing the eastern Mediterranean Sea Box 2 shows the ground coverage of (b). (Credit: US NASA). (b) Photograph STS41G-35-94 taken by astronauts with handheld Hasselblad 500 EL/M 70-mm camera on October 7, 1984 (Credit: US NASA)

patterns. Among them a striking feature is a spiral train consisting of four interconnected spirals. The average horizontal length scale of spirals is 5 km and the average separate distance is 10 km. All four spirals rotate cyclonically.

It is believed that the complex streak patterns including spirals result from surface signature of differential roughness, which depends on the straining and damping of capillary and short gravity waves. On the other hand, the surfactant and associated surface films dissipate capillaries and the shortest gravity waves. Published values for film thicknesses required to dampen capillary waves and produce a visible effect are of the order of 0.01–0.1 $\mu$m. On a smooth surface, the sun appears as a single reflection. While on a roughed surface, the sunglitter is formed by the reflection from broadly distributed wave facets. The glitter pattern depends on the wave slope statistics [Cox and Munk, 1954]. This is heavily biased towards capillaries and short gravity waves, which are damped by surfactants. Differentially smooth lines will be dark in the outer glitter, where reflection into the camera requires large RMS slopes and bright in the inner glitter. The shortest capillary wavelength greatly exceeds the wavelength of light, thus the glitter is governed by geometric optics [Munk *et al.*, 2000].

Microlayer surfactants on the ocean surface originate mainly from two natural sources: the biological activities of plankton and fish, and petroleum oil seepage [MacDonald *et al.*, 1993] and two man-made sources: polluted river discharges and oil spilling (see Figure 3.3(b). Natural biogenic surface films are found in all parts of the oceans [Peltzer *et al.*, 1992], but by no means are they evenly distributed. There is a relatively high level of occurrence in marginal seas and a relatively low level in the Southern Hemisphere [Bresciano *et al.*, 1998]. The films are detected visually at light winds. Off the Norwegian coast, Espedal *et al.* [1998] noted 40% of surfactant coverage for 2.5 m $s^{-1}$ winds, but only 5% for 5–10 m $s^{-1}$ winds. The coverage is below 1% for winds exceeding 12.5 m $s^{-1}$. Thus, sunglitter, surfactants and low wind conditions comprise three favorable conditions for spirals to be imaged by visible band sensors.

### 3.2.2. *ERS-1 SAR images*

As documented in Section 1.4 in Chapter 1, the intensity of radar return signal from the ocean surface depends on the roughness parameterized by wavenumber spectral density of the ocean surface wave field, which satisfies the Bragg resonance scattering condition. The X band radar resonates with capillaries (wavelengths 1.8–3.2 cm for 60° to 30° angles of incidence), C band with very short gravity waves (3.3–8.7 cm) and L band with short gravity waves (14–25 cm). All these short waves are damped by the surface film, which is advected by the ocean surface current [Apel *et al.*, 1998; Gasparovic *et al.*, 1988; Lyzenga and Marmorino, 1998]. Thus, as active sensors, SARs are capable of detecting spirals on the sea no matter of sunglitter.

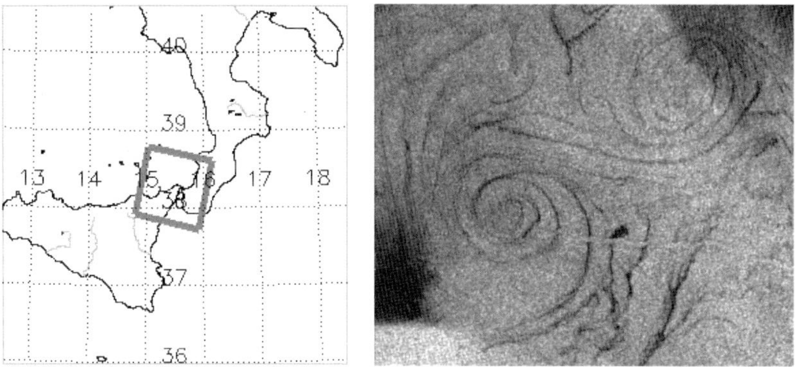


(a) (b)

Fig. 3.2 (a) Map of the Sicily Island Box at the center shows the coverage of ERS-1 SAR image 2853 in (b). (b) Zoomed spiral pair imagery on ERS-1 SAR image 2835 of Sicily Island Imaging time was 09:41 UTC on September 12, 1993 (Credit: ESA)

Figure 3.2 shows the location and zoomed spiral imagery on ERS-1 SAR image (frame number 2835) taken at 09:41 UTC on September 12, 1993. The image is centered at 38°16′N 15°29′E in the Mediterranean Sea north of the Sicily Island. The zoomed image with dimensions of 15 by 13 km shows an eye-glass-like spiral pair. From dark streamlines, one can see that both spirals rotate anticlockwise. The average horizontal length scale of spirals is 4 km and the separate distance is 8 km, which are close to the case shown in Figure 3.1(b).

Figure 3.3 shows the ERS-1 SAR image (frame number 0891) taken at 18:54 UTC on October 12, 1993. The image is centered at 44°45′N 49°03′E in the Caspian Sea south of the Volga River Estuary. The most remarkable features on this image are the mushroom-like features, which may be treated as an initial stage of spiral evolution. The Volga River flows passing through developed agricultural area, thus the river water carries a heavy load of pollutants originating from fertilizers washed out from agricultural fields and municipal plants. The pollutants serve as nutrients for the marine organisms, which produce biogenic surfactants. They are advected by the current field associated with eddies and thus act as tracers as shown in the image.

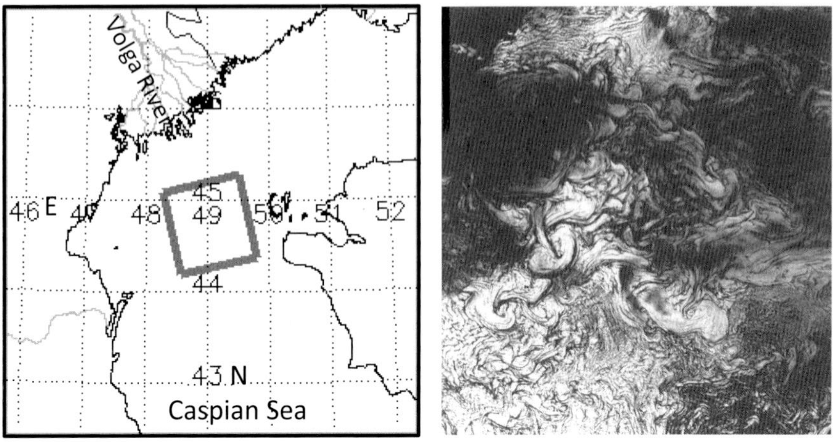


(a) (b)

Fig. 3.3 (a) Map of the northern Caspian Sea Box at the center shows the coverage of ERS-1 SAR image 0891 in (b). (b) ERS-1 SAR image of the Caspian Sea south of the Volga River Estuary Imaging time was 18:54 UTC on October 12, 1993 (Credit: ESA)

## 3.3. Formation Mechanism: The Stuart Solution

As mentioned in Section 3.1, several theories may explain the formation of spiral on the sea. Here, we brief the Stuart solution [Munk *et al.*, 2000; Stuart, 1967].

Taking the simplest case of a non-rotating, incompressible equal-density fluid,

$$\frac{D\boldsymbol{v}}{Dt} = -\rho^{-1}\nabla p \tag{3.3.1}$$

and

$$\nabla \cdot \boldsymbol{v} = 0, \tag{3.3.2}$$

these can be combined to yield the conservation of vorticity

$$\frac{D\zeta}{Dt} = \left(\frac{\partial}{\partial t} + \mathbf{v}\cdot\nabla\right)\zeta = 0 \tag{3.3.3}$$

with

$$\zeta = \frac{\partial v}{\partial x} - \frac{\partial u}{\partial y} = -\nabla^2\psi, u = \frac{\partial \psi}{\partial y}, v = -\frac{\partial \psi}{\partial x}, \tag{3.3.4}$$

where $\psi$ is the stream function.

An exact steady-state solution ($D\zeta/Dt = 0$) by Stuart [1967]

$$\psi(x, y) = -k^{-1}U \log q, \zeta = +kU\left(1 - \alpha^2\right)q^{-2}, \tag{3.3.5}$$

$$q = \cosh ky - \alpha \cos kx \tag{3.3.6}$$

yields the well-known cat's-eye configuration (Figure 3.4). The stagnation streamlines (the braids) separate the closed interior circulation (the core) from the only slightly perturbed exterior streamlines. $\alpha$ is an arbitrary parameter, $0 \leq \alpha \leq 1$. For $\alpha = 0$,

$$u(y) = -U \tanh ky, v = 0, \zeta = kU/\cosh^2 ky \tag{3.3.7}$$

going from $u(\pm\infty) = \mp U$ ($U$ is positive). Most of the current shear is between $ky = \pm 1$. As $\alpha \to 1$, we have $q \to 0, \zeta \to \infty$ at the core center, yielding concentrated point vortices.

We use the following numerical magnitudes for scale estimates:

$$k = 3 \times 10^{-4}\mathrm{m}^{-1}, \quad \zeta_0 = 10^{-4}\mathrm{s}^{-1},$$

where $\zeta_0 = kU$ is the vorticity at $y = 0$ for $\alpha = 0$. This implies a distance $2\pi/k \approx 20\,\mathrm{km}$ between stagnation points and a time scale of $10^{-4c}\,\mathrm{s}^{-1}$ for winding up the spirals (or about one day per revolution). The initial shear zone of width $2k^{-1} = 7\,\mathrm{km}$ separates the velocity differential of $\pm U = \pm 0.3\,\mathrm{ms}^{-1}$.

We now convert to dimensionless coordinates, referring distances to $k^{-1}$ and times to $\zeta_0^{-1}$. At the stagnation points, $x = \pm\pi, y = 0$

$$q = 1 + \alpha, \quad \zeta = (1 - \alpha)/(1 + \alpha), \tag{3.3.8}$$

whereas at the core center $x = 0$, $y = 0$,

$$q = 1 - \alpha, \quad \zeta = (1 + \alpha)/(1 - \alpha). \tag{3.3.9}$$

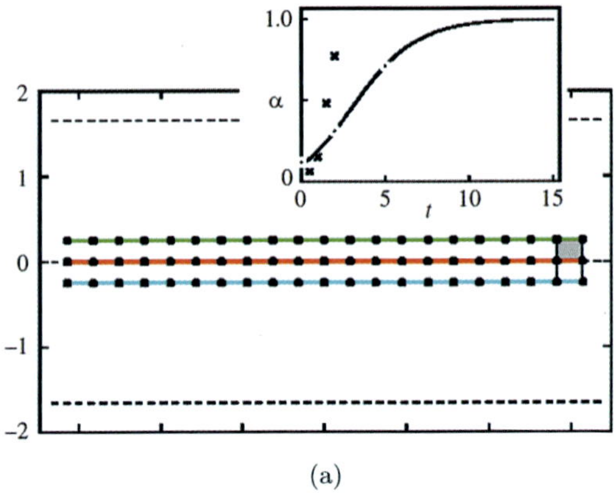


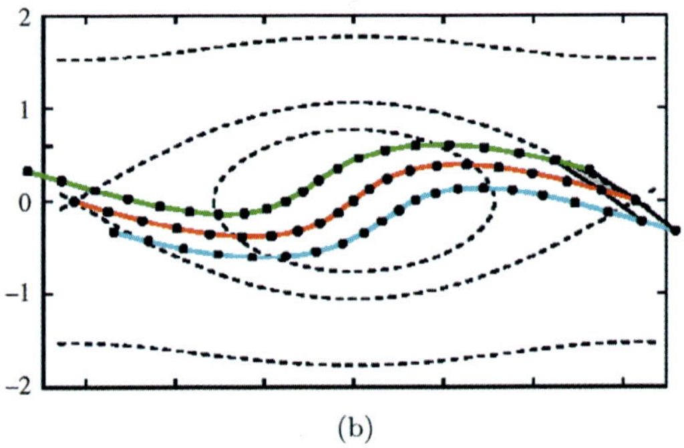


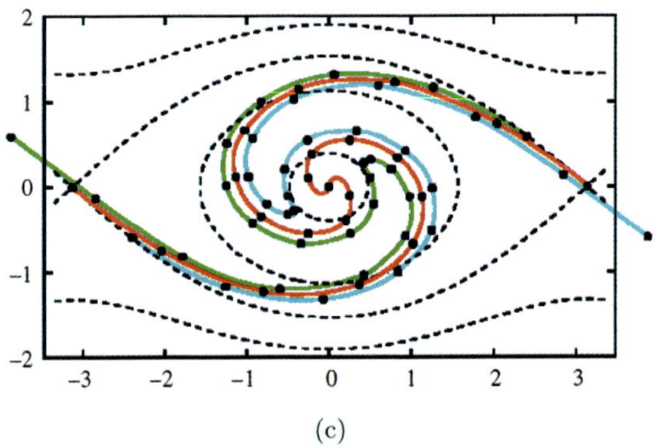


Fig. 3.4 Particle distribution in a Stuart vortex particles are initially placed between stagnation points $x = \pm\pi$ along the three lines $y = 0, \pm 0.25$. Subsequent positions are computed for $\alpha(t)$ determined by the incipient stability calculations of Corcos and Sherman [1976]. Particles are concentrated along the widening braids and injected into the core. The deformation of a rectangle formed by four particles (including the fixed stagnation point) is shown. (a) $t = 0, \alpha = 0.12$; (b) $t = 2, \alpha = 0.31$; (c) $t = 5, \alpha = 0.71$ [Munk *et al.*, 2000]

The tilt $\theta$ of the streamlines at the stagnation points is given by

$$\tan\theta = \pm\sqrt{\alpha} = \sqrt{(\zeta-1)/(\zeta+1)}$$

[Milne-Thomson, 1960]. The local inclination of the stagnation streamlines (the 'braids') is given by

$$\sin\theta = \sqrt{\frac{\alpha}{1+\alpha}}\sin(x/2). \tag{3.3.10}$$

The velocity component and the rate of strain along the braid are

$$V_{\text{braid}} = 2\sqrt{\frac{\alpha}{1+\alpha}}\cos(x/2), \tag{3.3.11}$$

$$\gamma = \sqrt{\frac{\alpha}{1+\alpha}}\sin(x/2)\left[1-\frac{\alpha}{1+\alpha}\sin(x/2)\right]^{1/2}, \tag{3.3.12}$$

with the principal strain axis directed along the braids.

Certain integral properties follow readily. The integrated vorticity per unit wavelength $2\pi/k$ is given by the circulation

$$\Gamma = 4\pi U/k \tag{3.3.13}$$

with a fraction

$$\Gamma_c/\Gamma = 4\pi^{-1}\tan^{-1}\sqrt{\alpha} \tag{3.3.14}$$

within the stagnation streamlines (the core circulation).

In the two-dimensional flow, the vorticity can be redistributed but not creatable, yet the appearance of core vorticity $(1+\alpha)/(1-\alpha)$ exceeds unit vorticity of the initial shear flow. Away from the concentrated cores the flow field depends on the integrated vorticity $\Gamma$, not on its distribution or maximum value. From the above analysis, one can see that the Stuart formalism can provide a clear dynamic framework for exploration of the spiral formation process with growing shear instability.

## 3.4. Symmetry Broken in Favor of Cyclonic Rotation

As mentioned in Section 3.1, the distinction of spirals from submesoscale vortex train is that spirals appear in the form of vortexes with single polarity, cyclonic overwhelmingly. This phenomenon is termed by "breaking symmetry". In order to explain the dominance of cyclonic vortices, previous investigators have raised several scenarios [Hoskins and Bretherton, 1972; MacVean and Woods, 1980; Munk *et al.*, 2000; Woods *et al.*, 1977] and developed numerical simulations [Métais *et al.*, 1995]. Among them, Munk *et al.* [2000] provided the simplest, easy-understandable explanation. It follows from the Rayleigh criterion of stability (extended to include Coriolis acceleration) that cyclonic circular vortices are stable and anticyclonic vortices are unstable. This leads to an "inertial instability" criterion $Ro < -1$, which goes back to Pedley [1969]. There is ample numerical evidence for instabilities that impede the development of anticyclonic vortices but have no such effect on cyclonic vortices. [Lesieur *et al.*, 1991; Potylitsin and Peltier, 1998]. The distinction is equally clear in laboratory experiments [Bidokhti and Tritton, 1992]. Kloosterziel [1990] observed that "it is virtually impossible to create anticyclonic vortices ... by simply stirring the fluid locally. But cyclonic vortices are easily created in this way! In the anticyclonic case stirring ... leads to turbulent motion and the generation of waves whereas by cyclonic stirring a well-defined smooth vortex forms".

## 3.5. Three-Dimensional Structure Derived from Cruise Observations

### 3.5.1. *SAR image of spirals in ECS*

Figure 3.5 shows an ERS-1 SAR image (frame number 3087–3105) taken at 02:26 UTC on July 23, 1994. The image is centered at 25°20′N 121°56′E in the ECS north of the Taiwan Island. The top half of image shows two southwest-northeast aligned, cyclonic spirals. The average horizontal length scale of spirals is about 50 km and the separate distance is 80 km, which are much greater than that in the cases shown in Figures 3.1–3.3. The large dark area in the lower right-hand section of the image shows an upwelling area northeast of Taiwan, which is induced by interaction of the intruded Kuroshio current with the shoaling bottom topography.

Later, in summer 2003, Qiao *et al.* [2005] carried out cruise observation with Conductivity-Temperature-Depth profilers (CTDs)

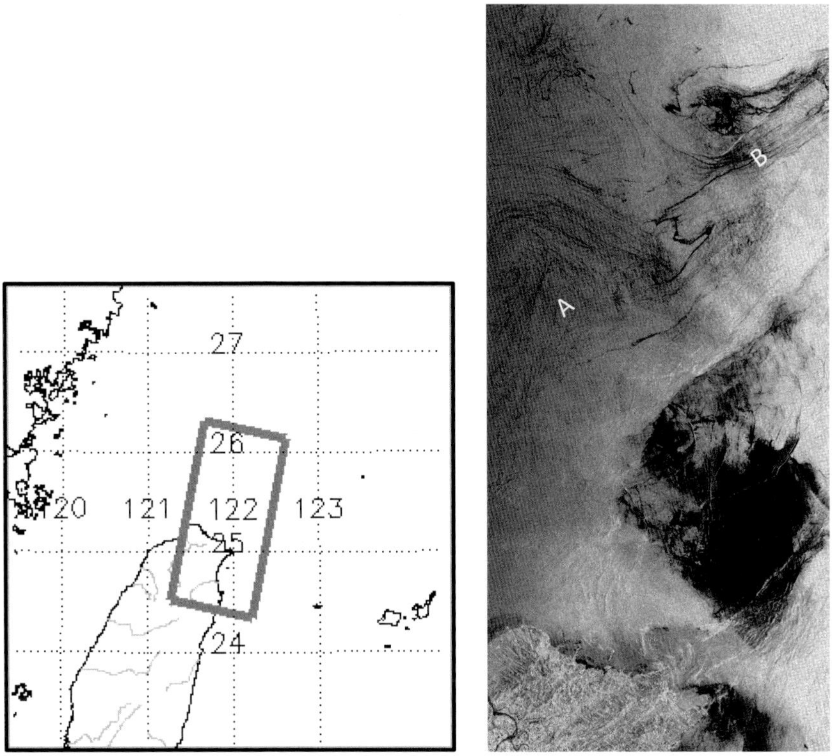


(a) (b)

Fig. 3.5 (a) Map of the southern ECS and (b) ERS-1 SAR image 3087–3105 north of the Taiwan Island Box at the center of (a) shows the coverage of the image. Imaging time was 02:26 UTC on July 23, 1994 (Credit: ESA)

and a moored Acoustic Doppler Current Profiler (ADCP) on the southern ECS. Their observation area covered spiral B shown in Figure 3.5(b) and downstream spirals beyond the SAR image as shown in Figure 3.6. Thus, we obtain the best chance to gain insight into the three-dimensional structure of spirals.

### 3.5.2. *Study area and cruise observation*

The ECS is a marginal sea located in the northwest Pacific. It borders on the East China coast on the west side, and separates from the Pacific by the Ryukyu Islands and Kyushu Island on the east side.

Its water body connects with the Yellow Sea in the north, with the Sea of Japan through the Korea Strait and Tsushima Strait in the northeast, and with the South China Sea through the Taiwan Strait in the southwest. The total water area of the sea is 770,000 km$^2$. The bottom topography is characterized by a wide continental shelf with an average depth of about 300 m extended from the East China coast, and a southwest–northeast running deep trough, Okinawa Trough, with a maximum depth of 2719 m on the south and east sides. The Kuroshio, a western boundary current of the Pacific, flows northeastward through the Okinawa Trough all the year round.

The Kuroshio plays a dominant role in the ECS circulation system [Guan, 1985; Guo *et al.*, 2003; Yuan *et al.*, 2001; Zheng and Klemas, 1982]. Owing to interaction of the Kuroshio with the shelf water and bottom topographic effects, the mesoscale and sub-mesoscale processes in the ECS are active. The typical examples include mesoscale eddies [Zheng *et al.*, 1990], spirals, the shear waves [Zheng and Xiu, 1991], upwelling [Hsu *et al.*, 1995], and the branching of the Kuroshio [Hsueh *et al.*, 1993].

In summer 2003, a cruise observation project for investigating the fine structure of the continental circulation was carried in the southern ECS from July 30 to August 10. The observation area is located at 25.5°–27.5°N 122.0°–124.5°E as shown as a shaded box in Figure 3.6. During the cruise observation, the studied area was scanned by CTDs along seven transects from the bottom to the surface with a horizontal resolution of about 35 km. At station C shown in an inset of Figure 3.6, a moored shallow water ADCP (150 kHz) was employed to measure the horizontal current velocities at depths from 16 to 104 m with a vertical resolution of 4 m.

### 3.5.3. *Water temperature distribution*

Cruise-observed water temperature distribution at depth of 50 m is shown as a color-coded image in the inset of Figure 3.6. One can see a sandwich pattern, which shows two high temperature regions with the highest temperature of 28°C, one is on the shallow water side, the other on the deep water side, and a low temperature region with

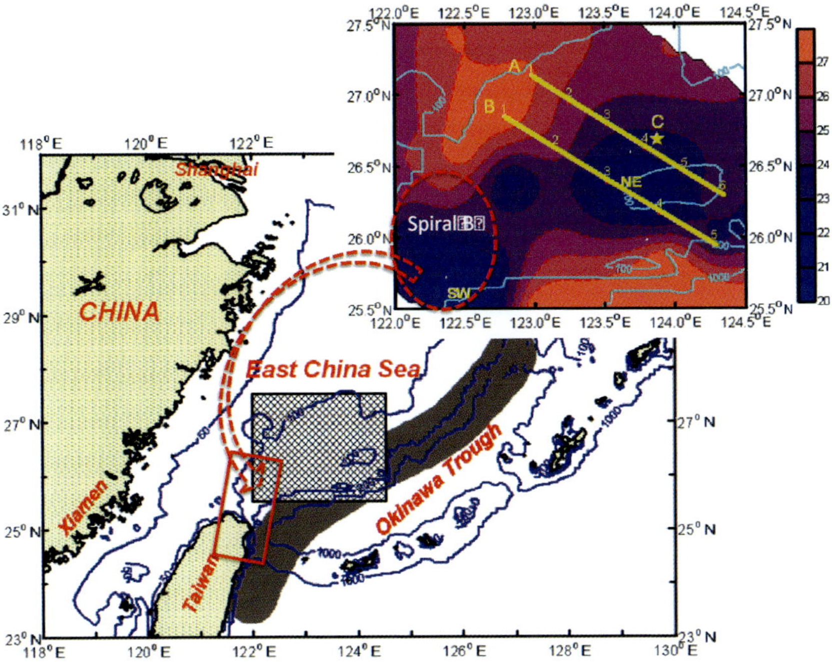

Fig. 3.6 Cruise observation area and the temperature distribution at 50 m (inset) A and B are two CTD transects, and C is the ADCP mooring station. Dark belt shows the Kuroshio path flowing through the Okinawa Trough. Cold eddy SW corresponds to spiral B on ERS-1 SAR image 3087–3105 (Figure 3.5) covering the area in red box [Qiao *et al.*, 2005]

the lowest temperature of 22°C between the two. The temperature difference between the highest and the lowest temperatures is as high as 6°C. Within the low temperature region, there are three low temperature centers. They are centered at 25.8°N 122.5°E in the southwest of the investigation area (marked by SW on the image), 26.4°N 123.9°E (marked by NE) and 26.4°N 122.9°E (in the middle), respectively. The two larger centers are in the size of about 80 by 70 km (with 24°C isotherm as a boundary). The average separate distance is about 80 km.

The vertical thermal structure of low temperature center NE, which was measured along transects A and B, is shown in Figure 3.7. One can see that on the shallow water side represented by stations

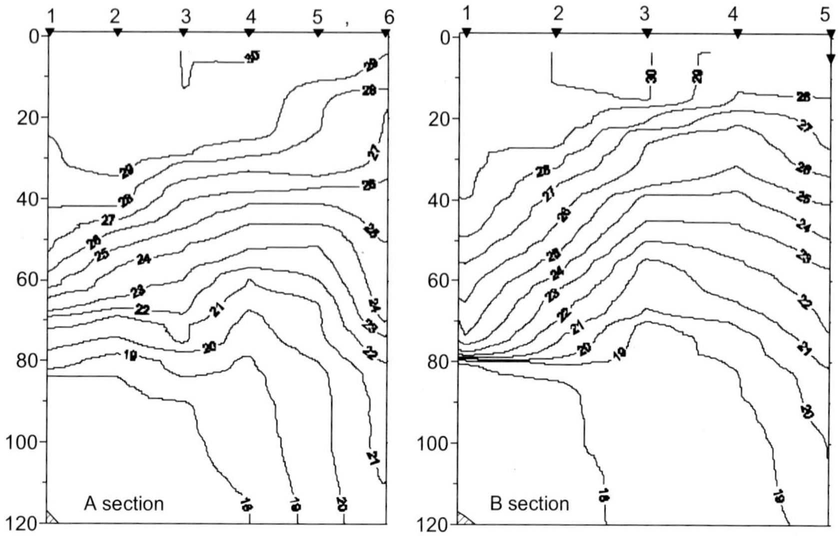


Fig. 3.7 Vertical distributions of water temperatures measured along transects A and B

A1 and A2, the ocean stratifies into three layers: an upper mixed layer with a uniform temperature of 29°C occupies the upper 30 m layer, a thermocline with temperature decreasing from 29°C to 18°C occupies the middle layer from 30 to 84 m, and a bottom layer with temperature of 18°C occupies the lower layer from 84 m to the bottom. The isotherms measured at all other stations show a protruding pattern. At station B4, isotherms of 25°C, 26°C, 27°C and 28°C are elevated as high as 35 to 40 m compared with water temperatures on the shallow water side. These thermal features imply that the low temperature center NE, in fact, is a cold core vortex [Parker, 1971; Cheney and Richardson, 1976].

### 3.5.4. *Horizontal current velocity*

During the cruise, a shallow water type ADCP was moored at 28.1908°N 125.3109°E to measure the horizontal current velocity vectors for 24 h. The mooring station C marked in the inset of Figure 3.6 is located near the outer edge of cold core vortex NE.

The water depth is 110 m. The measurements were taken from 16 to 104 m with a vertical interval of 4 m (total 23 layers). The sampling time interval is 5 mins. The observed horizontal velocity vectors are shown in Figure 3.8(a).

In order to extract the vertical structure of the mean current from the observations, the raw data are processed by the following procedures. First, the observed horizontal current velocity vectors are analyzed into the zonal component$u$ and the meridional component $v$. Then each component is further analyzed into

$$u(t, z) = u(z) + \tilde{u}(t, z) + \delta(t, z), \tag{3.5.1}$$

where $u(t, z)$ is the observed velocity, $u(z)$ is the mean current velocity, $\tilde{u}(t, z)$ is the tidal current velocity and $\delta(t, z)$ is a residual value. Using the fast fourier transform (FFT) method, the tidal current velocity is approximately determined as a sum of all components with the periods from 6 to 25 h. The mean current velocity is approximately calculated from

$$\overline{u(z)} = \frac{1}{T}\int_0^T [u(t, z) - \tilde{u}(t, z)]dt, \tag{3.5.2}$$

where $T$ is the observation period (24 h). The results are shown in Figure 3.8(b).

One can see that the amplitudes of daily averaged horizontal velocity vectors reach a maximum value of 30 cm s$^{-1}$ at the depth of 16 m, the $u$ component crosses zero at the depth between 40 and 44 m, and the vectors rotate with the depth, implying a $\beta$-spiral-like feature [Stommel and Schott, 1977; Schott and Stommel, 1978]. From Figure 3.8(b), the vertical velocity component $w$ can further be calculated using the following equation [Huang, 1988]

$$w = \frac{f\rho_o}{g}\frac{u^2 + v^2}{\rho_z}\frac{d\theta}{dz}, \tag{3.5.3}$$

where $f$ is the Coriolis parameter, $\rho_o$ is the water density at the sea surface, $g$ is the gravitational acceleration, $u$ and $v$ are the zonal and meridional velocity components, respectively, $\rho_z = d\rho(z)/dz$, $z$ is the

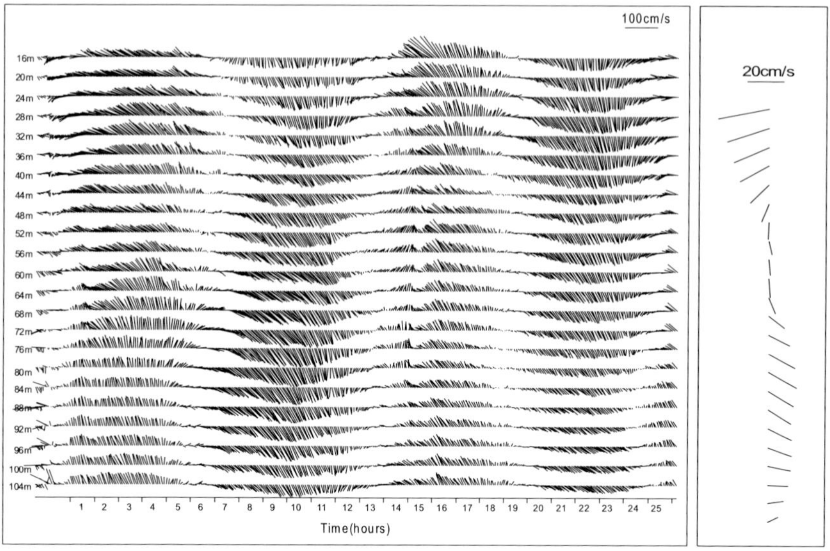


(a) (b)

Fig. 3.8 (a) ADCP-observed horizontal velocity vectors at station C and (b) daily average horizontal velocity vectors from 16 to 104 m [Qiao *et al.*, 2005]

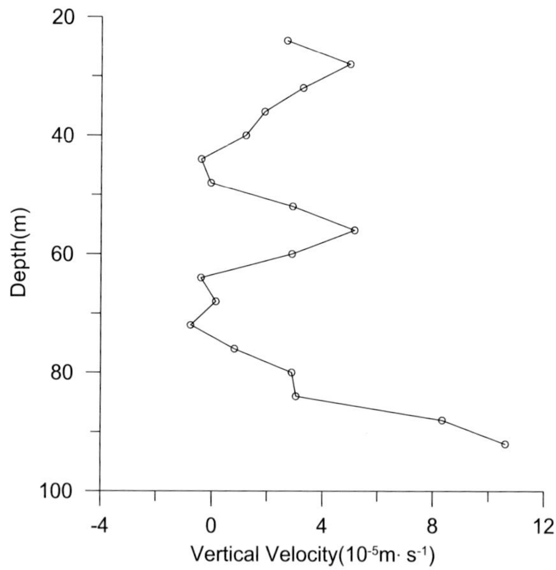


Fig. 3.9 Estimated vertical velocities at station C based on Eq. (3.5.3) in the text [Qiao *et al.*, 2005]

vertical coordinate axis upward positive, and $\theta$ is defined as

$$\theta = \tan^{-1}(v/u). \tag{3.5.4}$$

The calculated results of $w$ are shown in Figure 3.9. One can see that $w$ takes its maximum values $5 \times 10^{-5}$ and $10 \times 10^{-5}\,\mathrm{m\,s^{-1}}$ at the surface layer and bottom layer, respectively. At the middle layer, $w$ is variable. But, at the majority of depths, $w$ is positive, further confirming the low temperature center as a cold core vortex.

## 3.6. Summary

Spirals showing up on space shuttle photographs and satellite SAR images are a type of sub-mesoscale ocean dynamic processes. From observed cases, we determine that the characteristic horizontal scale on the sea surface is in the order of 5–50 km. They appear in pairs or as vortex train with the characteristic separate distance (wavelength) in the order of 10–80 km.

The analysis results of the cruise data taken on the continental shelf of the southern ECS in summer 2003 reveal a chain consisting of three low temperature centers, which were distributed along the continental slope side front of the Kuroshio. The location of the cold center SW is corresponding to a spiral on ERS-1 SAR image 3087–3105. Based on this connection, we interpret that the three low temperature centers comprise a spiral train. The ocean current data measured at downstream cold center NE indicate that it has $\beta$-spiral horizontal velocity structure, implying that spirals on satellite images are indeed spirals in the ocean interior.

Spirals appear in the form of vortexes with single polarity, cyclonic overwhelmingly, *i.e.*, breaking symmetry. Cruise observations further confirm the existence of this phenomenon.

# Chapter 4

# Ocean Near-Inertial Oscillations

## 4.1. Introduction

The near-inertial motion in the ocean is a specific feature of the oscillations with frequencies near the local inertial frequency $f$ ($= 2\Omega \sin\theta$, the Coriolis parameter, where $\Omega$ is the angular velocity of the earth rotation, and $\theta$ is the local latitude). The estimated average velocity amplitude of near-inertial oscillation (NIO) in the global oceans is about $10\,\mathrm{cm\,s^{-1}}$, greater in the mid-latitude areas than that in the tropical and polar areas [Chaigneau *et al.*, 2008; Park *et al.*, 2005]. Extreme cases of strong NIOs could have the velocity amplitude as high as $2.0\,\mathrm{m\,s^{-1}}$ [Teague *et al.*, 2007]. With such universal distribution and strong intensity, the NIO constitutes a significant component of the upper ocean dynamics and may have remarkable influence on variation of sea surface level and turbulent mixing [Sun *et al.*, 2015].

A major feature of NIO motion is that the two velocity components, $u$ and $v$, are close in magnitude, thus, the trajectories of water particles are nearly circular, rotating clockwise (counter-clockwise) in the northern (southern) hemisphere. In other words, the vortexes induced by the NIOs have single polarity, anti-cyclonic (cyclonic) in the northern (southern) hemisphere. Interestingly, this feature distinguishes the NIOs from sub-mesoscale vortex train examined in Chapter 2 and spirals in Chapter 3. The vortex train may appear in the form of vortex pairs with both polarities, cyclonic and anti-cyclonic, and spirals appear in the form of vortexes with single polarity, cyclonic overwhelmingly, in both hemispheres.

Storms and hurricanes (typhoons in the East Asia) that pour a huge amount of energy into the ocean for a very short duration are the most important trigger mechanism of NIOs. Thus, the performance of the NIO is controlled by the parameters of the preceding atmospheric event including wind speed, moving speed and spatial scale [Shay and Elsberry, 1987; Xie *et al.*, 2009; Zheng *et al.*, 2006a]. In addition to external forcing, ocean conditions, such as stratification and background mean flow, also have strong influence on NIO propagation and evolution. The characteristics of NIOs in a mean flow with or without geostrophic flow shear have been analyzed under the theory of linear waves [Kunze, 1985; Mooers, 1975; Orvik and Mork, 1995]. The Doppler shift is supposed to happen due to the wave–current interaction, and the background vorticity modifies the frequency band of NIOs [Sun *et al.*, 2011].

The SCS, the largest marginal sea of the western Pacific, has average 15 typhoon visits annually, therefore, has frequent NIO occurrences. Meanwhile, the SCS has complex topography, variable hydrographic structure and circulation [Hu *et al.*, 2000] due to the alternating monsoon winds, causing the NIO to have variable characteristics. In recent years, more and more NIO events in the SCS have been reported [Chen *et al.*, 2013; Chu *et al.*, 2000; Sun *et al.*, 2011, 2012; Zhang *et al.*, 2005, 2014]. The parametric sub-harmonic instability (PSI) is believed to be an important reason for the NIO generation or intensification at specific latitudes [Sun *et al.*, 2011]. The nonlinear wave–wave interaction between NIO and internal tides may restrain the NIO growth and accelerate its damping [Guan *et al.*, 2014]. So far the mooring equipped with current profilers or current meters has been the most important method for NIO observation, due to its advantages in detecting and resolving NIO processes for long-term and continuous sampling. Numerical modeling has been also used to simulate NIO events in the SCS [Kuo *et al.*, 2011].

On the Atlantic side, the Gulf of Mexico (GOM) is a pathway of tropical hurricanes. The intensity of tropical hurricanes usually much exceeds that of mid-latitude storms. For example, the peak wind speed of 1998 Hurricane Georges reached 67.0 m s$^{-1}$ [Zheng *et al.*, 2006a]. Thus, this chapter will use the GOM as a test area to demonstrate the dynamical features and satellite SAR imaging of NIOs induced by hurricanes and storms.

## 4.2. NIO Signals from Mooring Observations

### 4.2.1. *Test area and field program*

The GOM is a semi-closed, sub-tropical ocean area. The bottom topography is characterized by a deep basin with an average depth of 2000 m and extended continental shelves on the northern, eastern and southern sides. The most famous feature of regional oceanography is the Loop Current, which originates from the Caribbean Sea and enters the GOM through the Yucatan Channel. In the eastern basin, the current takes the loop shape and flows out of the GOM through the Straits of Florida. Diurnal tide is the major tidal component along the most coast of the GOM. The GOM is located at the pathway of tropical storms and hurricanes. A typical example is Hurricane Katrina in 2005. It was the costliest natural disaster and one of the five deadliest hurricanes in the history of the US. At least 1,245 people died in the hurricane and subsequent floods. Total property damage was estimated at $108 billion (2005 USD). The test site is the northern continental shelf of the GOM near the Mississippi River Delta as shown in Figure 4.1. The SAR imaging area is on the western side covering water depths from 50 to 3000 m. Two mooring stations of the field program were located in the DeSoto Canyon centered at 29°N 87.5°W on the eastern side.

The data taken by the Field Program of the DeSoto Canyon Eddy Intrusion Study during the period from 1996 to 1999 are used for the test [Minerals Management Service, 1999]. Two mooring stations, C1 and C3 marked in Figure 4.1, are used to represent the near-inertial response at the shelf-break to the hurricane. Mooring station C1 on the shallow water side is located at 29°35.210′N 87°20.972′W with the depth of 100 m, the instrument-measured depth of 95 m and the offshore distance of 80 km. Mooring station C3 on the deep water side is located at 29°00.191′N 87°21.387′W with the depth of

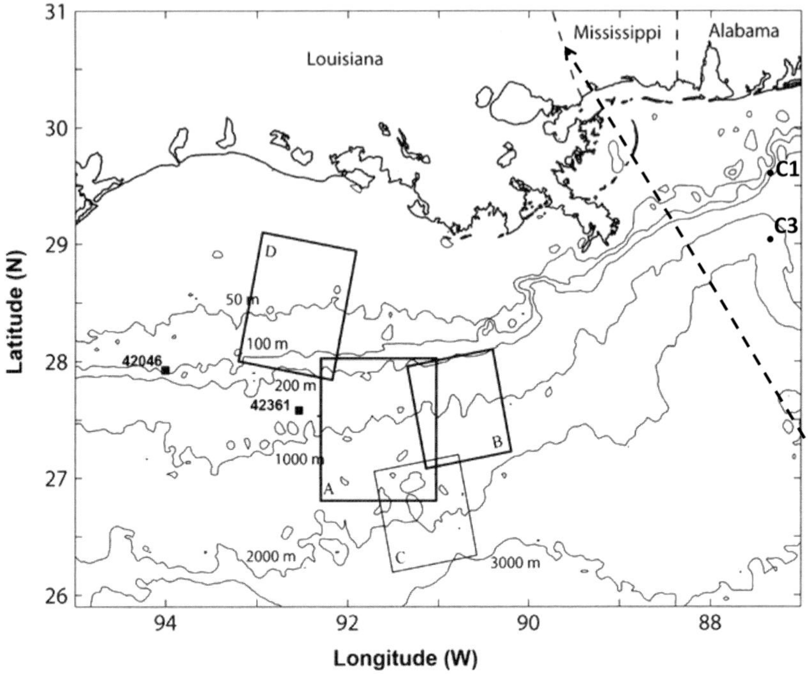


Fig. 4.1 Map of the test area. Boxes represent coverage of four SAR images containing NIO signatures Station 42361 is the location of a platform for measuring current data. C1 and C3 were mooring stations. Long dashed arrow represents the ground track of Hurricane Georges in 1998 [Li *et al.*, 2013]

1300 m, the maximum instrument-measured depth of 1290 m and the offshore distance of 150 km. One can see that the mooring stations were located near the path of 1998 Hurricane Georges.

### 4.2.2. *Ocean current time series*

#### 4.2.2.1. *Mooring station* C1

Time series of ocean current velocity profile measured at mooring station C1 from August 12 (Julian day 224) to December 8 (day 342), 1998 are shown in Figures 4.2(a) and (b). The temporal resolution of the records is 30 min. Ripples along the curves include signals of NIOs with periods near one day. One can see that the NIOs always existed during the season and in the whole water column from the surface to the bottom, though sometimes they strengthened and sometimes weakened. The oscillation peak-to-peak amplitudes are on the order of $30\,\mathrm{cm\,s^{-1}}$. The records also show two obvious oscillation strengthening processes, happened at the beginning and the end of September 1998 (around days 245 and 270), respectively.

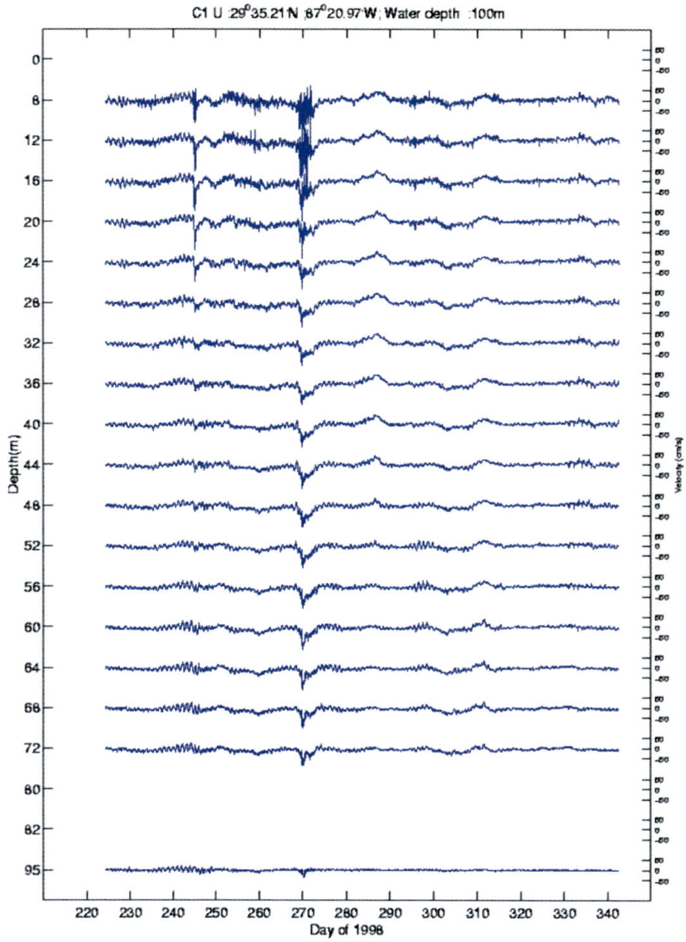


(a)

Fig. 4.2 (a) Time series of zonal component of ocean current velocity profile measured at mooring station C1 from August 12 (day 224) to December 8 (day 342), 1998 [Zheng *et al.*, 2006]. (b) Same as Figure 4.3(a), but for meridional component [Zheng *et al.*, 2006]

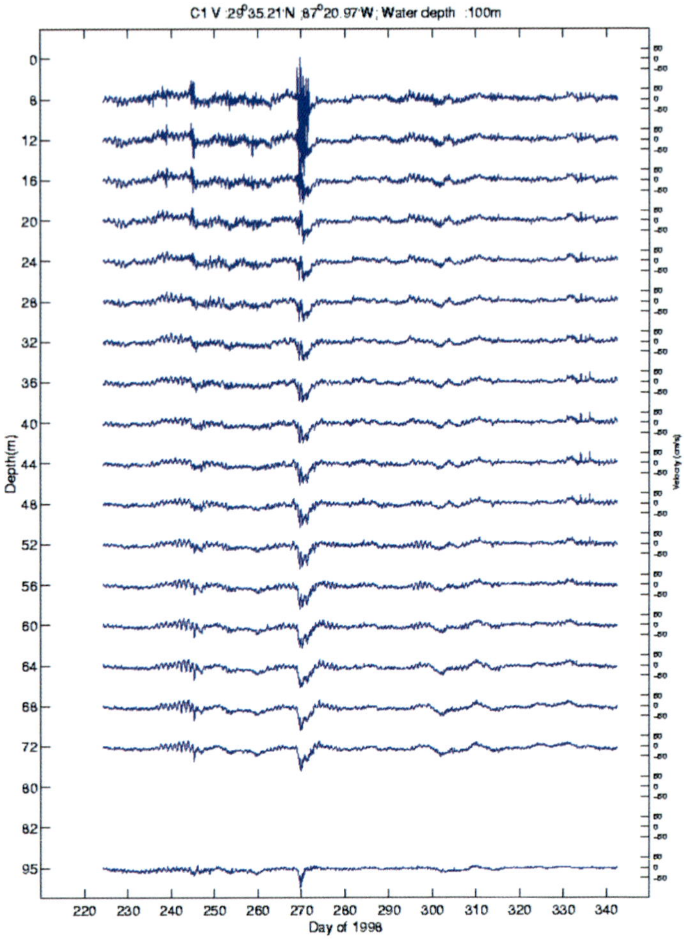


(b)

Fig. 4.2 (*Continued*)

In particular, during the second event, the oscillation amplitudes in the upper surface layer increased almost one order of magnitude. Comparing with hurricane records, we find that those two periods corresponded to the passages of Hurricane Earl and Hurricane Georges, respectively.

In order to show the evolution process and vertical propagation of the ocean current response clearly, we calculate the ocean current energy density from the current velocity components shown in Figures 4.2(a) and (b). Assuming a cosine oscillation function, the average kinematic energy density over a period is

$$\overline{E_k} = \frac{1}{4}\rho A^2, \tag{4.2.1}$$

where $\rho$ is the water density, and $A$ is the velocity amplitude of a fluid parcel. For discrete data, Eq. (4.2.1) has an approximate form

$$\overline{E_k} = \frac{\rho}{4}\left[\frac{\sum_{m=-l}^{l}(u^2, v^2)_{i+m}}{2l+1}\right], \tag{4.2.2}$$

where $u$ and $v$ are zonal and meridional components of the ocean current velocity, respectively, $i + m(= 0, 1, 2, \ldots)$ is a time series number of a data point, $2l + 1$ is a number of data points over the oscillation period, one day in our case, implying that $l$ equals to 24.

A time-depth near-inertial energy image from day 266 to day 276 corresponding to an extended period of the passage of Hurricane Georges is shown in Figure 4.3. One can see that the image has two peaks. The major one is located above 16 m, and the secondary one is centered at 52 m with a range from 36 to 72 m.

Comparing with the vertical thermal structure [Zheng *et al.*, 2006], we found that the depths of the two peaks are closely associated with the thermocline. The major peak extends to the depth of 16 m, which is the bottom of mixed layer just atop the thermocline. The secondary peak sets up at 52 m, which is located in the lower layer below the thermocline. In between the two there is a transition layer from 16 to 36 m corresponding to the thermocline itself, implying that there is inertial energy propagating downward from the mixed layer to the lower layer through the thermocline.

#### 4.2.2.2. *Mooring Station* C3

Time series of ocean current velocity profile time series measured at mooring station C3 from August 12 to December 8, 1998 are shown in Figures 4.4(a) and (b). The temporal resolution of the records also is 30 min. One can see that inertial oscillations with the average

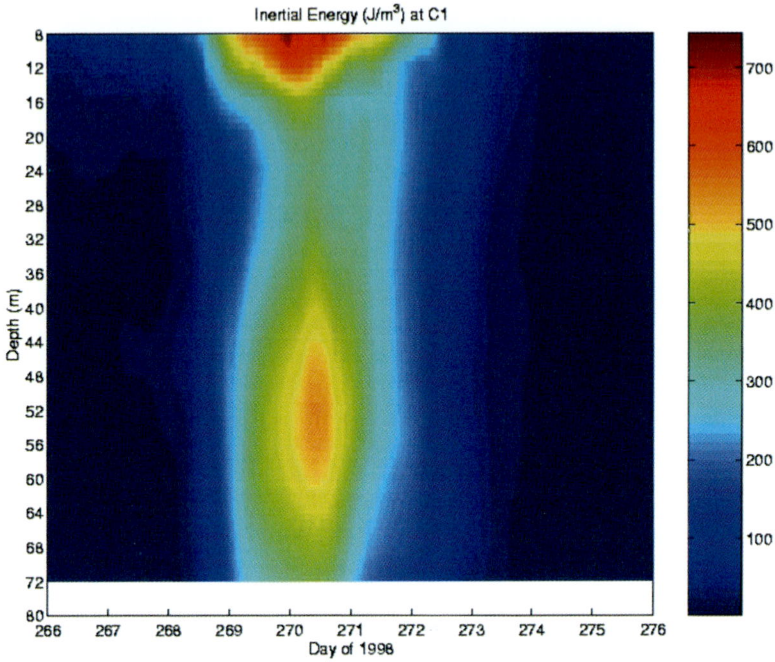


Fig. 4.3 Time series of the near-inertial energy calculated from the current velocity components shown in Figures 4.2(a) and (b)

period of about one day always existed during the season and in the whole water column. The oscillation peak-to-peak amplitude is slightly larger than that of station C1, implying that the deep water side of the shelf-break has gained more energy from hurricanes. The data sets also show two obvious strengthening processes of NIO happened at the beginning and the end of September 1998, corresponding to the passages of Hurricane Earl and Hurricane Georges, respectively.

The time-depth near-inertial energy image from day 266 to day 276 is shown in Figure 4.5. One can see only one peak above 72 m. The maximum depth of the peak is 50 m.

Comparing with the vertical thermal structure, we found that the depth of 50 m is the local mixed layer depth. There is not the secondary peak below the thermocline to 72 m. The peak set up from the upper surface layer on day 269, and reached its maximum intensity and maximum depth on day 270. After that the depth of the peak became shallower quickly on day 271. This major peak lasted for two days. Following this major peak, three minor peaks were consequential in the following five days.

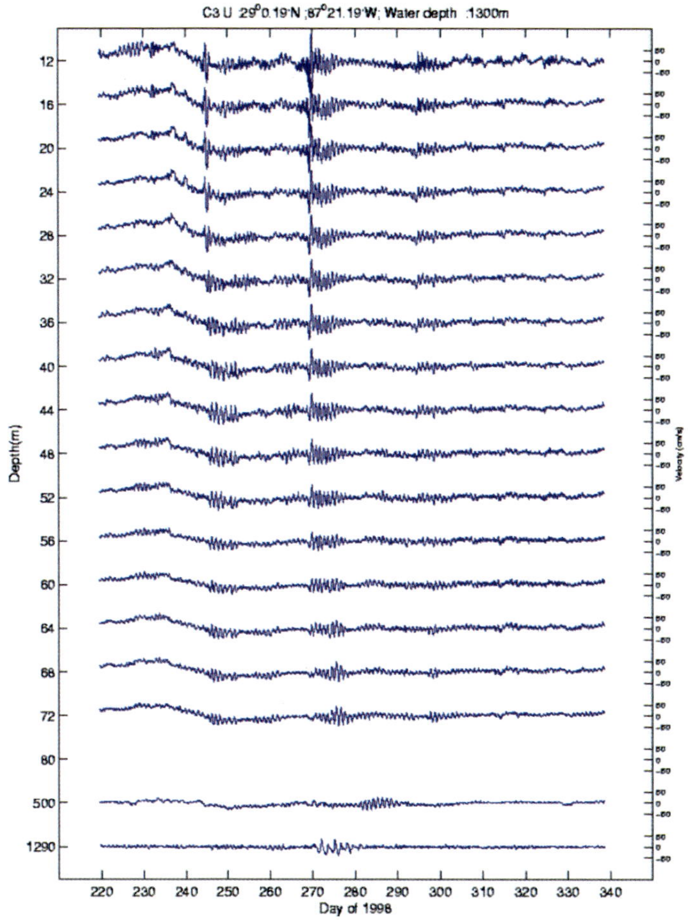


(a)

Fig. 4.4 (a) Time series of zonal component of ocean current velocity profile measured at mooring station C3 from August 8 (day 220) to December 5 (day 339), 1998 [Zheng *et al.*, 2006]. (b) Same as Figure 4.6(a), but for meridional component

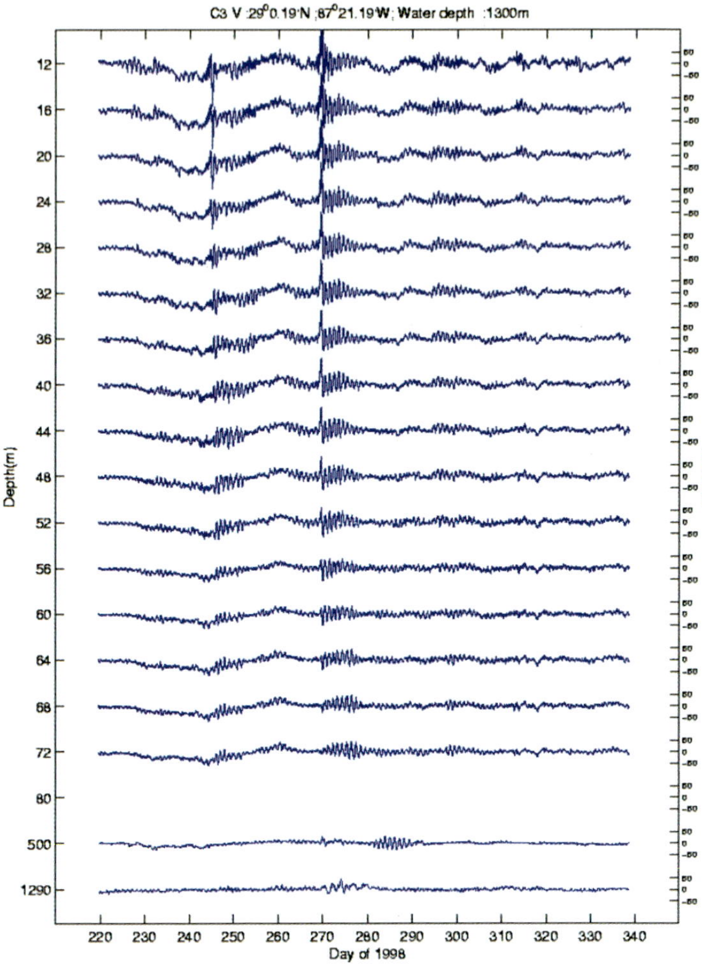


(b)

Fig. 4.4 (*Continued*)

## 4.3. Determination of NIO Frequencies

Previous investigations indicated that the frequency of hurricane-induced NIO would shift above the local inertial frequency, so-called "the blue shift" [Gill, 1982; Price, 1981; Shay and Elsberry, 1987]. We use field observations to determine NIO frequencies. The empirical

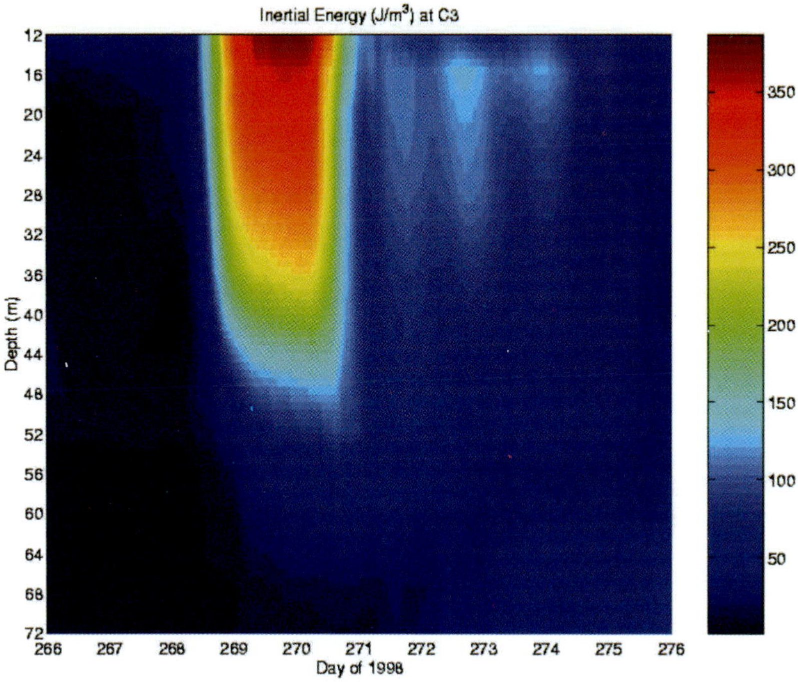


Fig. 4.5 Time series of the near-inertial energy calculated from the current velocity components shown in Figures 4.4(a) and (b) [Zheng *et al.*, 2006]

mode decomposition (EMD) method [Huang *et al.*, 1998] is first used as a band-pass filter to extract the NIO signals from ocean current data. Then the band-pass-filtered data sets are least-square fitted to an empirical function

$$u = -a[\sec\text{h}^2(t) + G(t)\, e^{-i\sigma t}] + be^{-i(\sigma t-\theta)}, \qquad (4.3.1)$$

where $a$, $b$, $G(\mathrm{t})$, $\sigma$ and $\theta$ are adjustable.

Using this method, the frequency of NIO, $\sigma$, can be determined. The best fitting of band-pass-filtered mooring current velocities to the empirical function (4.3.1) shown in Figure 4.6 yields the average NIO frequencies at C1 and C3 are $1.06f$ and $1.03f$, respectively [Zheng *et al.*, 2006a].

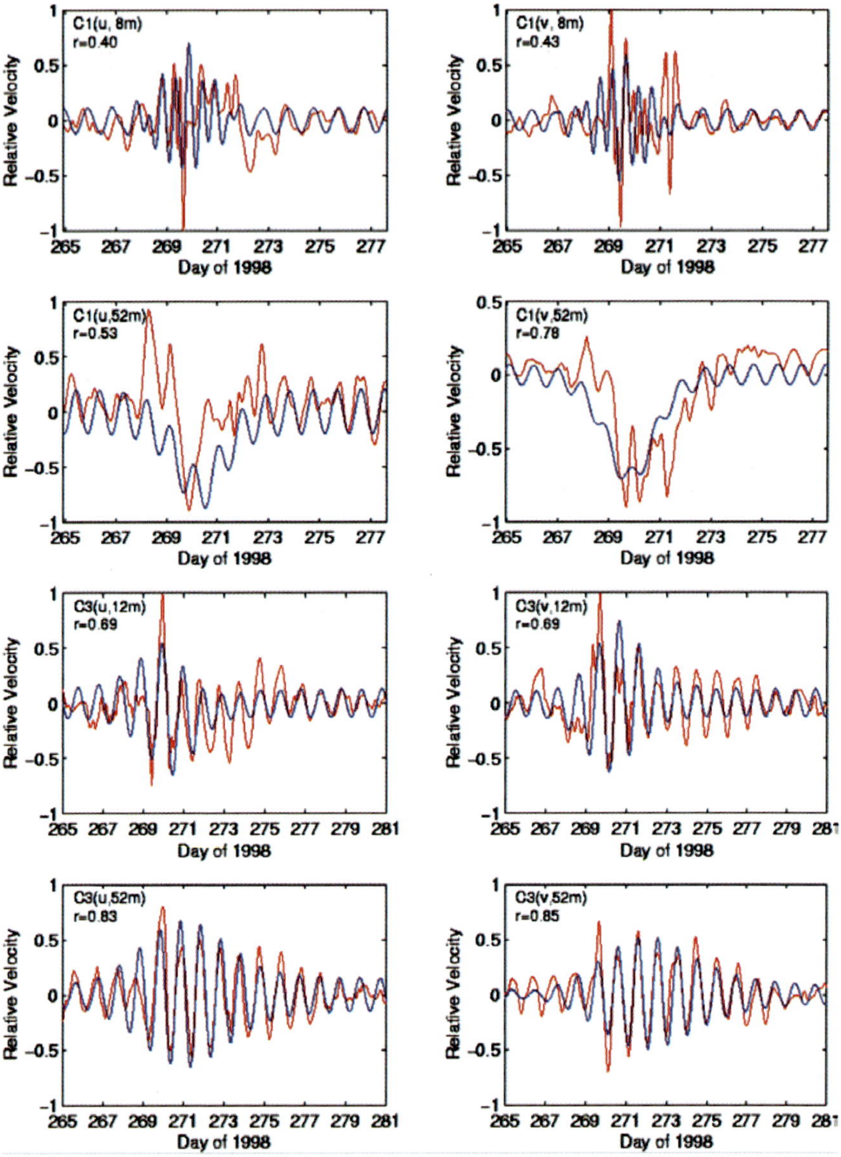


Fig. 4.6 Band-pass-filtered mooring current velocities fitted to the empirical function. Code Cn(U, M) represents mooring stations (C1 and C3), current components ($u$ and $v$) and the depth of M m

## 4.4. Interpretation of SAR Images of NIOs

### 4.4.1. *SAR imaging of NIOs*

Oil slicks on the ocean surface resulting from spills from ships, oil and gas platforms and natural seeps from the ocean bottom are a kind of marine pollutions. On the other hand, they are moved by winds or ocean currents, thus, may serve as a tracer for ocean currents. It is feasible, therefore, to extract the data and information of ocean currents from oil slick trajectories through the dynamic analysis.

As mentioned in Chapter 3, microlayer surfactants and surface films generated by oil slicks on the ocean surface damp short wavelength sea surface waves and make the ocean surface smooth. In this case, the incident radar pulses are reflected forward, and a little radar power is backscattered to the receiving antenna. Thus, oil slicks would appear as dark streaks or dark patches with complex patterns on SAR images, depending on quantities and types of oil slicks as well as the ocean surface dynamic environment.

The test area, GOM, is rich in oil and gas resources and provides about 23% of the US gas and 30% of its oil production. In the GOM there are approximately 3,800 fixed platforms and over 60,000 km of pipelines. Oil slicks from the platforms and pipeline leakage happen often. A typical example was the accident that drilling rig Deepwater Horizon exploded and fired on April 20, 2010. Following explosion and sinking of the oilrig, a sea-floor oil gusher flowed for 87 days, until it was capped on July 15, 2010. The US Government estimated the total discharge at 4.9 million barrels into the GOM. There are also many active natural oil seep sites in the GOM verified by submersible sampling [MacDonald *et al.*, 1996], geophysical data [Roberts, 2006] and satellite observations [Garcia-Pineda *et al.*, 2009; Hu *et al.*, 2009]. Figures 4.7–4.10 show four satellite SAR images of oil slicks in the GOM acquired in different years as listed in Table 4.1 [Li *et al.*, 2013]. Four boxes in Figure 4.1 show their ground coverage distributed on the western side of the test area covering water depths from 50 to 3000 m. The circular dark patterns on the SAR images are interpreted as the trajectories of oil slicks caused by the ocean NIOs [Li *et al.*, 2013].

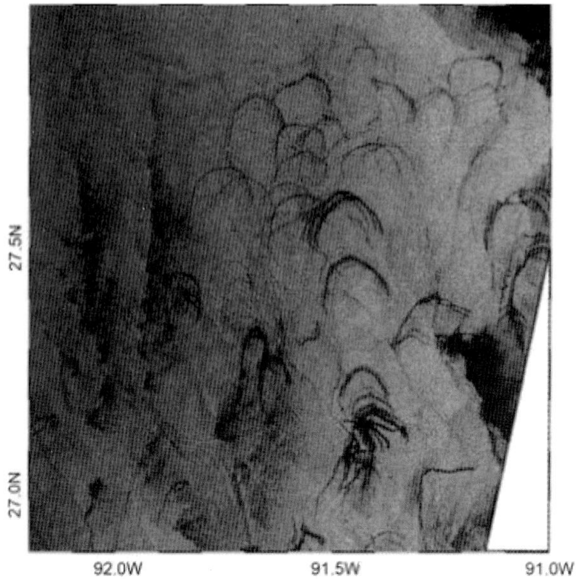


Fig. 4.7 RADARSAT-1 SAR image of oil seepage in the GOM Circular dark patterns are interpreted as the trajectories of oil seepage caused by the ocean NIOs. Imaging time was 12:12:22 UTC, June 12, 1997. Image © CSA

### 4.4.2. *Verification of NIOs*

Interpretation of circular dark patterns on the SAR images as the trajectories of oil slicks caused by the ocean NIOs is verified through rotary spectrum analysis, a proven technique to show the inherent rotational tendency of a flow field (winds or currents) in a clockwise or counterclockwise direction [O'Brien and Pillsbury, 1974]. The current velocity measurements using an acoustic Doppler current profiler (ADCP) fixed on oil platform at station 42361 were taken from May 1 to 31, 2006. This period coincided with imaging time of Figure 4.10. The ADCP was installed at subsurface looking downward and covered between 51 and 803 m below the surface. The current data were collected at a time interval of 10 min. The rotary spectrum calculated from the most near-surface (51 m) ADCP measurements is shown in Figure 4.11. One can see that the clockwise rotary spectrum energy (solid line) is overwhelmingly larger than that of the counterclockwise rotary spectrum (dashed

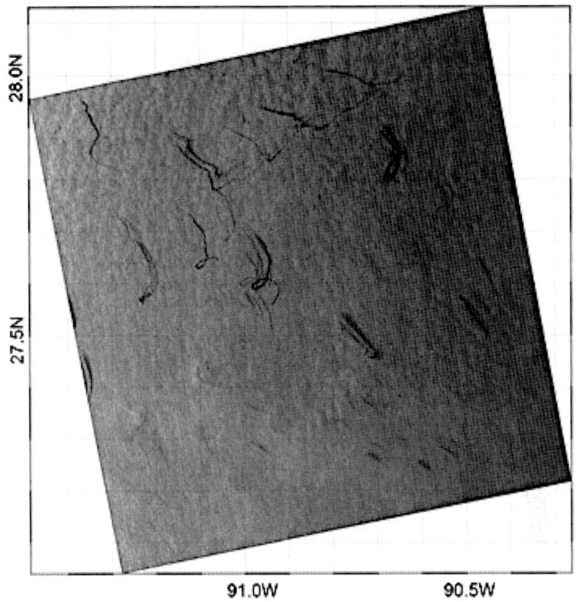


Fig. 4.8 Same as Figure 4.7, but at 00:50:20 UTC, November 6, 1997. Image ©CSA

line) at various frequencies. There is a very small group of peaks at around the 12-h period, which indicates a semi-diurnal tidal oscillation. On the other hand, there are four large clockwise rotary peaks at $T_1 = 23.79$, $T_2 = 24.88$, $T_3 = 25.90$ and $T_4 = 29.0$ h. Station 42361 is located at 28.40°N 92.60°W, where the inertial period is 25.17 h. Thus, the frequencies of four large rotary spectral peaks are $\omega_1 = 1.06f$, $\omega_2 = 1.01f$, $\omega_3 = 0.97f$ and $\omega_4 = 0.88f$, respectively. Among the four frequencies, the first two, $\omega_1$ and $\omega_2$, satisfy the dispersion relation (4.4.1), the other two not. Compared to the NIO frequencies derived from the current data measured at stations C1 and C3 in Section 4.3, we identify that $\omega_1$ should be a NIO component generated in the shallow water, and $\omega_2$ in the deep water. Component $\omega_3$ might originate from a little southern latitude south of 27.5°N, and $\omega_4$ seems to represent a little low frequency process, such as the mesoscale eddy or the GOM Loop Current.

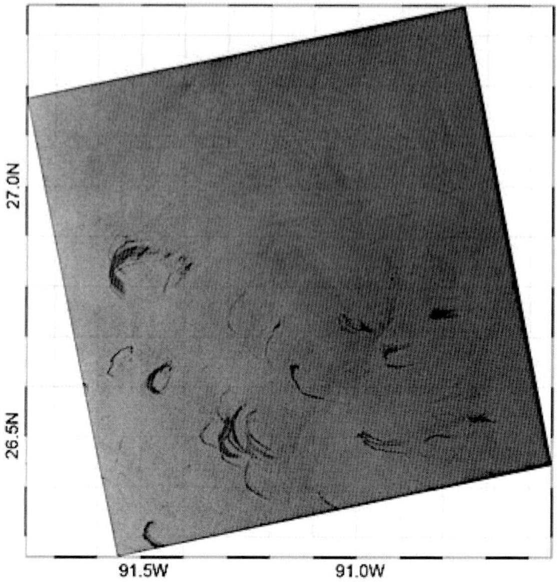


Fig. 4.9 Same as Figure 4.7, but at 00:00:40 UTC, May 16, 2004. Image © CSA

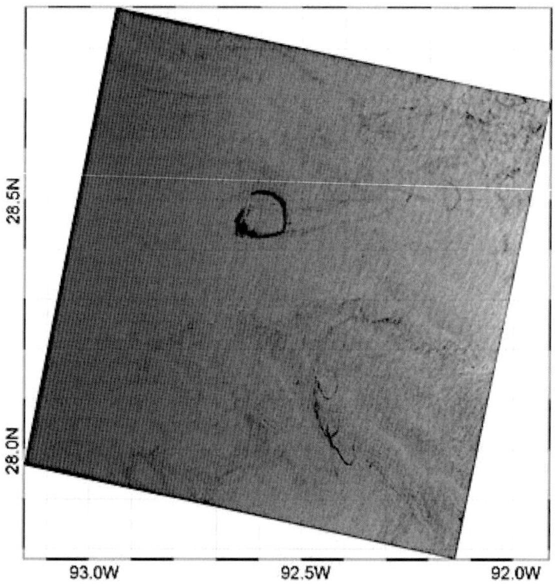


Fig. 4.10 Same as Figure 4.7, but at 12:14:20 UTC, May 13, 2006. Image © CSA

Table 4.1 Specifications of RADARSAT-1 SAR images used in this chapter

| Image | Acquisition Time/Date | Imaging Mode | Polarization | Spatial Resolution (m) | Swath (km) |
|---|---|---|---|---|---|
| Figure 4.7 | 12:12:22 UTC June 12, 1997 | ScanSAR Narrow | HH | 50 | 300 |
| Figure 4.8 | 00:50:20 UTC Nov. 6, 1997 | Standard | HH | 25 | 100 |
| Figure 4.9 | 00:00:40 UTC May 16, 2004 | Standard | HH | 25 | 100 |
| Figure 4.10 | 12:14:20 UTC May 13, 2006 | Standard | HH | 25 | 100 |

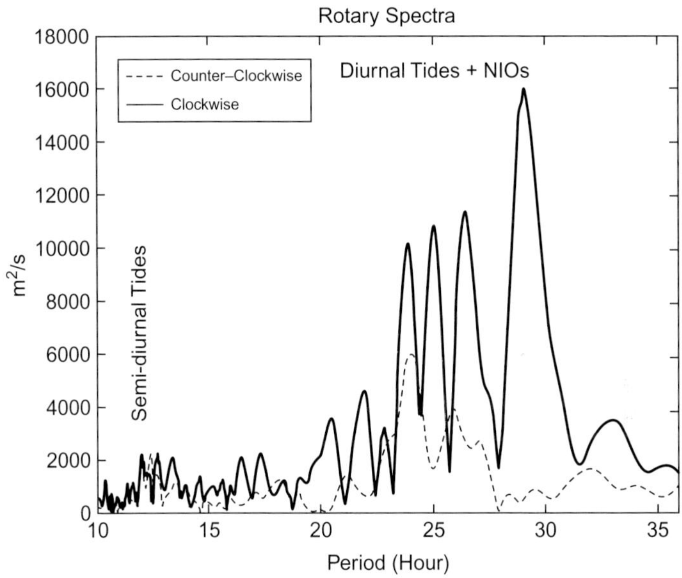


Fig. 4.11 Rotary spectrum of the ADCP current measurements at Station 42361. Four large peaks are at 23.79, 24.88, 25.90 and 29.0 h, respectively [Li *et al.*, 2013]

It is worth noting that the local tidal frequency is very close to that of NIOs. But, tidal currents in the deep ocean are generally small (on the order of a few $\text{cm}^{-1}$). The tides along the GOM coast are micro-tides. Thus, it is reasonable to assume that the NIO is the dominant factor that governs the oil slick motion observed by

SAR. The rotary spectrum analysis has further demonstrated that the motion is indeed dominated by clockwise (anti-cyclonic) rotation, which is a major characteristic of NIOs.

### 4.4.3. *Dynamic interpretation of SAR images of NIOs*

#### 4.4.3.1. *Dispersion relation*

The NIOs are generally viewed as a degenerated or special case of Poincaré waves. If we choose the $x$-axis in the wave propagation direction, the dispersion relation for a Poincaré wave is [Gill, 1982]

$$\omega^2 = f^2 + K^2 R_0^2 f^2, \tag{4.4.1}$$

where $\omega$ is the wave frequency, $K$ is the wave number and $f$ is the Coriolis parameter $f = 2\Omega \sin\phi$, here, $\Omega$ is the earth's angular velocity and $\phi$ is the latitude, $R_0$ is the Rossby radius of deformation

$$R_0 = \sqrt{gH}/f, \tag{4.4.2}$$

where $g$ is the gravitational acceleration and $H$ is the mean water depth.

The velocity components $u$ and $v$ for a Poincaré wave are expressed as

$$u = \frac{\omega\eta_0}{KH}\cos(Kx - \omega t), \tag{4.4.3}$$

and

$$v = \frac{f\eta_0}{KH}\sin(Kx - \omega t), \tag{4.4.4}$$

where $\eta_0$ is the wave amplitude. The motion trajectory derived from Eqs. (4.4.3)–(4.4.4) is an ellipse, and the ratio of the two axes is $\omega/f$. From Eq. (4.4.1), we have $\omega > f$. For higher frequency Poincaré waves ($\omega \gg f$), the elliptical trajectory would be long and approaches rectilinear. However, in the ocean, a Poincaré wave happens mostly at the lowest energy frequency [Gill, 1982]. For this lowest possible frequency Poincaré wave, the second term containing $g$ in Eq. (4.4.1) is negligible (gravity is not important). Then we have $\omega \approx f$. Therefore, $u_{\max} \approx v_{\max}$ and the trajectory is a circle (the near-inertial circle). In this case, the motion is denoted as near-inertial motion with a near-inertial frequency close to the Coriolis parameter $f$, which is only a function of latitude $\phi$. The radius of the near-inertial circle

$$R = \frac{U}{f}, \tag{4.4.5}$$

where $U$ is the initial velocity set up by an external force, such as the wind. The Coriolis force causes the oil slicks on the ocean surface to veer to the right (clockwise) in the northern hemisphere, and to the left (counter-clockwise) in the southern.

4.4.3.2. *Derivation of the initial velocity*

The analysis and calculations here and in the next two subsections are based on the assumption that tidal currents are negligibly small compared to that of NIOs. Figure 4.7 shows a group of oil slick semi-circular features. The radiuses of these open semi-circles are similar in size. We focus on a small area centered at about 27.5°N, 91.3°W. The water depth at this location is about 500 m. The NIO has a period $T = 25.92$ h. The measured radiuses of the near-circular patterns are between 4.0 and 6.5 km. Therefore, from (4.4.5), we calculate the rotation velocity, $U = fR$, as between 0.27 and $0.44\,\mathrm{m\,s^{-1}}$. Meanwhile, all the water and oil particles were moving in the same near-circular fashion, indicating no major ambient current or net current.

In Figure 4.10, there are two oil slicks apparently showing the NIO, a big circle to the north and a group of spiral-like circles. The big nearly complete circle shows what appears to be a full cycle of the NIO. We measure the radius of the big circle is 6.0 km. The inertial oscillation period in that area (28.40°N, 92.60°W) is 25.17 h, thus, the rotating velocity would be $0.42\,\mathrm{m\,s^{-1}}$.

4.4.3.3. *Determination of the NIO frequency*

From Eq. (4.4.1), we have $\omega > f$. From Eqs. (4.4.3)–(4.4.4), we know that the trajectories of the water and oil particles should be an ellipse with a major axis $a$ in the zonal direction and a manor axis $b$ in the meridional direction ($a > b$). Meanwhile, $a$ and $b$ satisfy the relation

$$\frac{a}{b} = \frac{\omega}{f}. \tag{4.4.6}$$

From the big circle in Figure 4.10, we measure $a/b = 1.06$. Thus, we obtain $\omega = 1.06f$ which is just corresponding to the clockwise rotary peak at $T_1 = 23.79\,\text{h}$ in Figure 4.11.

4.4.3.4. *Determination of the NIO wavelength*

The wavelength is a fundamental parameter of the wave motion. However, it is impossible to be determined by the mooring measurements shown in Figures 4.3 and 4.4. Fortunately, this important parameter can be directly measured from the oil slick patterns on the SAR images.

From Eq. (4.4.3), we have $u = 0$ for $x = R$ and $t = T/4$, *i.e.*,

$$\cos(Kx - \omega t)\Big|_{\substack{x=R \\ t=T/4}} = 0. \tag{4.4.7}$$

From Eq. (4.4.7), we have

$$KR - \frac{\omega T}{4} = \pi/4. \tag{4.4.8}$$

Then we obtain the wavelength

$$\lambda = 2R. \tag{4.4.9}$$

From the circular oil slick patterns on the SAR image shown in Figure 4.7, the radiuses of near-circular patterns are between 4.0 and 6.5 km. Therefore, from (4.4.9), we calculate that the NIO wavelengths are between 8 and 13 km. Similarly, in the case of Figure 4.10, we obtain that the NIO wavelength is 12 km. If we consider the effects of tidal currents, these numbers should be larger. For example, if the tidal current velocities are 10% of the NIO velocities, the above wavelengths should be 20% larger than the calculated values and so on, because the reverse rotation of the tidal currents.

#### 4.4.3.5. *Determination of the NIO phase speed*

The dispersion relation of linear waves is

$$c = \frac{\omega}{K}, \tag{4.4.10}$$

where $c$ is the phase speed of wave propagation. From the calculation results obtained in Sections 4.4.3.3–4.4.3.4, we derive that the phase speeds of NIO propagation are between 0.09 and $0.14\,\mathrm{m\,s^{-1}}$ in the case of Figure 4.7 and $0.15\,\mathrm{m\,s^{-1}}$ in the case of Figure 4.10, respectively. These values are within a reasonable range.

#### 4.4.3.6. *Estimate of the mean ambient flow velocity*

In Figure 4.10, there is a group of spiral-like circles, similar to Figure 4.8. The average radius of the three southern spiral-like near-inertial circles is about 2.8 km. Circle trains indicate that there is an ambient current, so that the trajectory appears to be translated by the mean current. The spacing distance between two co-phase points over two subsequent oil slick circles is measured to be about 14 km. The NIO period at this latitude (28.00°N) is 25.50 h. Thus, the mean ambient flow velocity is about $0.15\,\mathrm{m\,s^{-1}}$.

In summary, the interpretation results of SAR imagery of oil slicks on the ocean surface obtained in Sections 4.4.3.1–4.4.3.6 indicate that the SAR images are unique data resource for the NIO research. From the SAR images, almost all dynamic parameters of NIO waves, including the wavelength, the frequency (with blue shift or red shift), the phase speed, the initial rotation velocity and the mean ambient flow velocity, can be directly calculated or estimated.

# Chapter 5

# Estuary Outflow Jet

## 5.1. Introduction

Estuary plumes or estuary outflow jets are formed by outflows of lower salinity water into the coastal ocean [Garvine, 1987]. This chapter will use the Delaware Bay plumes as examples to examine their dynamical features, which have been investigated by using field observations since early 1990s. Garvine [1991] discovered that lighter water outflow from Delaware Bay constitutes a source of the Delaware coastal current. Münchow *et al.* [1992] studied subtidal currents at the Delaware Bay mouth region and found a narrow, baroclinic, subtidal jet exiting on the right looking seaward. Münchow and Garvine [1993a] further described the presence of the jet. Sanders and Garvine [1996] observed a twin front feature of the Delaware Bay plumes. These previous investigations have revealed general features of the Delaware Bay plumes. However, the fine structure, detailed dynamic features and seasonal variability of the plumes have not completely been understood yet, due to limited spatial resolutions and a limited repeatable period of field observations.

Satellite images may serve as an important information source for detecting the fine structure of estuary plumes. Changes in watercolor, differences of water temperature and salinity of plumes from surrounding waters are detectable by various remote sensors onboard satellites [Clemente-Colón, 2001; Garvine, 1984]. Among these sensors, SAR has unique advantages to be used for detecting estuary plumes. Besides all weather and all day detecting abilities, SAR images are characterized by high spatial resolutions and the sensitivity to the ocean dynamic processes. As reviewed in Section 1.2.3 and listed in Table 1.1, the state-of-art resolution of satellite SAR images may reach about 1 m, which is suitable for observation of two-dimensional structure of ocean processes with a length scale larger than 10 m. In principle, satellite SAR also has a potential to indirectly detect ocean water salinity variation [Clemente-Colón, 2001] as well as ocean current variability [Zheng *et al.*, 1998a, 2001, 2004], which are major features of estuary plumes. In this chapter, we use RADARSAT-1 and ERS-2 satellite SAR images to detect the Delaware Bay plumes. Through image morphological interpretation and data analysis, we obtain the plume fine structure on the $x-y$ plane, dynamical features of the plumes along with some insight into their seasonal variability.

## 5.2. Theoretical Model of a Jet SAR Image

### 5.2.1. *Two-dimensional jet*

Formally, a two-dimensional jet is a special flow formed by an efflux of fluid from a long and narrow orifice into the same type of fluid as the jet itself, as illustrated in Figure 5.1 [Kundu, 1990]. Dynamically, the jet is controlled by two conditions. One is the momentum flux

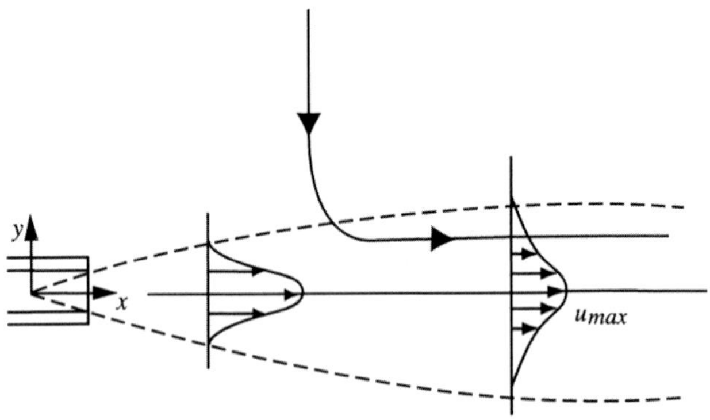


Fig. 5.1 Scheme of a two-dimensional jet flow [Zheng *et al.*, 2004]

preservation, i.e.,

$$M = \rho \int u^2 dy = \text{independent of } x, \tag{5.2.1}$$

where $\rho$ is the density of fluid, $u$ is the $x$-component of velocity and the coordinate system is defined as that in Figure 5.1.

Another condition is a similarity solution of the stream function in the form of

$$\psi = a x^{1/3} \tanh \eta, \tag{5.2.2}$$

where

$$a = \left( \frac{9 A_H M}{2\rho} \right)^{1/3}. \tag{5.2.3}$$

Here $A_H$ is the horizontal eddy viscosity, and

$$\eta = \frac{y}{b x^{2/3}} \tag{5.2.4}$$

in which

$$b = \left( \frac{48 A_H^2 \rho}{M} \right)^{\frac{1}{3}}. \tag{5.2.5}$$

From Eq. (5.2.2) we derive the velocity components of the jet. The $x$-component or axial velocity is given by

$$u = u_{\text{max}} \text{sech}^2 \eta, \tag{5.2.6}$$

where $u_{\text{max}}$ is the maximum axial velocity at the center of the jet, given by

$$u_{\text{max}} = \frac{a}{b} x^{-\frac{1}{3}} = \left( \frac{3M^2}{32\rho^2 \nu} \right)^{1/3} x^{-1/3}. \tag{5.2.7}$$

The $y$-component or transverse velocity is given by

$$v = v_{\text{max}} (2\eta \text{sech}^2 \eta - \tanh \eta), \tag{5.2.8}$$

where $v_{\text{max}}$ is the maximum transversal velocity given by

$$v_{\text{max}} = \frac{a}{3} x^{-2/3}. \tag{5.2.9}$$

These equations imply that the dynamics of an ideal two-dimensional jet are characterized by (1) a jet axis that is kept unchanged, (2) a jet width that increases as $x^{2/3}$, (3) an $x$-component velocity that decreases as $x^{-1/3}$, (4) a $y$-component that decreases as $x^{-2/3}$ and (5) point $x = 0$ is a singular point.

### 5.2.2. *Jet modulation to wave spectra*

As mentioned in Section 5.1, we use RADARSAT and ERS-2 satellite SAR images to detect the Delaware Bay plumes. The two SARs operate at C band (wavelength of 5.7 cm), of which the corresponding ocean surface Bragg waves are in the gravity-capillary wave band. Thus, the wavenumber spectral density of the high frequency ocean surface wave field is written in the form of

$$\psi_g(k) = m_3^{-1}\left[m\left(\frac{u_*}{c}\right)^2 - 4\gamma k^2\omega^{-1} - \left(S_{\alpha\beta}\frac{\partial \mathrm{U}_\beta}{\partial x_\alpha}\right)\omega^{-1}\right]k^{-4}, \tag{1.4.10}$$

where $m$ and $m_3$ are coefficients, $u*$ is the friction velocity, $c$ is the wave phase speed, $\gamma$ is the viscosity of seawater, $\omega$ is the angular frequency of the ocean surface waves, $k$ is the wavenumber of ocean surface waves and the scalar product of the excess momentum flux tensor and the modification velocity tensor is given by

$$\left(S_{\alpha\beta}\frac{\partial U_\beta}{\partial x_\alpha}\right) = \left[\frac{\partial u}{\partial x}\cos^2\phi + \left(\frac{\partial u}{\partial y} + \frac{\partial v}{\partial x}\right)\cos\phi\sin\phi + \frac{\partial v}{\partial y}\sin^2\phi\right]\Big/2, \tag{1.4.12}$$

where $u$ and $v$ are velocity components, and $\varphi$ is the wave direction. Assuming that both wind field and seawater viscosity are spatially homogeneous, Eq. (1.4.10) can be re-written as

$$\psi = \psi_0 + \psi_j \tag{5.2.10}$$

where $\psi_0$ represents a mean, homogeneous wave field and $\psi_j$ represents the variability of wave field induced by an ocean jet, which is in the form of

$$\psi_j = m_3^{-1}\left[-S_{\alpha\beta}\frac{\partial U_\beta}{\partial x_\alpha}\omega^{-1}\right]k^{-4}. \tag{5.2.11}$$

Table 5.1 Information and parameters of SAR images used for this chapter

| **Image Code** | **96813** | **97207** | **97514** |
|---|---|---|---|
| Satellite | RADARSAT-1 | RADARSAT-1 | ERS-2 |
| Orbit direction | Descending | Descending | Descending |
| Time, UTC | 11:07 August 13, 1996 | 11:15 February 7, 1997 | 15:43 May 14, 1997 |
| Beam mode | SWB | SWB | Standard |
| Resolution, m | 100 × 100 | 100 × 100 | 30 × 26.3 |
| Swath, km | 450 | 450 | 100 |
| Tidal phase | Maximum flood | Maximum flood | Early flood |
| Wind speed, m $s^{-1}$ | 9.2 | 3.9 | 5.1 |
| Wind direction | 180° | 335° | 185° |
| Wave height, m | 1.86 | 0.53 | 0.89 |
| Dominant wave period, *s* | 5.88 | 9.09 | 8.33 |
| Air temperature, °C | 23.5 | N/A | 13.4 |
| Water temperature, °C | 22.7 | 5.9 | 12.1 |

Here, $U_\beta$ may contain two major components: the long ocean wave–current field and the jet current field. In our case, three SAR images used for the analysis were taken at very low sea states with the wave heights of 1.86, 0.55 and 0.89 m as listed in Table 5.1. Therefore, it is an acceptable approximation if we suppose that the effects of the ocean wave–current field on SAR imaging are ignored. In this case, the two-dimensional jet comprises a dominant component of the spatially variable current $U_\beta$. Substituting (5.2.6) and (5.2.8) into (5.2.11) yields

$$\psi_j = -m_3^{-1}\omega^{-1}k^{-1}\cdot$$
$$\left[\begin{array}{l} -\dfrac{a}{6b}x^{-\frac{4}{3}}\left(\cos^2\phi + 2x^{-1}y\cos\phi\sin\phi - \sin^2\phi\right)\operatorname{sech}^2\left(\dfrac{y}{bx^{2/3}}\right) \\ +\dfrac{2a}{3b^2}x^{-2}\left(y\cos^2\phi - \dfrac{2}{3}x\cos\phi\sin\phi + \dfrac{2}{3}x^{-1}y\cos\phi\sin\phi - \sin^2\phi\right) \\ \cdot\operatorname{sech}^2\left(\dfrac{y}{bx^{2/3}}\right)\tanh\left(\dfrac{y}{bx^{2/3}}\right) + \dfrac{a}{9}x^{-5/3}\cos\phi\sin\phi\tanh\left(\dfrac{y}{bx^{2/3}}\right) \end{array}\right]. \tag{5.2.12}$$

Substituting (5.2.12) into (1.4.1) yields the backscatter cross section per unit area induced by the two-dimensional jet

$$\sigma_0(\theta)_{ij,jet} = -16\pi k_0^4 |g_{ij}(\theta)|^2 m_3^{-1}\omega^{-1}k^{-1}.$$
$$\left[ -\frac{a}{6b}x^{-\frac{4}{3}}\left(\cos^2\phi + 2x^{-1}y\cos\phi\sin\phi - \sin^2\phi\right)\operatorname{sech}^2\left(\frac{y}{bx^{2/3}}\right) \right.$$
$$+\frac{2a}{3b^2}x^{-2}\left(y\cos^2\phi - \frac{2}{3}x\cos\phi\sin\phi + \frac{2}{3}x^{-1}y\cos\phi\sin\phi - \sin^2\phi\right).$$
$$\left. \cdot\operatorname{sech}^2\left(\frac{y}{bx^{2/3}}\right)\tanh\left(\frac{y}{bx^{2/3}}\right) + \frac{a}{9}x^{-5/3}\cos\phi\sin\phi\tanh\left(\frac{y}{bx^{2/3}}\right) \right] \tag{5.2.13}$$

This is a theoretical model for radar imagery of two-dimensional jet. One can see that the imagery is not only dependent on the structure of jet current field, but also on the wind direction (In the case of gravity–capillary waves, the wave direction is consistent with the sea surface wind direction).

In order to simplify the analysis without loss of physics and the major features of the jet radar images, we derived an approximate form of Eq. (5.2.13) from an assumption that the $y$ component velocity, $v$, is ignored in comparison with the $x$-component (In fact, this condition stands in most of cases). Thus, we have

$$\sigma_0(\theta)_{ij,\text{jet}} = -16\pi k_0^4 |g_{ij}(\theta)|^2 m_3^{-1}\omega^{-1}k^{-1}$$
$$\times\left\{\frac{u_{\max}}{6}x^{-1}\left[\cos^2\phi + \left(4\eta\cos\phi - \frac{6x^{1/3}}{b}\sin\phi\right)\right.\right.$$
$$\left.\left.\times\cos\phi\tanh\eta\right]\operatorname{sech}^2\eta\right\}, \tag{5.2.14}$$

where $\eta$ is given by Eq. (5.2.4). This theoretical model predicts that the jet radar image has three major features. First, the jet current structure, $\operatorname{sech}^2\eta$, constitutes a determinant factor for its radar image. Second, the image intensity depends on the maximum $x$-component of velocity and decreases as $x^{-1}$ along the jet axis.

Third, the imagery intensity is sensitive to the wind direction with respect to the jet axis. The jet may even become invisible on radar images for wind directions $\phi = \pi/2,\ 3\pi/2$.

In order to examine effects of the wind direction on radar imaging of the jet, we further define function $G(\eta, \phi)$ as

$$G(\eta,\phi) = \left[\cos^2\phi + \left(4\eta\cos\phi - \frac{6x^{1/3}}{b}\sin\phi\right)\cos\phi\tanh\eta\right]\operatorname{sech}^2\eta. \tag{5.2.15}$$

The physical meaning of $G(\eta,\ \phi)$ is the dimensionless radar backscatter section of a two-dimensional jet. We plotted the curves of $G(\eta,\ \phi)$ versus $\eta$ for $x = b^3$ and for typical values of $\phi$, which are shown in Figure 5.2.

These curves represent theoretical models of normalized radar image intensity (or gray level) patterns along the transverse section through a point on the jet axis at $x = b^3$ for various wind directions. One can see that in all cases, $G$ varies obviously with $\eta$ and has only one positive peak, which is accompanied by one negative peak (two negative peaks for $\phi = 0,\ \pi$). This feature implies that the jet has a self-radar-image-contrast-enhancement effect. The case for $\phi = 0$, $\pi$ is the only case, in which the $G(\eta,\ \phi)$ versus $\eta$ curve is symmetric about the $x$ axis ($\eta = 0$). In all other cases, the positive peaks shift off the $x$ axis either rightward (for $\phi = \pi/4$) or leftward (for $\phi = 3\pi/4$, $5\pi/4$ and $7\pi/4$), implying that the locations of peak intensity values of the jet radar image would not coincide with the jet axis, which is the location of the maximum $x$-component velocity.

## 5.3. Satellite SAR Images and Field Data

The SAR images used for this study were taken by RADARSAT-1 and ERS-2 satellites (see Section 1.2.3 and Table 1.1 for details). Table 5.1 lists main information and parameters of the images, which were acquired during a period from summer 1996 to spring 1997.

We use cruise data measured by Sanders and Garvine [1996] to determine the salinity distribution and current field structure of a low salinity estuary plume off the Delaware Bay. The original

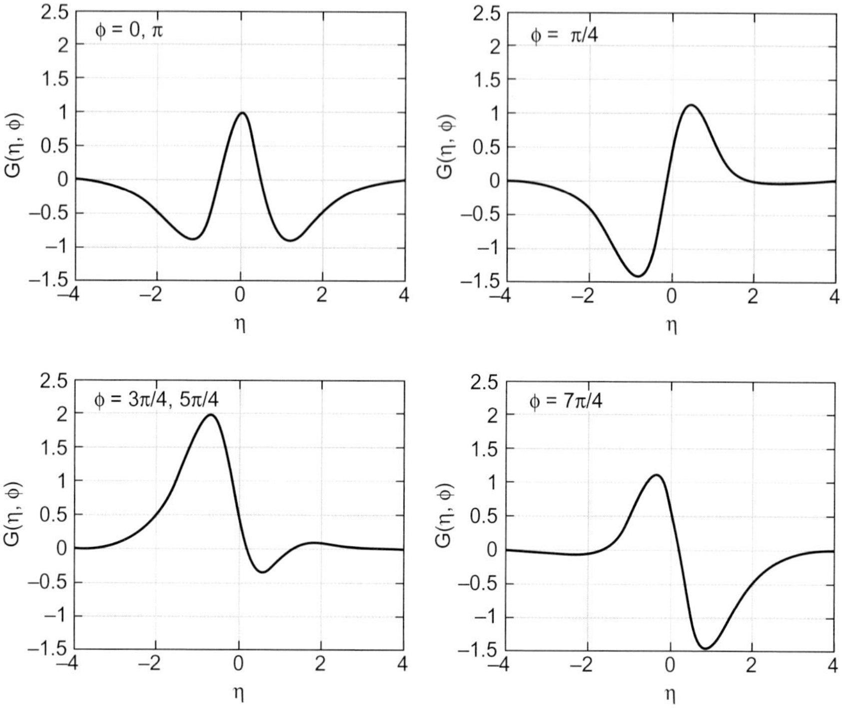


Fig. 5.2 Theoretical models of normalized radar image intensity (or gray level) structure along the transverse section through a point on the jet axis for $x = b^3$. The wind direction with respect to the jet axis is given at the left-upper corner in each case [Zheng *et al.*, 2004]

data were measured with the conductivity-temperature-depth (CTD) profiler along a transect line labeled AA in Figure 5.3 on May 28–29 and June 25, 1992. We used density anomaly $\rho_t$ data derived from the measurements to calculate the current field structure. Assume that the $y$-component of velocity is ignored again, the following approximate relation stands

$$\rho_t = -\alpha u(y), \tag{5.3.1}$$

where $\alpha$ is a constant to be determined, and $u(y)$ represents the $x$-component of velocity. This means that using density anomaly data, we may determine a normalized form of function of $u(y)$ used for dynamical analysis even though $\alpha$ is unknown.

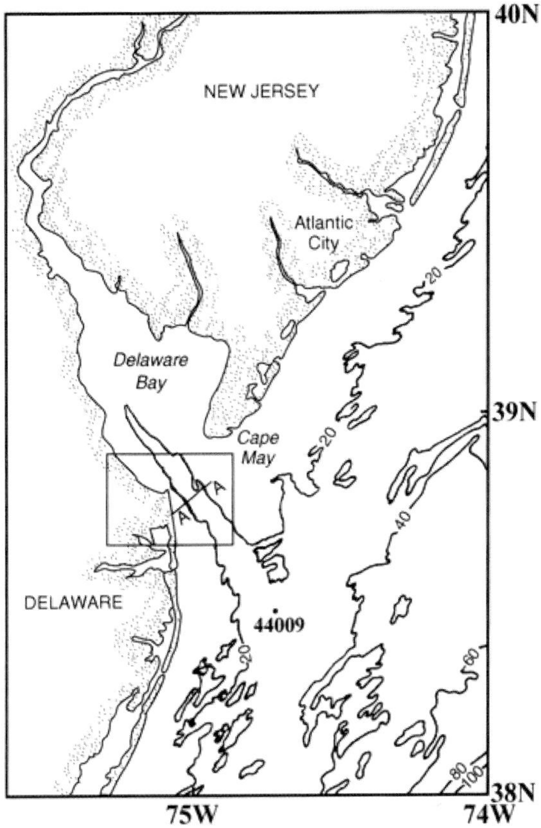


Fig. 5.3 Map of the Delaware Bay area A rectangular box marked at the bay mouth shows the study area, in which transect line AA shows the locations of field data measured by Sanders and Garvine [1996] as shown in Figure 5.4. Point 44009 shows a location of the NOAA Delaware Bay Buoy, from which we obtained simultaneous field measurements at SAR imaging time [Zheng *et al.*, 2004]

From the above analysis, one can see that winds simultaneously measured at SAR imaging time are necessary data for SAR image interpretation. Fortunately, there is a NOAA Delaware Bay Buoy (ID: 44009, location: 38°27′49″N 74°42′07″W) available nearby the study area (15 km southeast). We use continuous winds measured by the buoy at a height of 5 m above the sea level to represent the sea surface winds in the plume region. We also use the ocean surface wave data measured by the Buoy for the sea state analysis.

The tides are another important dynamical element, which may have influence on the SAR imaging of the estuary plumes. We consulted data available at the website http://tidesonline.nos.noaa.gov/ (Center for Operational Oceanographic Products and Services, Tides Online, NOAA National Ocean Service, Silver Spring, Maryland, 2004, hereinafter referred to as Center for Operational Oceanographic Products and Services, 2004) to determine the tide phase at the SAR imaging time.

## 5.4. Interpretations of SAR Images of Delaware Bay

### 5.4.1. *Study area*

The Delaware Bay is located on the east coast of the US as shown in Figure 5.3. The total surface area of the bay is about 1864 km$^2$. The south–north length of the bay is about 75 km, and the east–west breadth varies from 18 km at the mouth from Cape Henlopen, Delaware to Cape May, New Jersey to a maximum of 45 km at the broadest point [Polis and Kupferman, 1973; Zheng *et al.*, 1993a]. The Delaware Bay plumes originate at the bay mouth and extend offshore along a deep channel, which has a maximum channel depth of over 20 m, and the width of 3 km at the bay mouth, up to about 20 km downstream onto the continental shelf [Sanders and Garvine, 1996; Wong and Münchow, 1995].

The Delaware Bay plumes, in fact, are an outflow of low-salinity estuarine water from the bay. The measurements along transect line AA marked in Figure 5.3 by Sanders and Garvine [1996] indicated the salinity inside the plume is 2–3 psu lower than that of surrounding shelf water.

From the measurements, they determined that the width of the plume is about 4 km, and the depth is about 5 m. They also found that the plume has a twin front structure. The onshore and offshore fronts lie at about 1 and 5 km from the Delaware coast, respectively. Using the density anomaly data [Sanders and Garvine, 1996] and relation (5.3.1), we calculated the current field structure along transect AA as shown in Figure 5.4. One can see that we may treat the current field of the plume into two jet current fields, which

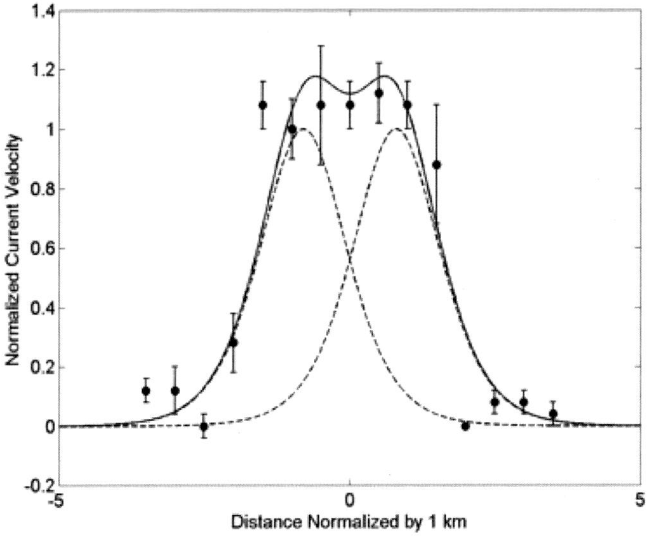


Fig. 5.4 Twin jet structure model of the Delaware Bay plume current field. Data points are derived from the density anomaly data measured along transect AA in Figure 5.3 [Sanders and Garvine, 1996, Figure 4]. Dashed line shows two model jet current fields symmetric about the jet axis or central point. Solid line shows a composite of the two jet current fields [Zheng *et al.*, 2004]

are symmetric about the central point. We call this kind of the plume current field a twin jet structure.

The field-based investigations have revealed basic features of the physical oceanography in the study area. Owing to the limited temporal and spatial resolutions of field measurements, however, the physical nature, dynamic features and the fine structure of the estuary plumes with a spatial resolution up to 10 m still remain unclear. Motivation of this effort is to fill these gaps with the data and information extracted from high resolution SAR images. Following are the results obtained by interpretations of three cases, which occurred in summer, winter and spring, respectively.

### 5.4.2. *Summer case*

In summer, June through August, southerly winds prevail in the region. Southerly alongshore winds produce an Ekman offshore

Fig. 5.5 RADARSAT-1 SWB image covering the Delaware Bay mouth region taken at 11:07 UTC on August 13, 1996. White line represents transect along which gray level values are taken for comparison to the theoretical model [Zheng *et al.*, 2004]

surface drift. This Ekman divergent flow is compensated by an onshore deeper flow that brings colder water to the surface forming coastal upwelling. Therefore, the local summer is also called "upwelling favorable season" [Münchow and Garvine, 1993b]. The river discharge is minimum in summer and reaches maximum in spring. Figure 5.5 shows a RADARSAT-1 image covering the Delaware Bay mouth region taken at 11:07 UTC on August 13, 1996. We code the image by identification number 96813 and list the major specifications in Table 5.1. The time series vector wind records measured by NOAA Delaware Bay Buoy 44009 gave the wind speed as $9.2\,\mathrm{m\,s^{-1}}$ and the wind direction as 180° at the SAR imaging time. The tide phase was at maximum flood.

The bright area on the lower left of the image is land, part of the coastal region of southern Delaware. To the right, there is a light-bright, plume-like pattern. Based on the known general features of the Delaware Estuary Plumes derived from field measurements, as shown by Münchow and Garvine [1993b, Figure 9] and Sanders and Garvine [1996, Figure 5], we interpret this imagery pattern as a signature of the Delaware Estuary Plume. The surrounding dark areas are the coastal waters, the Delaware Bay on the upper left, the Atlantic Ocean on the right, and the Rehoboth Bay on the lower left. The main features of the image are drawn on the map shown in Figure 5.6. One can see that the axis of plume takes an angle of 23° with the coastline, which is close to the direction of the deep channel (30° with the coastline), but turns right 7° toward alongshore direction. The plume has twin image fronts corresponding to the twin jet structure as shown in Figure 5.4. Along the axis, we divide the plume into three sections: the source section characterized by an almost unchanged transverse width of 3.0 km, the jet section characterized by a gradually expanding width with a maximum value of 7.7 km, and the dispersion section characterized by a gradually shrinking width and vague imagery. The observed total length of the plume is 21.3 km, of which the length of the jet section is about 11.0 km.

Comparing the imagery fronts in Figures 5.5 and 5.6 to the envelope curves of two-dimensional jet in Figure 5.1, one can see there is a similarity between the two. In order to prove their physical identity, however, we need further evidence. For this reason, we perform three tests to be discussed next.

#### 5.4.2.1. *Transverse structure of the imagery front*

If we suppose that a linear function was used for conversion of the radar backscatter section to the gray level on the SAR image, then it is reasonable to use a normalized gray level curve to represent the spatial structure of dimensionless radar backscatter section. For the two-dimensional jet, we have derived the theoretical model of dimensionless radar backscatter section as Eq. (5.2.15) and plotted the curves for typical wind directions as shown in Figure 5.2. On the other hand, we take transect of the imagery front along the

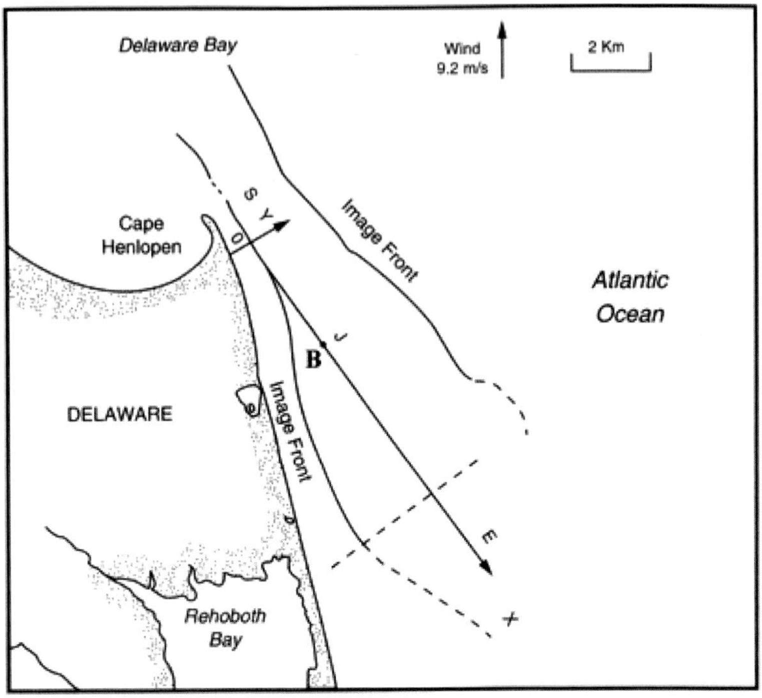


Fig. 5.6 Interpretation map of Figure 5.5 S, J and E represent the source, jet, and extension sections of the jet, respectively. B is the point we used for estimating the axial velocity [Zheng *et al.*, 2004]

white line marked in Figure 5.5. We plot the normalized gray level values measured along transect (data points) and the theoretical model curve for $\phi = 3\pi/4$ in Figure 5.7. One can see that the model curve fits the data points quite well. Both of them are characterized by a positive peak shifting off the $x$-axis rightward, which is a unique response of SAR imaging to the wind direction derived from this chapter. This means that the ocean process that generates the imagery front has the transverse structure identical or similar to the two-dimensional jet.

#### 5.4.2.2. *Spreading of the imagery front*

From the image shown in Figure 5.5, one can see that the imagery front spreads within the jet section. We measured the spreading

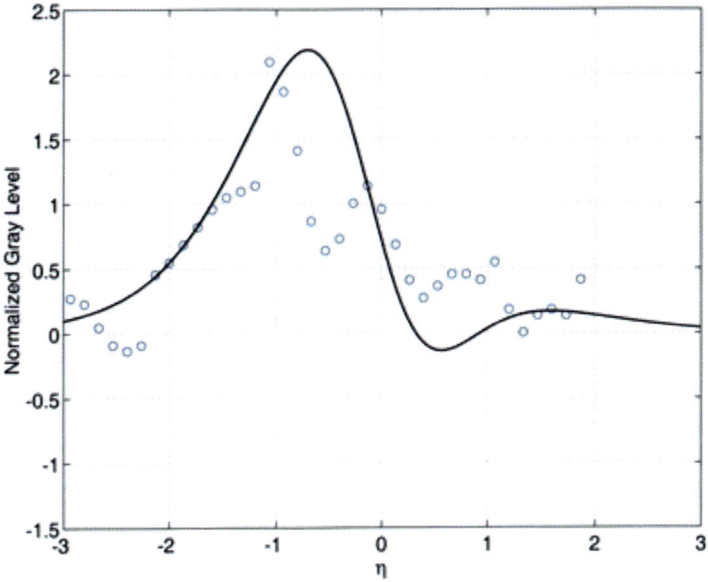


Fig. 5.7 Comparison of normalized gray level values measured along transect marked in Figure 5.5 (data points) and the theoretical model curve for $\phi = 3\pi/4$ [Zheng *et al.*, 2004]

process of the imagery front representing a streamline with respect to the axis, which is an extension line of the imagery front in the source section. We plotted the measured data as solid points in Figure 5.8. The solid straight line represents a linear relation

$$y = -0.40x \tag{5.4.1}$$

This relation represents width spreading of a turbulent jet [Kundu, 1990]. The dashed line represents a relation

$$y = -0.72x^{2/3} \tag{5.4.2}$$

This relation represents width spreading of a laminar jet [Kundu, 1990]. One can see that from 0 to 6 km, the turbulent jet gives an excellent description of width spreading of the plume, and beyond that the laminar jet does better job than the turbulent jet.

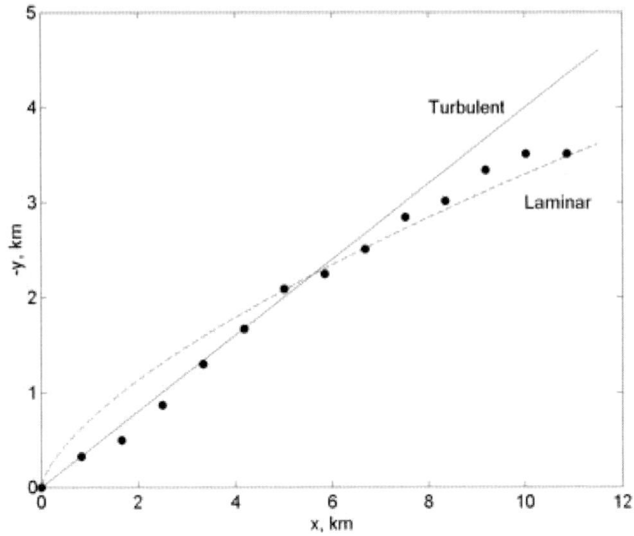


Fig. 5.8 Comparison of the spreading process of the imagery front with respect to the axis to theoretical models. Data points are taken from J section on Figure 5.6. The solid line represents width spreading of a turbulent jet. The dashed line represents width spreading of a laminar jet [Zheng *et al.*, 2004]

### 5.4.2.3. *Estimates of axial velocity and Reynolds number*

From theories of the two-dimensional jet, we derive a theoretical model for the axial velocity. Substituting (5.2.6) into (5.2.1) and computing the integration yield

$$M = \frac{4}{3} b\rho x^{2/3} u_{\max}^2. \tag{5.4.3}$$

Substituting (5.4.3) into (5.2.7) yields

$$u_{\max} = \frac{6A_H}{b^2 x^{1/3}}. \tag{5.4.4}$$

This relation indicates that if the eddy viscosity $A_H$ is reasonably chosen, and parameter $b$ is determined from the SAR image, the maximum axial velocity at a given point on the $x$-axis can be calculated. In our case, from the SAR image shown in Figure 5.5, we have the position of the peak gray level line of the SAR imagery $y = -2.1\,\text{km}$ for $x = 5.0\,\text{km}$ (point B in Figure 5.6). From Figure 5.7

we know that the position of the peak gray level corresponds to $\eta = -0.676$. Substituting these values into Eq. (5.2.4) yields $b = 1.06\,\text{km}^{1/3}$. We choose $A_H$ as $2 \times 10^2\,\text{m}^2\,\text{s}^{-1}$, a typical value for the upper ocean [Kundu, 1990]. From (5.4.4), we obtain the axial velocity at B

$$u_{\max} = 6 \times 10^{-1}\text{m}\,\text{s}^{-1}. \tag{5.4.5}$$

This value is comparable with field measurements. For example, Wong and Münchow [1995] gave ADCP observations collected in the same region and a regular distribution of the largest noise (= Raw data – Mean – Tide) vectors with amplitudes of the order of 60 cm $\text{s}^{-1}$ and downstream orientations at the deep channel [see Wong and Münchow, 1995, Figure 17(c)]. We interpret that the noise vectors represent the outflows generated by Delaware Estuary Plumes, thus, it is not surprising that our estimate for the jet axial velocity is consistent with their field measurements. This consistence implies that the two-dimensional jet is a good model for describing the behavior of this estuary plume.

Furthermore, from the data points shown in Figure 5.8, we determine the length scale of the plume as 6 km. Therefore, we obtain that an estimate of the Reynolds number $R_e (= UL/A_H)$ is 18, implying a relatively strong jet.

### 5.4.3. *Winter case*

In local winter, December–February, northerly winds prevail. Northerly alongshore winds produce an Ekman onshore surface drift. Therefore, the local winter is also called “downwelling favorable season” [Münchow and Garvine, 1993b]. The river discharges in winter are below the peak phase. Figure 5.9 shows a RADARSAT-1 SWB image covering the Delaware Bay mouth region taken at 11:15 UTC on February 7, 1997. We code the image by identification number 97207 and list the major specifications in Table 5.1. Buoy 44009 gave the wind speed as $3.9\,\text{m}\,\text{s}^{-1}$ and the wind direction as 335° at the SAR imaging time. The tide phase was at maximum flood.

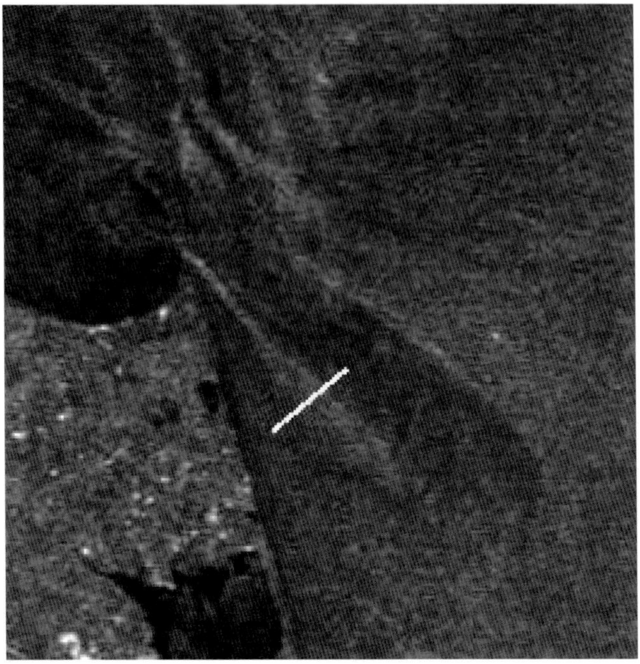

Fig. 5.9 RADARSAT-1 SWB image covering the Delaware Bay mouth region taken at 11:15 UTC on February 7, 1996. White line represents transect, along which gray level values are taken for comparison to the theoretical model [Zheng *et al.*, 2004]

On the image, the bright area on the lower left is the coastal region of southern Delaware. Just to the right, there is a light-bright, plume-like imagery, which is interpreted as a signature of the Delaware Estuary Plume. Figure 5.10 provides an interpretation map of the image. The axis of the plume takes an angle of 26° with the coastline, which is found 3° more offshore than that in summer case. The axial direction of the plume is also close to the direction of the deep channel (30° with the coastline), but turns right 4° facing downstream direction. The plume also has twin image fronts similar to the twin jet structure as shown in Figure 5.4. Again, we divide the plume along the axis into three sections: the source section characterized by an almost unchanged transverse width of 3.8 km, the jet section characterized by a gradually expanding width with a

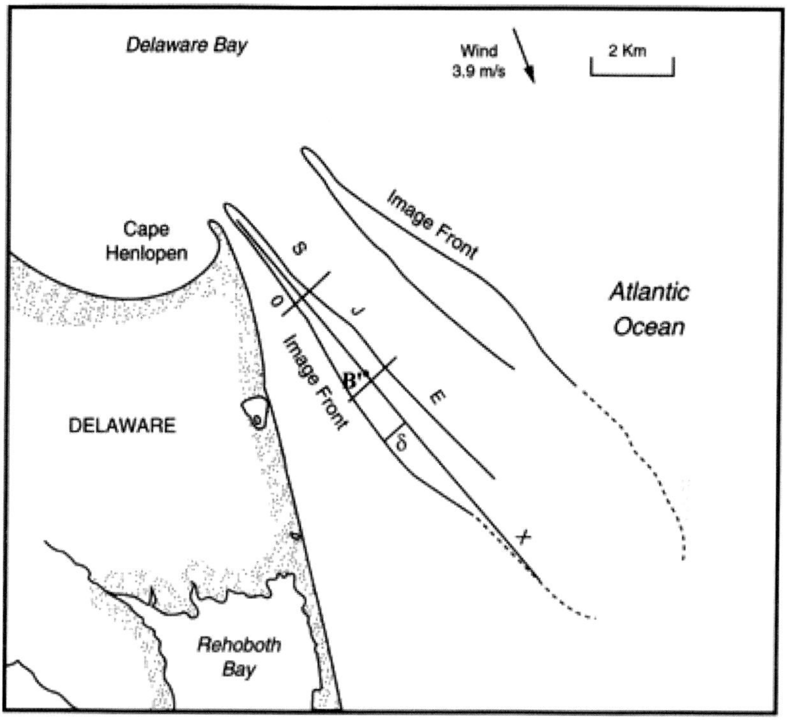


Fig. 5.10 Interpretation map of Figure 5.9 S, J and E represent the source, jet and extension sections of the jet, respectively. B′ is the point we used for estimating the axial velocity [Zheng *et al.*, 2004]

maximum value of 6.6 km, and the dispersion section characterized by a gradually shrinking width and vague imagery. The total length of the observed plume feature is 19.7 km, in which the length of the jet section is 4.0 km.

#### 5.4.3.1. *Transverse structure of the imagery front*

In Figure 5.11, we plot the normalized gray level values measured along the white line transect marked in Figure 5.9 along with the theoretical model curve for $\phi = 0$. One can see that the model curve fits the data points quite well. Two curves are characterized by a positive peak on the $x-$axis. This verifies again that the ocean

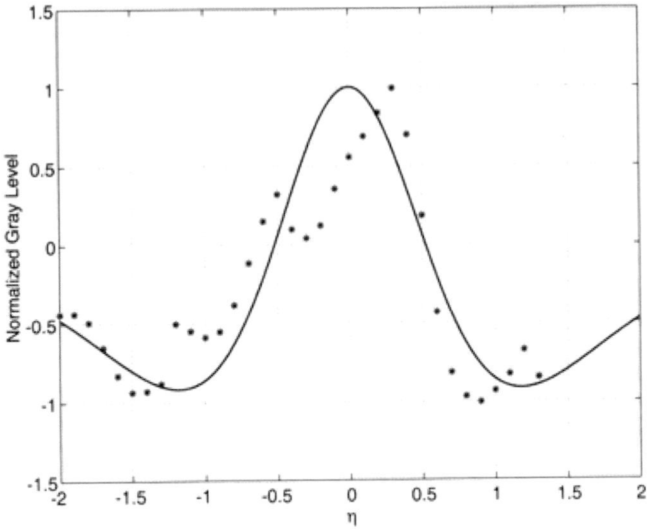


Fig. 5.11 Comparison of normalized gray level values measured along transect marked in Figure 5.9 (data points) and the theoretical model curve for $\phi = 0$ [Zheng *et al.*, 2004]

process that generates the imagery front has the transverse structure identical to the two-dimensional jet.

#### 5.4.3.2. *Spreading of the imagery front*

From the image shown in Figure 5.9, one can see that the front spreads once again within the jet section. We measure the front spreading with respect to the axis and plot these values in Figure 5.12. The solid curve in the figure represents a relation given by

$$\delta = 0.334x^{2/3}. \tag{5.4.6}$$

This equation represents the width spreading of a laminar jet [Kundu, 1990].

#### 5.4.3.3. *Estimates of axial velocity and Reynolds number*

We can also estimate the maximum axial velocity at a given point on the $x$-axis. For example, for point B′ in Figure 5.10, we have

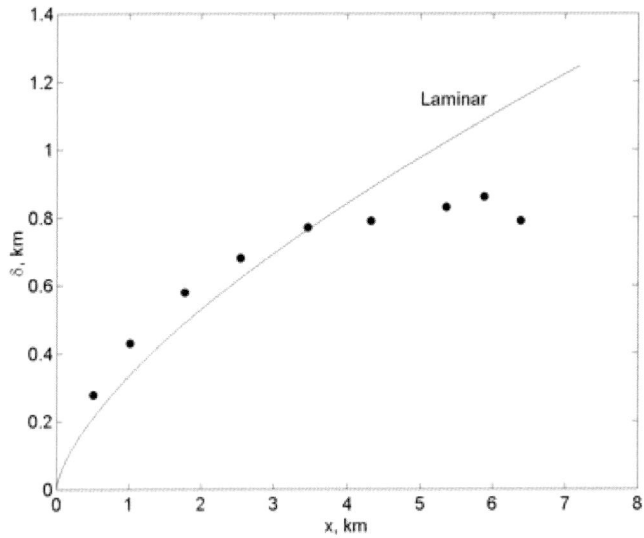


Fig. 5.12 Comparison of the spreading process of the imagery front with respect to the axis to theoretical model. Data points are taken from J section in Figure 5.10. The curve represents width spreading of a laminar jet [Zheng *et al.*, 2004]

$y = -0.77\,\text{km}$ for $x = 3.5\,\text{km}$, implying that $b\eta = 0.33$. For this relatively weak plume, we assume that the observed position of the boundary corresponds to $\eta = -0.2$. Substituting these values into Eq. (5.2.4) yields $b = 1.65\,\text{km}^{1/3}$. From Eq. (5.4.4) and choosing $A_H$ as $2 \times 10^2\,\text{m}^2\,\text{s}^{-1}$ again, we obtain an axial velocity at B′ of

$$u_{\max} = 4 \times 10^{-1}\text{m}\,\text{s}^{-1}$$

Furthermore, from the data plotted in Figure 5.12 we determine the length scale of the plume as 3.5 km. Therefore, we obtain an estimate of the Reynolds number $R_e$ of 3, implying a relatively weak jet.

### 5.4.4. *Spring case*

In local spring, March–May, the winds are in a transit period from winter to summer. The river discharge is in the peak phase, particularly in April and May. Figure 5.13 shows an ERS-2 SAR

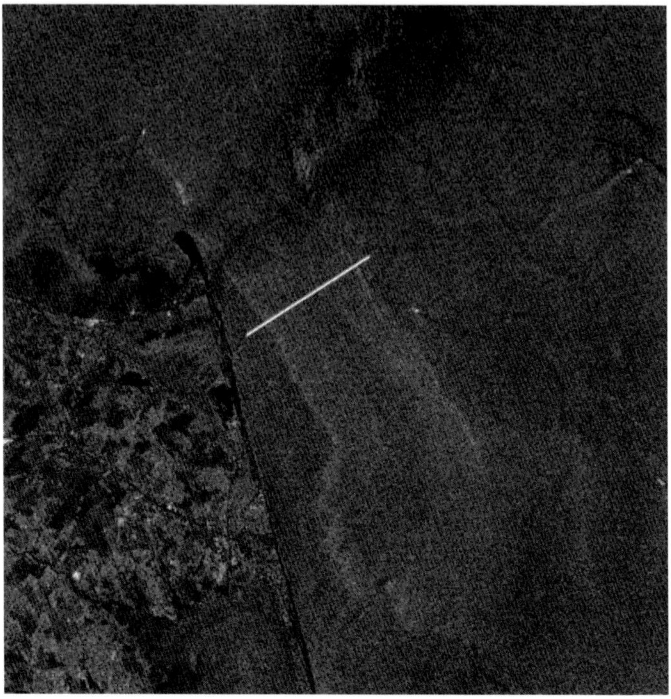

Fig. 5.13 ERS-2 Standard image covering the Delaware Bay mouth region taken at 15:43 UTC on May 14, 1997. White line represents transect, along which gray level values are taken [Zheng *et al.*, 2004]

image of the Delaware Bay mouth region taken at 15:43 UTC on May 14, 1997. We code the image by identification number 97514 and list the major specifications in Table 5.1. Buoy 44009 registered a wind speed of 5.1 $\mathrm{m\,s^{-1}}$ and a wind direction of 185° at the SAR imaging time. The tide phase was at the early flood.

The bright area on the lower left side of the image is the coastal region of southern Delaware. To the right, there is a light-bright, plume-like feature, which is interpreted as a signature of the Delaware Estuary plume. The plume has twin image fronts corresponding to the twin jet structure shown in Figure 5.4. The total length of the plume is about 23.1 km. According to the imagery features, the plume can be divided into only two sections: (1) a width-unchanged section characterized by generally parallel image fronts and an almost unchanged transverse width of 3.2 km and (2) a width-changed section characterized by a sharp width expansion to a maximum value of 5.3 km, and a gradually shrinking width as well as a less defined imaging of the front. If we assume that the plume axis is parallel to the fronts of the width-unchanged section, the axis angle would be 17° relative to the coastline, which is an additional 7° shoreward when compared to the summer case.

Unlike the summer and winter cases, image 97514 shows the plume as an integrated body with relatively uniform bright tunes. This may be explained by spring being the peak period for fresh water discharge in the Delaware Bay. The salinity of seawater in the plume, therefore, takes the lowest values of the year, which may reach as low as 26 psu or as much as 6 psu lower than surrounding continental shelf water [Münchow and Garvine, 1993a]. In this case, the salinity signals are intensive enough to overwhelm the current modulation signals. Figure 5.14 shows a curve representing the radar return intensity signals measured along transect marked by the white line in Figure 5.13. One can see that the twin fronts located on the

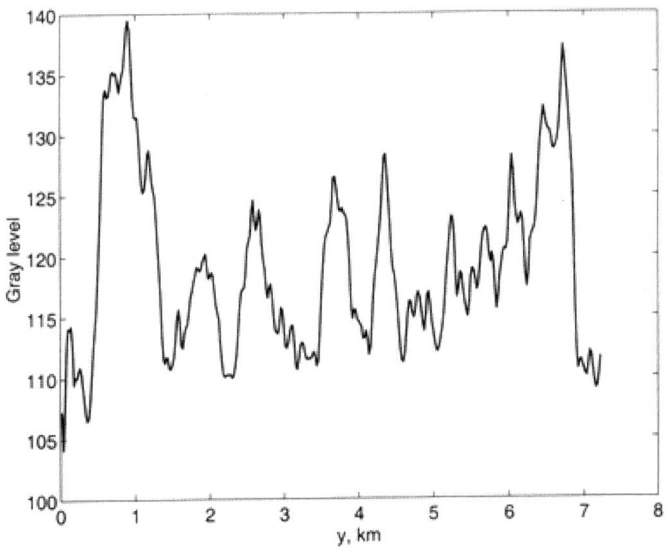


Fig. 5.14 Curve of gray level values taken along the white transect marked in Figure 5.13. [Zheng *et al.*, 2004]

two sides of the plume still generate strong radar return signals, but it is hard to determine the boundaries of the two jets within the plume. Hence, we are unable to extract further information related to the jet dynamics from this image.

## 5.5. Conclusions and Discussion

From the physics of radar imaging, ocean gravity–capillary wave spectra, and analytical solutions of a fluid jet, we derive theoretical model of a radar image for an ocean jet. If the effects of the ocean waves are ignored, the theoretical model indicates that the jet radar image intensity depends strongly on the jet current structure and the sea surface wind direction with respect to the jet axis. The peak value of image intensity is dependent on the maximum axial velocity, while the position of the peak with respect to the axis depends on the wind direction. In all cases, the jet SAR image has a single peak.

We develop a twin jet physical model to describe the transverse current structure of the Delaware Bay plumes. The analysis indicates that this model not only conforms field measurements and also explains the twin SAR imagery fronts.

Interpretations of SAR images of the Delaware Bay plumes from summer 1996 to spring 1997 indicate that in all seasons, low salinity plumes appear as bright, jet-like patterns on the SAR images. As listed in Table 5.2, measurements give the axial length of the plumes at about 20 km and the transverse width varying from 3 to 8 km. In summer and winter, the plumes can be divided into three sections along the axis: the source, jet and dispersion sections. The typical axial velocity is estimated at 0.6 and $0.4\,\mathrm{m\,s^{-1}}$, and the Reynolds number at 18 and 3, respectively. In spring, salinity signals overwhelm the current modulation signals, thus, the plume appears as an integrated body with relatively uniform bright tunes on the SAR image.

On the basis of the spreading of SAR imagery fronts, which, in fact, represent the streamlines, we determine the physical natures of the plume jets. In summer, the jet behaves like a turbulent jet off the source region down to about 5 km downstream, and like a laminar

Table 5.2 Seasonal variability of Delaware Bay plumes derived from SAR image interpretation

| **Date**<br>**Tide**[a] | **Summer**<br>**August 13, 1996**<br>**Maximum Flood** | **Winter**<br>**February 7, 1997**<br>**Maximum Flood** | **Spring**<br>**May 14, 1997**<br>**Early Flood** |
|---|---|---|---|
| Wind speed[a], $m\,s^{-1}$ | 9.2 | 3.9 | 5.1 |
| Wind dir[a]. with jet | 135° | 0° | 135° |
| Plume width: source region, km | 3.0 | 3.6 | 3.9 |
| Plume width: maximum, km | 7.7 | 6.6 | 6.6 |
| Axial dir. with coastline | 23° | 26° | 17° |
| Axial length of plume, km | 21.3 | 19.7 | 23.1 |
| Length of source and jet, km | 11.0 | 8.0 | 13.7 |
| Axial velocity, $m\,s^{-1}$ | 0.6 | 0.4 | N/A |
| Reynolds number | 18 | 3 | N/A |
| Nature of jet | Turbulent/laminar | Laminar | N/A |

[a]Field data

jet from about 5 km down to about 10 km downstream. In winter, the jet behaves like a laminar jet off the source region down to about 4 km downstream.

We note that the three analyzed cases of Delaware Bay plumes occurred at maximum flood tide or early flood tide of the Delaware Bay entrance [Flater *et al.*, 1997]. This implies that countercurrents against the general tidal inflows form the plumes. Although, this argument seems hard to be explained, it is in fact highly possible. Li [2001] presented a three-dimensional solution for tides in channels with arbitrary lateral depth variation. One of his conclusions is that the time of flood (ebb) in deep water lags that in shallow water, and the time of flood (ebb) on the surface lags that at the bottom. This conclusion provides a dynamic basis for our interpretations for the Delaware Bay plumes, which occur on the surface layer of a narrow and deep channel. It is the longest flood time lag to maintain the existence of the Delaware Bay plumes as a countercurrent. Furthermore, we examine ADCP observations collected in the same region by Wong and Münchow [1995] and find a regular distribution of the largest noise vectors with amplitudes of the order of $60\,\mathrm{cm\,s^{-1}}$ and downstream orientations at the deep channel [see Wong and Münchow, 1995, Figure 17(c)]. This seems to further indicate that outflow at the deep channel may always exist without regard to the tide stage.

The case studies demonstrate that the methodology developed in this chapter can be used to establish the relationship between SAR signals and flow field. Therefore, there should be potential applications to a wide range of problems.

## Chapter 6

# Ocean Internal Waves: I. Dynamics and Parameter Extraction

### 6.1. Introduction

Ocean internal waves (IWs) are generated by vertical oscillations of seawater occurring within the ocean, in contrast to ocean surface waves occurring on its surface. To sustain an oscillation, the restoring force is a necessary condition. For ocean surface waves, the restoring force is the gravity force, thus, ocean surface waves are also called ocean surface gravity waves. For IWs, the restoring force is the reduced gravity force or the buoyance force, which is depending on the vertical density difference due to changes in temperature and/or salinity of seawater. This implies that IWs may only occur in the vertical stratified ocean. If the density changes continuously, IWs can propagate horizontally and vertically through seawater. IWs, also called internal gravity waves and many other names, go by the fluid stratification, generation mechanism and influence of external forces. If propagating horizontally along an interface, where the density rapidly increases with the water depth that is called the thermocline, they are specifically called interfacial (internal) waves. If the amplitudes of interfacial waves are large enough so that the nonlinearity becomes important, they are called internal solitary waves or internal solitons, which can be described by solutions of the Korteweg and de Vries (KdV) equation [Korteweg and de Vries, 1895]. If generated by flow over bottom topography, they are called lee waves. If generated by tidal flow over submarine ridges or the continental shelf with frequencies of semidiurnal or diurnal tides, they are called semi-diurnal internal tides or diurnal internal tides, respectively, which are also referred to as baroclinic tides. If their frequencies are comparable to the earth rotational frequency so that the Coriolis effect becomes important in their dynamics, they are called near-inertial oscillations (NIO), which are usually generated by sudden wind events [Zhao *et al.*, 2004].

IWs travel at much slower speed than surface waves, as a result of the density difference between the upper and lower water layers being much less than that between air and water. Their wavelengths may reach tens of kilometers rather than in meters as for surface waves, and their wave amplitudes can be tens of meters, even reach greater than one hundred meters. The frequency range of IWs is limited to below the Brunt–Väisälä frequency and near the local inertial frequency ($N(z) \geq \omega \sim f$). All the above-mentioned IWs belong to sub-mesoscale processes as defined in Section 2.1.1 and shown in Figure 2.1. They are distinguished from large-scale waves, such as Rossby waves propagating in the open ocean.

IWs play an important role in ocean mixing, acoustic signal propagation, sediment re-suspension and cross-shore pollutant transport. The research of IWs may be a great help to better understand upper ocean dynamics and the ocean interior structure. Their breaking is an effective mechanism for mixing nutrient-rich water from the bottom to the euphotic layer. These waves may also play an important role in maintaining the large-scale, deep circulation by providing downward mixing of heat. In particular, the huge power of large amplitude IWs is a non-negligible factor, which may have critical impacts on submarine navigation, coastal engineering and oil exploration [Zhao, 2005].

IWs are one of the earliest ocean phenomena detected in SAR images [Apel and Gonzalez, 1983; Elachi and Apel, 1976; Fu and Holt, 1984]. In order to understand the radar imaging mechanisms of ocean IWs, critical efforts were made in early 1980s, such as the Joint Canada–U.S. Ocean Wave Investigation Project (JOWIP) in 1983 [Hughes and Dawson, 1988] and the SAR Internal Wave

Signature Experiment (SARSEX) in 1984 [Gasparovic *et al.*, 1988]. The results obtained from these projects are broadly used as the basis for interpretation and application of ocean SAR images taken by follow-up missions. Liu *et al.* [1998] used ERS-1 SAR images to analyze IW evolution process in the East and South China Seas (SCSs). Hsu and Liu [2000] determined the group velocity of ocean IWs and observed the nonlinear wave–wave interaction shown in RADARSAT-1 ScanSAR wide images. SAR images were also used for studies of generation mechanisms [Yuan *et al.*, 2006; Zhao *et al.*, 2004; Zheng *et al.*, 2008b], polarity change [Zhao *et al.*, 2003] and statistics of IWs in SCS [Wang *et al.*, 2012; Zheng *et al.*, 2007].

## 6.2. IW Dynamics

### 6.2.1. *Linear theories*

An equation that governs the motions of IWs can be written in the form of [Apel, 1995]

$$\frac{\partial^2}{\partial t^2}\nabla^2 w + \left(N^2\nabla_h^2 + f^2\frac{\partial^2}{\partial z^2}\right) w = 0, \tag{6.2.1}$$

where $w$ is the vertical velocity, $f$ is the Coriolis parameter, $\nabla$ is the Laplacian, $\nabla_h$ is the horizontal Laplacian and $N$ is the Brunt–Väisälä frequency or the buoyancy frequency, which is a decisive factor for IW occurrence defined as

$$N^2 = -\frac{g}{\rho}\frac{d\rho}{dz}, \tag{6.2.2}$$

where $g$ is the gravitational acceleration and $\rho$ is the potential density of seawater. Assuming that (6.2.1) has a plane wave solution

$$w = w_0 \exp[i(kx + ly + mz - \omega t)], \tag{6.2.3}$$

where $w_0$ is the wave amplitude, $k$, $l$ and $m$ are the wavenumbers of $x$-, $y$- and $z$-components, respectively, and $\omega$ is the angular frequency. Substituting (6.2.3) into (6.2.1) yields the dispersion relation for IWs in a rotating, stratified fluid

$$\omega^2 = \frac{N^2(k^2+l^2) + f^2m^2}{k^2 + l^2+m^2}. \tag{6.2.4}$$

In the finite boundary case, we further assume that the wave solution of the vertical velocity to be the product of a horizontally traveling plane wave and a vertical eigenfunction $\phi(z)$, which must meet the boundary conditions at the surface and the bottom, *i.e.*,

$$w(\boldsymbol{x}, t) = \phi(z)\exp[i(kx + ly + mz - \omega t)]. \tag{6.2.5}$$

Substituting (6.2.5) into (6.2.1) yields an eigen equation

$$\frac{\partial^2\phi}{\partial z^2} + (k^2 + l^2)\left[\frac{N^2(z) - \omega^2}{\omega^2 - f^2}\right]\phi = 0, \tag{6.2.6}$$

where the quantity multiplying $\phi$ is the square of the effective vertical wavenumber $m^2$, which varies with $z$ because of the variable buoyancy frequency $N$. From (6.2.6), one can see the following special cases for frequency relations:

(1) When $\omega^2 = N^2$, so that $m = 0$. This is a condition termed as cutoff.
(2) When $\omega^2 = f^2$, so that $m \rightarrow \infty$. This is a condition called a resonance.
(3) For $\omega > N$ or $\omega < f$, the vertical wavenumber is imaginary, and the solutions are evanescent.
(4) For $f < \omega < N$, the solutions are oscillatory with forms determined by the functional behavior of $N(z)$. In other words, the stratified ocean constitutes a band-pass waveguide for IWs, and allows propagation only between the two characteristic frequencies $f$ and $N$.

Since $N \gg f$ in the ocean, IWs are most nearly like the buoyancy oscillations. When $\omega \rightarrow f$, however, IWs take on characteristics of inertial oscillations [Apel, 1995].

For an arbitrary form of $N(z)$, (6.2.6) does not always yield an analytical solution. Hence the solution is usually accomplished numerically using observed profiles of $N$. The case of $N =$ constant is solvable in terms of sinusoidal functions under the boundary conditions

$$\phi(-H) = 0 \quad \text{and} \quad \phi(0) = 0$$

at the bottom and the surface, respectively. The complete solutions are then written in terms of summations over the normal modes

$$w(\boldsymbol{x}, t) = \sum_{n=-1}^{-\infty} w_{0n} \sin m_n(z + H) \cos[(kx + ly - \omega_n t)],$$

for

$$m_n = n\pi/H, \quad n = -1, -2, -3, \ldots, \tag{6.2.7}$$

where the eigenfrequencies are

$$\omega_n^2(\boldsymbol{k}) = \frac{N^2(k^2 + l^2)H^2 + f^2(n\pi)^2}{(k^2 + l^2)H^2 + (n\pi)^2} \tag{6.2.8}$$

The first three orthogonal eigenfunctions for vertical motion of IWs in the ocean with a constant-$N$ buoyancy profile are shown in Figure 6.1.

As an example, Figure 6.2 shows a case of near linear IWs, which were observed in the Lombok Strait of Indonesia by an EK500 Echosounder operating at a frequency of 38 kHz during the 2005 Indonesian throughflow cruise. The echogram clearly shows IWs with a wavelength of ~1.8 km. The IW amplitude (peak to trough) exceeds 100 m.

### 6.2.2. *Nonlinear theories*

#### 6.2.2.1. *The KdV equation*

John Scott Russell made the first documented observation of surface solitary waves in 1834. A solitary wave was induced by a boat moving in a canal and propagated more than one mile while keeping its shape. Russell termed this type of wave motion "solitary wave", mainly because it stands alone and apart from other types of oscillatory wave motion. Later, Korteweg and de Vries [1895] derived a famous nonlinear equation (KdV equation) to describe the behavior of

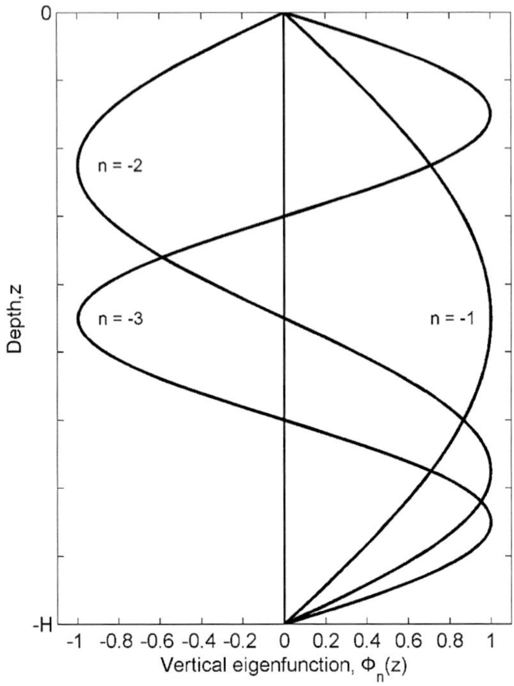


Fig. 6.1 First three orthogonal eigenfunctions for vertical motion of IWs in the ocean with a constant-$N$ buoyancy profile

solitary waves. Since then, the oceanographers have documented that the behavior of a single soliton in a two-layer model ocean can be well described by KdV equation in a dimensional form of [Osborne and Burch, 1980; Ostrovsky and Stepanyants, 1989; Zheng *et al.*, 2001a]

$$\frac{\partial \eta}{\partial t} + C\frac{\partial \eta}{\partial x} + \alpha\eta\frac{\partial \eta}{\partial x} + \beta\frac{\partial^3 \eta}{\partial x^3} = 0, \tag{6.2.9}$$

where $\eta(x, t)$ is the interface displacement between the two layers from the unperturbed level induced by the single soliton, and

$$C = \left[\frac{g(\rho_2 - \rho_1)h_1 h_2}{\rho_2 h_1 + \rho_1 h_2}\right]^{1/2} \tag{6.2.10}$$

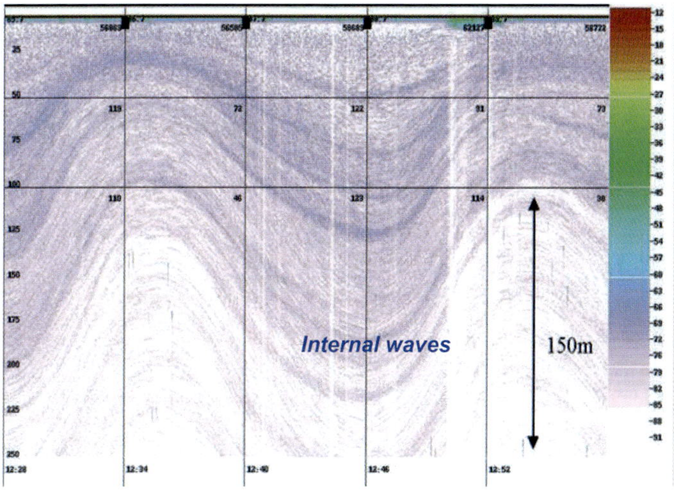


(a)

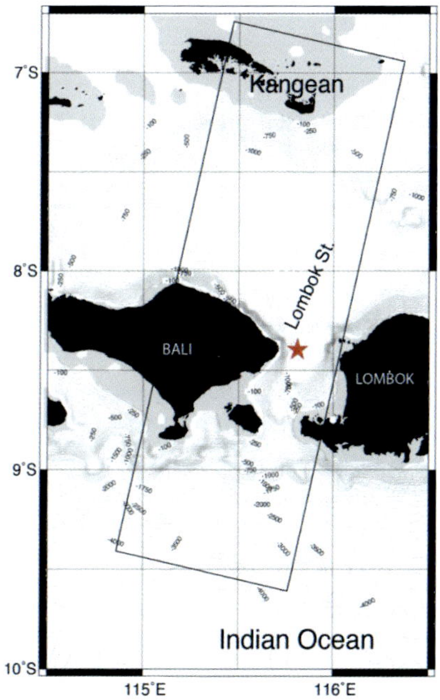


(b)

Fig. 6.2 A case of near linear IWs. The acoustical echogram of the water column (a) measured by an EK500 Echosounder onboard a standing R/V during the 2005 Indonesian throughflow cruise in Lombok Strait. (b) Vertical axis is depth in m. Horizontal axis is time. Color codes represent relative backscatter strength in dB [Susanto *et al.*, 2005] which is defined as the linear phase speed,

$$\alpha = \frac{3}{2}\frac{C}{h_1 h_2}\frac{\rho_2 h_1^2 - \rho_1 h_2^2}{\rho_2 h_1 + \rho_1 h_2} \tag{6.2.11}$$

which is defined as the nonlinear coefficient, and

$$\beta = \frac{C h_1 h_2}{6}\frac{\rho_1 h_1 + \rho_2 h_2}{\rho_2 h_1 + \rho_1 h_2} \tag{6.2.12}$$

which is a measure of dispersion. Here, $h_1$ and $h_2$ are the thickness of the upper and lower layers, respectively, and their water densities are $\rho_1$ and $\rho_2$, which are uniform within the layer as shown in Figure 6.3.

Equation (6.2.10) has a soliton solution in the form of

$$\eta = \eta_0 \,\mathrm{sech}^2\left[\frac{(x - Vt)}{\Delta}\right], \tag{6.2.13}$$

where the soliton velocity

$$V = C + \frac{\alpha\eta_0}{3} \tag{6.2.14}$$

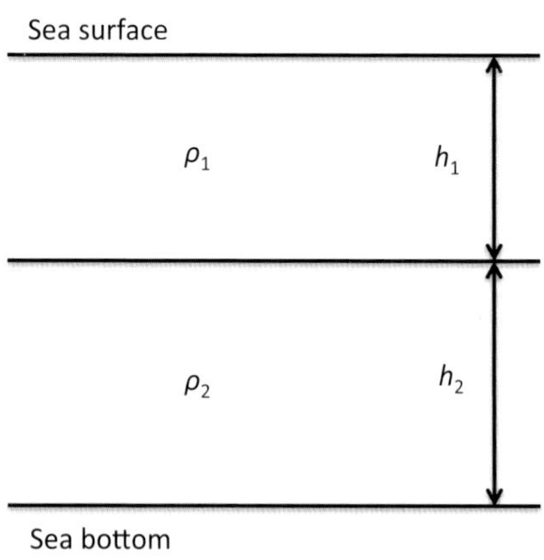


Fig. 6.3 Two-layer model ocean

and

$$\Delta = \left(\frac{12\beta}{\alpha\eta_0}\right)^{1/2} \tag{6.2.15}$$

which is defined as the characteristic half width of the soliton.

Besides the single soliton solution, there exists the dnoidal wave solution, *i.e.*, solitary wave packet solution, to the KdV equation [Gurevich and Pitaevskii, 1973]. Examples of solitary wave packets measured in the northern SCS is shown in Figure 6.4, which shows time series of vertical profile of water temperature at station $S$ located at 21°54′N, 117°11′E on April 23–24, 2001. One can see that there was a thermocline between 40 and 60 m and three depression solitary wave packets $P_1$, $P_2$ and $P_3$, which propagated along the thermocline and passed through the station at around 18:00 on April 23, 3:00 and 18:00 on April 24, 2001, respectively. The three packets consist of 2, 3 and 3 solitons, which show the same trend of decreased amplitudes from the leading soliton to the rear one. The separate distances between two solitons in $P_2$ and $P_3$ also show a decrease trend. These are remarkable features of the dnoidal solution.

#### 6.2.2.2. *The PKdV equation*

When an internal soliton propagates along shoaling beach or thermocline as shown in Figure 6.5, its behavior can be described by the perturbed KdV (PKdV) equation in the form of [Newell, 1985]

$$q_\tau + 6qq_\theta + q_{\theta\theta\theta} = -\frac{9}{4}\frac{D_\tau}{D}q, \tag{6.2.16}$$

where $\tau$ is a rescaled distance coordinate

$$\tau = \frac{1}{6}\int^x D^{1/2}dX, \tag{6.2.17}$$

$\theta$ is the retarded time

$$\theta = -t + \frac{1}{\varepsilon}\int^x \frac{dX}{D^{1/2}}, \tag{6.2.18}$$

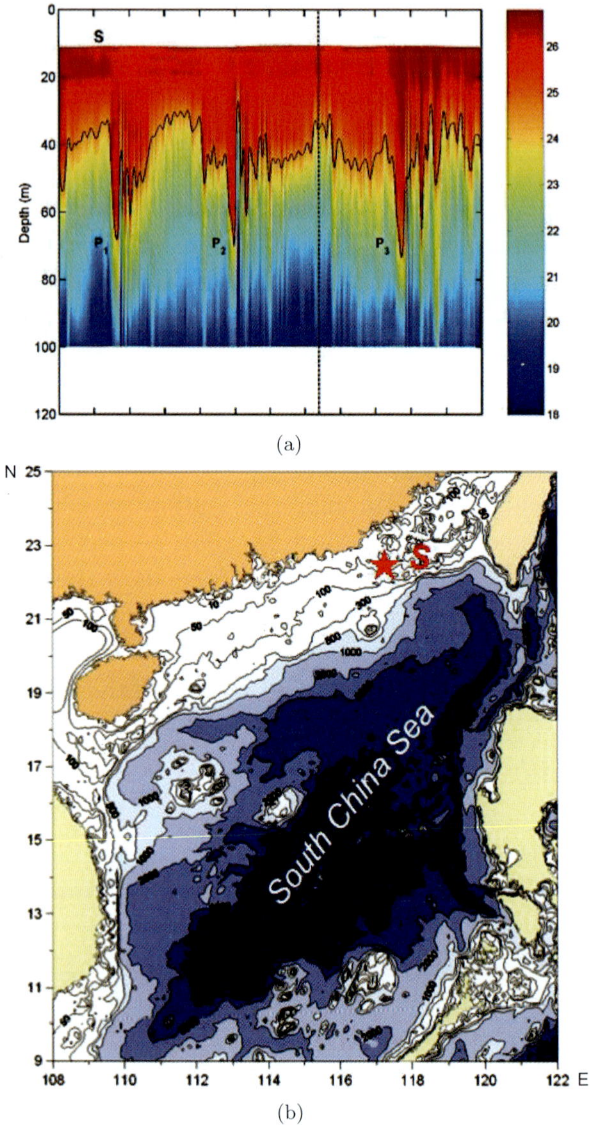


Fig. 6.4 (a) Three internal solitary wave packets ($P_1$–$P_3$) and (b) mooring station $S$ (120 m deep) in the northeastern SCS. The starting time of 36-h series of temperature profile was 12:00 UTC on April 23, 2001. Color codes are in °C. The fluctuating curve represents the isotherm of 24.5°C. The vertical dashed line represents the imaging time of the RADARSAT-1 SAR image shown in Figure 6.9(a) (Courtesy: Zhongxiang Zhao, University of Washington)

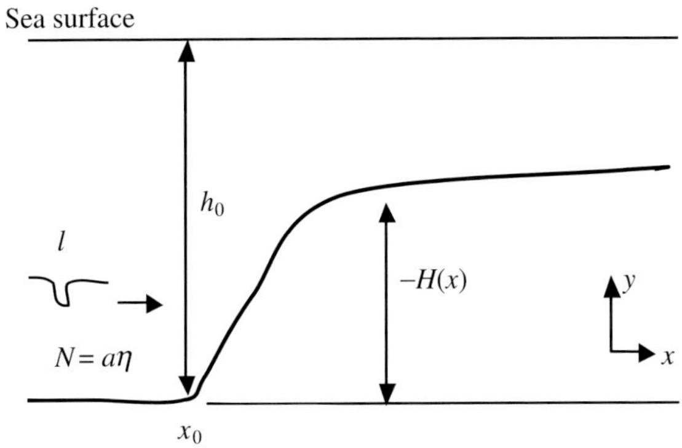


Fig. 6.5 Physics model for an internal soliton propagating along a shoaling thermocline [Zheng *et al.*, 2007]

where $t$ is the time, $D$ is the non-dimensional depth of the upper layer, $D = 1 + h$, in which $h = H(x)/h_0$, $X = \varepsilon x$ and the subscript represents partial differential with respect to that variable.

Solutions of (6.2.16) can be expressed as

$$q(\theta, \tau) = q^{(0)} + \sigma q^{(1)} + \cdots, \tag{6.2.19}$$

where

$$q^{(0)} = 2\eta^2 \mathrm{sech}^2 \eta(\theta - 4\eta^2\tau - \theta_0), \tag{6.2.20}$$

in which $\eta$ is the amplitude parameter, a slowly varying function of the distance, $\theta_0$ is the initial retarded time and

$$\begin{aligned} q^{(1)} = &\frac{\Gamma}{6\eta}\{-1 + \tanh \eta\xi + \mathrm{sech}^2\eta\xi(3 - 3\eta\xi \tanh \eta\xi \\ &+ 2\eta\xi - 2\eta^2\xi^2 \tanh \eta\xi)\} \\ &+ \frac{\Gamma}{2}(1 - \tanh \eta\xi)\,\mathrm{sech}^2\eta\xi, \end{aligned} \tag{6.2.21}$$

where

$$\Gamma = \frac{9}{4}\frac{D_\tau}{D}, \tag{6.2.22}$$

$\xi(= \theta - \bar{\theta})$ is a time variable in a coordinate system moving with the solitary wave, and $\sigma$ is the order of $\Gamma$ and satisfies $0 < \varepsilon \ll \sigma \ll 1$.

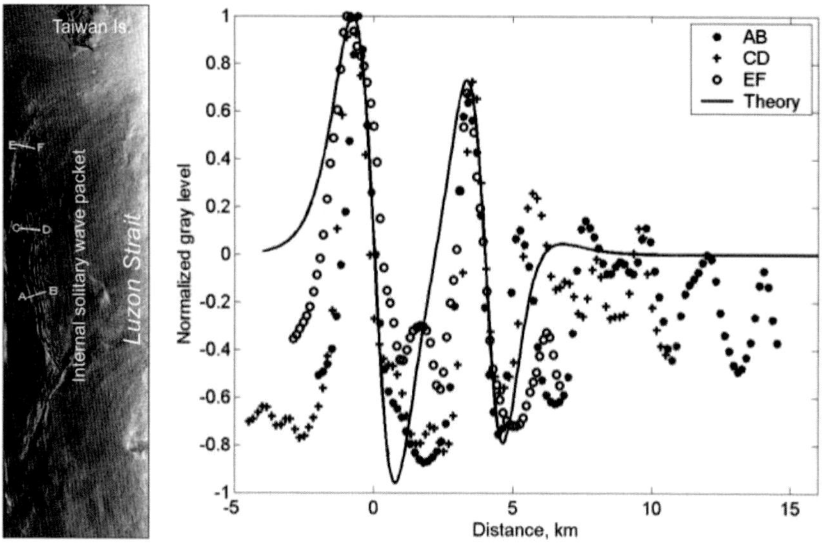


(a) (b)

Fig. 6.6 (a) ERS-1 SAR images of the Luzon Strait taken at 02:29 UTC on June 16, 1995 Dark area on the top is the southern tip of Taiwan Island. Grouped bright lines are imagery of solitary IW packets propagating westward. (b) A comparison of the PKdV solitary wave solutions (curve) to the SAR-observed ocean IW packets (data points) taken from three transects AB, CD and EF as marked on the image [Yuan *et al.*, 2006]

Solutions (6.2.20) and (6.2.21) represent a large soliton with a small oscillatory tail.

Figure 6.6(a) shows an example of a solitary wave packet imaged by ERS-1 SAR over the Luzon Strait at 02:29 UTC on June 16, 1995. The SAR swath of about 100 by 400 km scans the strait from the north to the south. The dark area on the top is the southern tip of Taiwan Island. One can see that on the west (left) side of the image, there is an IW packet led by a soliton with a crest line as long as 200 km followed by a group of small waves with an average wavelength of about 1 km and gradually shortened crest lines. The westward curvatures of the crest lines imply that the packet propagates westward [Yuan *et al.*, 2006].

In order to compare the structure of this packet with solutions to PKdV equations (6.2.20) and (6.2.21), the solutions are transformed into the form of a SAR image [Zheng *et al.*, 2001b]. The parameters are chosen as follows: $\Gamma = 0.3$, $\sigma = 0.3$, $\eta = 0.42$, $t = 0$, $\theta_0 = 0$, $\bar{\theta} = 8$, and the distance scale is 0.5 km. A normalized waveform is shown as a curve in Figure 6.6(b). The data points are gray level values taken from transects marked in Figure 6.6(a), corrected by SAR incidence angle and normalized by the maximum values. Here, the SAR system is assumed to act as a linear sensor, so that the gray level values of SAR imagery are used to represent the radar return intensity [Yuan *et al.*, 2006]. One can see that the solutions to the PKdV equation describe the major features of the observed solitary wave packet well.

To understand the ocean stratification conditions for occurrence of the above solitary wave packet, Figure 6.7 shows the horizontal distribution of monthly mean upper layer depth in June 1995 with a resolution of 2° latitude by 0.5° longitude along transect of 21°N from 116° to 122.5°E across the Luzon Strait. The upper layer depth is defined as the thickness from the sea surface to the depth of the 15°C isotherm [Zheng *et al.*, 2007], which is derived from the field data measured by conductivity-temperature-depth (CTD) profilers, the mechanical (MBT), expendable bathythermographs (XBT) and digital bathythermographs (DBT) [Boyer and Levitus, 1994].

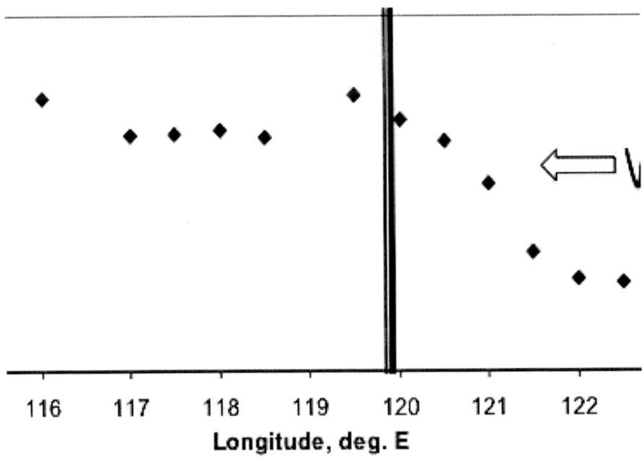


Fig. 6.7 Monthly mean upper layer depth along transect across the Luzon Strait in June 1995. Double vertical lines near 120°E represent the location of the leading soliton shown in Figure 6.6(a) [Zheng *et al.*, 2007]

Comparing Figure 6.7 to the physics model shown in Figure 6.5, one can see a similarity between the two, if the $x$-axis of Figure 6.7 is arranged from east to west, *i.e.*, in the reverse order of longitudes. This implies that the solutions to the PKdV equation stand in this case. The location of the leading soliton shown in Figure 6.6(a) with respect to the shoaling thermocline is marked in the figure. One can see that the leading soliton has climbed up to the shallow side of the upper layer. Recent investigations suggest that the initial disturbances of internal solitary waves in the Luzon Strait are generated by the internal tides, which are generated by the tidal waves propagating from the Pacific passing through the narrowing Luzon Strait [Zhao, 2014]. Thus, it is reasonable to interpret that the solitary wave packets shown in Figure 6.6(a) are products generated by the growth and fission of the initial disturbance in the form of a single soliton propagating along the shoaling thermocline.

The above analyses indicate that the solutions to the PKdV equation are of significance for understanding internal soliton evolution along a shoaling thermocline or shoaling bottom topography. In the next section, we will show that the single soliton solution to the KdV equation is a key for extracting the dynamical parameters of IWs from SAR images. Thus, nonlinear theories are even more suitable than linear theories for interpretation of behavior of IW packets.

## 6.3. SAR Imaging of IWs and Parameter Extraction

Form Section 1.4, we know that a radar image can be considered as a two-dimensional presentation of the intensity of radar return signals. The intensity depends on the two-dimensional wavenumber spectral density of the ocean surface wave field that satisfies the Bragg resonance scattering condition. On the other hand, variable background currents may modulate spatial distribution of spectral density of the ocean surface wave field as shown in Eqs. (1.4.10) and (1.4.11). In the case of IWs, elevation and depression of thermocline induced by IWs modulate sea surface capillary and short gravity waves into smoother divergence zone and rougher convergence zone. Thus, SAR images the rougher and smoother zones on the sea

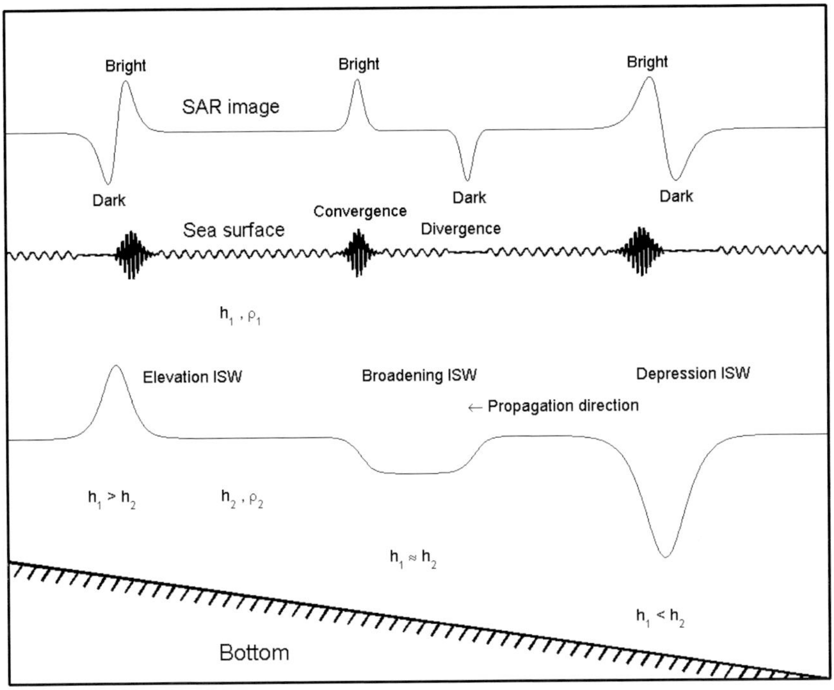


Fig. 6.8 Schematic diagram of SAR imaging of depression, broadening and elevation IWs (Courtesy: Zhongxiang. Zhao, University of Washington)

surface as brighter and darker stripes, respectively. Figure 6.8 is a schematic diagram of depression, broadening and elevation IWs, their convergence and divergence zones on the sea surface and different signatures in SAR image.

From Figure 6.8, one can see that a depression IW appears as a brighter stripe leading a darker stripe, an elevation IW appears as a darker stripe leading a brighter stripe and a broadening IW appears as a pair of separated darker and brighter stripes [Zhao, 2005]. The RADARSAT-1 SAR imaged all the three types of IWs in the northeastern SCS on April 24, 2001. The SAR image and interpretation map are shown in Figures 6.9(a) and (b), respectively.

From Figures 6.9(a) and (b), one can see that solitary wave packets $P_1$ and $P_2$, are characterized by a darker stripe leading a brighter stripe in SAR image, so that interpreted as elevation IWs

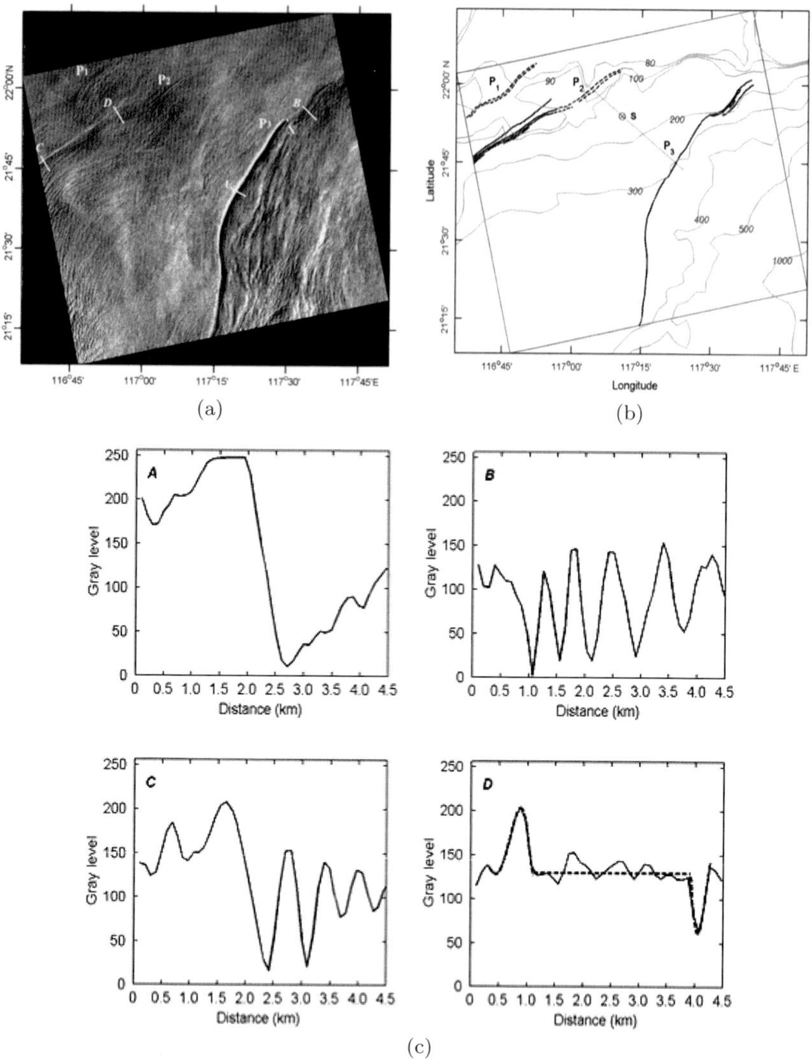


Fig. 6.9 (a) RADARSAT-1 SAR image taken at 10:00 UTC on April 24, 2001 (shown as the vertical dashed line in Figure 6.4(a)) over the northeastern SCS $P_1 - P_3$ are imagery of three neighboring IW packets, which were generated by three consecutive tidal cycles (Courtesy: Zhongxiang Zhao, University of Washington). (b) Interpretation map of (a) Solid frame represents the coverage of the SAR image. A mooring buoy is located at $S$ (21°54′N, 117°11′E). The numerals on isobaths are in m (Courtesy: Zhongxiang Zhao, University of Washington). (c) Gray level profiles (GLPs) along sections A, B, C and D in (a) dashed curve in D represents the smoothed gray level of a broadening IW (Courtesy: Zhongxiang Zhao, University of Washington)

shown as dashed curves in Figure 6.9(b). While $P_3$ is characterized by as a brighter stripe leading a darker stripe, so that interpreted as depression IWs shown as solid curves in Figure 6.9(b). The three solitary wave packets propagated northwestward and passed through the mooring station $S$ at around 18:00 on April 23, 3:00 and 18:00 on April 24, 2001, respectively, as shown by temperature profile time series in Figure 6.4(a).

Figure 6.9(c) shows four GLPs taken along sections A, B, C and D in Figure 6.9(a). From these gray level curves, one can see the detailed structure of IW packets. GLP A represents SAR imagery of a single depression soliton, of which the analytical solution will be derived in Section 6.4. GLP B represents a linear elevation wave packet. GLP C represents a depression solitary wave packet, *i.e.*, the dnoidal wave solution to the KdV equation [Gurevich and Pitaevskii, 1973]. GLP D represents a broadening depression IW.

On the basis of the above analysis results, we summarize the IW parameters that may be extracted directly from SAR image as follows:

(1) Exact occurrence time with accuracy up to one second.
(2) Exact occurrence location with accuracy up to one SAR imagery pixel.
(3) Number and geographic distribution of packets.
(4) Distance between two packets with accuracy up to one SAR imagery pixel.
(5) Length of crest line with accuracy up to one SAR imagery pixel.
(6) Polarity of packet.
(7) Curvature of crest line or propagation direction.
(8) Waveforms and detailed structure within a wave packet.
(9) Number of single waves or solitons in a wave packet.
(10) Wavelength (separate distance between two consecutive solitons) with accuracy up to one SAR imagery pixel.
(11) Relative amplitude of soliton and
(12) Information of interaction of two packets if colliding.

These data and information are unique and significant for operational applications and further dynamical analysis of IWs.

## 6.4. Indirectly Extracted Parameters: Soliton Characteristic Half Width and Amplitude

### 6.4.1. *Analytical solution of a single ocean internal soliton SAR Image*

In addition to the above parameters, the characteristic half width of an internal soliton is a key parameter, which determines the shape and the amplitude of the soliton. In this section, we present theories and methods for determining the characteristic half width of internal soliton from a SAR image [Zheng *et al.*, 2001b]. An analytical solution of a single ocean internal soliton SAR image from general theories of radar backscatter from the ocean surface will be derived from two-dimensional wavenumber spectral density of the ocean surface wave field (Eqs. (1.4.10) and (1.4.12) in Chapter 1). A single soliton solution to the KdV equation will be used to represent the solitary IW and calculate the current modulation, *i.e.*,

$$\eta = \eta_0 \operatorname{sech}^2\left[\frac{(x - Vt)}{\Delta}\right]. \tag{6.4.1}$$

The horizontal velocity of the water particles of the soliton in the upper layer is given by [Zheng *et al.*, 2001b]

$$u_1(x,t) = \left(\frac{C\eta_0}{h_1}\right) \operatorname{sech}^2\left[\frac{(x - Vt)}{\Delta}\right] \tag{6.4.2}$$

and that in the lower layer is given by

$$u_1(x,t) = \left(\frac{C\eta_0}{h_1}\right) \operatorname{sech}^2\left[\frac{(x - Vt)}{\Delta}\right]. \tag{6.4.3}$$

Equation (6.4.2) is the spatial function of the upper layer velocity induced by the single ocean internal soliton. Substituting (6.4.2) into (1.4.12) yields

$$S_{\alpha\beta}\frac{\partial U_\beta}{\partial x_\alpha} = -\left(\frac{C\eta_0 \cos^2\phi}{h_1\Delta}\right) \tanh\left[\frac{(x - Vt)}{\Delta}\right] \operatorname{sech}^2\left[\frac{(x - Vt)}{\Delta}\right]. \tag{6.4.4}$$

Substituting (6.4.4) into (1.4.10) and (1.4.1) yields

$$\sigma_0(\theta)_{ij} = 16\pi k_0^4 |g_{ij}(\theta)|^2 m_3^{-1}.$$
$$\left\{ m\left(\frac{u_*}{c}\right)^2 - 4\gamma k^2\omega^{-1} + \left(\frac{C\eta_0\cos^2\phi}{h_1\Delta}\right)\tanh\left[\frac{(x-Vt)}{\Delta}\right] \right.$$
$$\left. \times\ \mathrm{sech}^2\left[\frac{(x-Vt)}{\Delta}\right]\right\} k^{-4}. \quad (6.4.5)$$

The third term on the right represents the ocean IW radar image signature. This term can be rewritten

$$\sigma_{0\mathrm{IW}} = Q(k_0, g_{ij}, \theta)\left(\frac{C\eta_0\cos^2\phi}{h_1\Delta}\right)\tanh\left[\frac{(x-Vt)}{\Delta}\right]$$
$$\times\ \mathrm{sech}^2\left[\frac{(x-Vt)}{\Delta}\right], \quad (6.4.6)$$

where

$$Q(k_0, g_{ij}, \theta) = 16\pi k_0^4 |g_{ij}(\theta)|^2 m_3^{-1} k^{-4}. \quad (6.4.7)$$

For a satellite SAR, the frequency and antenna polarization are fixed, the incidence angle and dielectric constant change only a little within the spatial range of an ocean IW packet, which is of the order of 10–20 km. Therefore, $Q$ can be treated approximately as a constant.

Equation (6.4.6) is the analytical expression for the ocean IW SAR image we are looking for. The soliton solution to the KdV equation and the normalized form of (6.4.6) for $t = 0$ is shown in Figures 6.10(a) and (b), respectively. One can see the following two key points. (1) The internal soliton has a unique peak, but its SAR image has two peaks, one positive and one negative. (2) Both locations of the image peaks depart from the location of the soliton peak. In order to derive the soliton parameters from the data obtained from the SAR image on the basis of the solutions, suitable techniques and methods have been developed [Zheng *et al.*, 2001b].

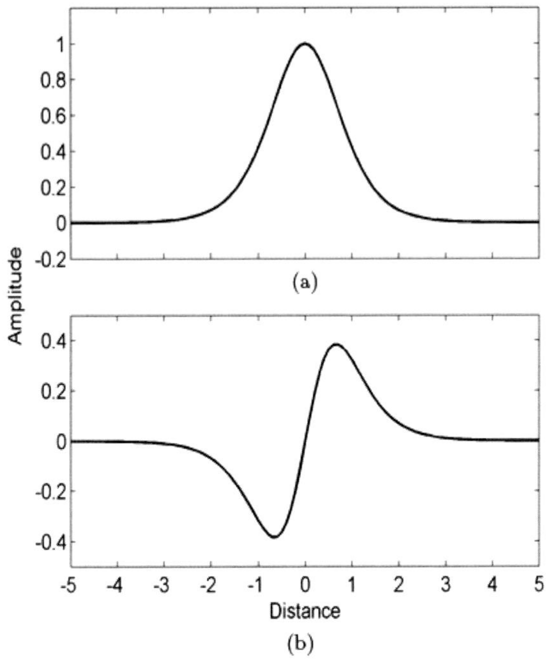


Fig. 6.10 Soliton solution of the KdV equation (a) and a dimensionless form of the analytical expression of an ocean internal soliton radar image (b) [Zheng *et al.*, 2001b]

## 6.4.2. *Determination of the characteristic half width*

As mentioned in the above section, the characteristic half width is a key parameter for an ocean internal soliton. If it is determined, we can calculate the soliton amplitude using Eq. (6.2.15) for known environmental parameters. In this section, we give two methods for determining the characteristic half width of ocean internal soliton using the SAR image.

### 6.4.2.1. *Curve fitting method*

Curve fitting is a straightforward method for deriving the characteristic half width of an internal soliton from the SAR image. Taking transect normally through an IW in the SAR image, we obtain a horizontal distribution curve of radar backscatter signature.

Normalizing the curve with the peak value and re-scaling the horizontal distances with a length scale to be adjusted, we obtain a normalized radar backscattering section curve of the internal soliton. Fitting this section curve with a normalized form of Eq. (6.4.6) for $t = 0$, *i.e.*,

$$\hat{\sigma}_{0\mathrm{IW}} = \tanh\left[\frac{x}{\Delta}\right]\mathrm{sech}^2\left[\frac{x}{\Delta}\right] \tag{6.4.8}$$

and adjusting length scale $\Delta$, we can find the best fit between the two. The length scale obtained for reaching the best fit is determined as the characteristic half width of the internal soliton.

6.4.2.2. *Peak-to-peak method*

In practice, it is a straightforward work to measure the locations of positive peak and negative peak in the radar backscattering section curve of an internal soliton. We define a horizontal distance between the two locations as a peak-to-peak distance $d_{p-p}$. The two peak locations satisfy

$$\frac{\partial\sigma_{0\mathrm{IW}}}{\partial x} = 0, \quad \text{for} \quad x = \pm\frac{1}{2}d_{p-p}. \tag{6.4.9}$$

From Eq. (6.4.6), we have

$$\mathrm{sech}^2\left(\frac{\frac{1}{2}d_{p-p} - Vt}{\Delta}\right) - 2\tanh^2\left(\frac{\frac{1}{2}d_{p-p} - Vt}{\Delta}\right) = 0. \tag{6.4.10}$$

Solving Eq. (6.4.10), we obtain

$$\frac{1}{2}d_{p-p} = \ln\left(\frac{\sqrt{2}+\sqrt{6}}{2}\right)\Delta, \quad \text{for} \quad t = 0. \tag{6.4.11}$$

Thus, the horizontal distance between the two peak locations

$$d_{p-p} = 2\ln\left(\frac{\sqrt{2}+\sqrt{6}}{2}\right)\Delta \approx 1.32\,\Delta. \tag{6.4.12}$$

The characteristic half width of the internal soliton, therefore, can be determined with this quantitative relationship, when the $p-p$ distance is measured from the SAR image.

### 6.4.3. *Validation*

#### 6.4.3.1. *Test 1: ERS-1 SAR image 1994 of the Portuguese continental shelf*

In August 1994, an ocean IW field observation experiment simultaneously with ERS-1 SAR overpasses was performed on the Portuguese Continental Shelf [da Silva *et al.*, 1998]. During the experiment, a moored buoy was deployed at 40°59.92′N 9°14.54′W to record the thermocline oscillations with a thermistor chain. The water depth is 149 m. The buoy recorded two internal solitary wave packets nearly simultaneously (1.5 h early) with the SAR overpasses. The case observed on August 8, 1994, for which the data were published in Figure 7 of da Silva *et al.* [1998], is used for testing our methods.

Figure 6.11 shows curve fittings of the normalized form of our theoretical model (6.4.8) to the normalized SAR image intensity of three large-amplitude solitons in Figure 7(b) of da Silva *et al.* [1998]. For the three curve fittings, all correlation coefficients are better than 0.99. The characteristic half widths, *i.e.*, the non-dimensionalized scale of the three solitons obtained by the curve fitting, are listed in Table 6.1.

Meanwhile, we measured the peak-to-peak distances of the three solitons directly from the SAR intensity curve in Figure 7(b) of da Silva *et al.* [1998]. All these data are also listed in Table 6.1. The navigation chart shows that the water depth at the locations of the three solitons is 140 m. From the nearly simultaneous thermistor chain data record shown in Figure 7(c) of da Silva *et al.* [1998], we determine the upper layer depth as about 25 m. Considering that the SAR image was taken 1.5 h later during the ebb tide period, the thermocline depth should be greater. Therefore, we reasonably choose $h_1 = 30$ m and $h_2 = 110$ m, respectively. These two parameters are used to calculate the amplitudes of the internal

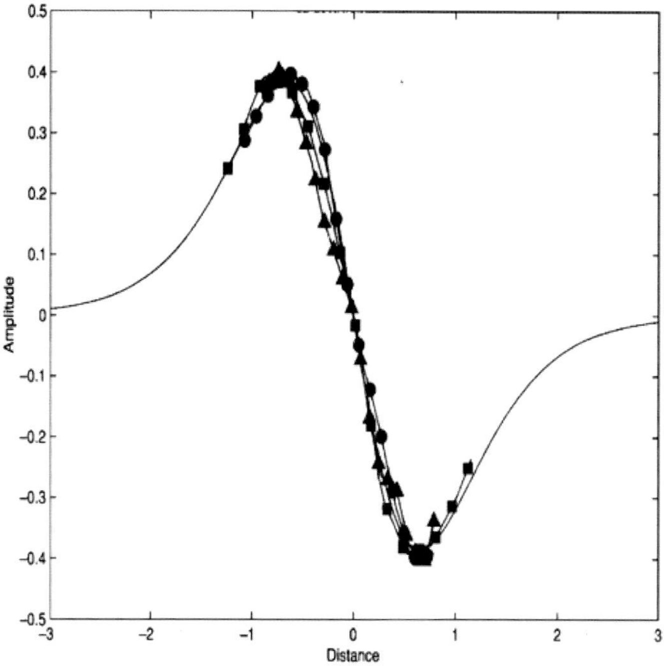


Fig. 6.11 Best fits of the normalized form of our theoretical model to the normalized SAR image intensity of three large-amplitude solitons in Figure 7(b) of da Silva *et al.* [1998]. The SAR image was acquired during the ocean IW field observation experiment simultaneously with ERS-1 satellite overpasses on the Portuguese Continental Shelf in August 1994 [Zheng *et al.*, 2001b]

Table 6.1 Characteristic half widths and amplitudes of ocean internal solitary waves on the Portuguese Continental Shelf in August 1994*

| No. | $\Delta_1$, km | $r$ | $\eta_{01}$, m | $d_{p\text{-}p}$, km | $\Delta_2$, km | $\eta_{02}$, m | $\eta_0$, m |
|---|---|---|---|---|---|---|---|
| IW11 | 0.10 | 0.997 | 18 | 0.11 | 0.08 | 26 | 23 |
| IW12 | 0.15 | 0.998 | 8 | 0.22 | 0.17 | 7 | 10 |
| IW13 | 0.18 | 0.996 | 6 | 0.24 | 0.18 | 6 | 5 |
| Mean | | | $11 \pm 5$ | | | $13 \pm 9$ | $13 \pm 8$ |

*$h_1 = 30$ m, $h_2 = 110$ m. $\Delta_1$, $r$ and $\eta_{01}$ represent the characteristic half width, correlation coefficient and amplitude determined with the curve fitting method, respectively. $\Delta_2$ and $\eta_{02}$ represent the same parameters obtained with the peak–peak method. $\eta_0$ is the measured amplitude

solitons with (6.2.15), using the characteristic half widths. The results are also listed in Table 6.1.

Statistically, the results derived from three different methods are comparable. Therefore, we list the mean and the standard deviations of three groups of soliton amplitudes in Table 6.1. One can see that the mean amplitudes are very close to each other. In particular, the results derived from the peak-to-peak method are almost the same as the field measurements.

#### 6.4.3.2. *Test 2: RADARSAT-1 SAR image 1998 of SCS*

Hsu and Liu [2000] observed ocean IWs on the northern continental shelf of SCS using RADARSAT-1 ScanSAR images. In an image acquired on May 4, 1998 [Figure 6 of Hsu and Liu, 2000], they found three IW packets near the Dongsha Islands. All three packets propagated westward from the SCS deep basin to shallow waters. Radar signals along transect across the three IW packets is shown in Figure 6.12 [Hsu and Liu, 2000]. The three packets of strong wave-like signals are coded IW21, IW22 and IW23, respectively. The leading wave of packet IW21 is an elevation IW, which is characterized by a negative (downward) peak leading a positive (upward) peak. The leading waves of packets IW22 and IW23 are depression IWs, which are characterized by positive peaks leading negative ones.

Two leading waves of the two depression IW packets are chosen for testing the methods because of their SAR signals much stronger than the background. First, theoretical model (6.4.8) is used to fit the normalized SAR image signals as shown in Figure 6.13. All correlation coefficients for two curve fittings are higher than 0.99. The characteristic half widths and amplitudes of ocean IWs obtained by curve fitting are 1 km and 24 m. Meanwhile, the peak-to-peak distances of the two solitons directly measured from the SAR signals shown in Figure 6.12 are 0.90 and 1.10 km, thus, the characteristic half widths are 0.70 and 0.80 km, respectively. From navigation charts, the average water depth at the locations of the two solitons is determined as 350 m. The field measurements indicate that the upper layer depth was about 150 m [Hsu and Liu, 2000]. The two parameters and the characteristic half widths determined above are

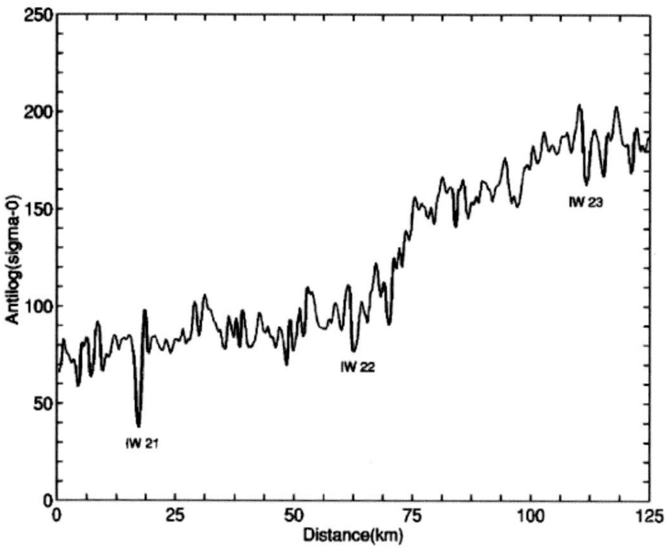


Fig. 6.12 Radar signals along transect across three packets of ocean IWs near the Dongsha Islands in the northern continental shelf of SCS taken from a RADARSAT-1 ScanSAR image acquired on May 4, 1998 [Hsu and Liu, 2000]. Three packets of strong wave-like signals are coded IW21, IW22 and IW23, respectively. Packet IW21 is an elevation IW, and the other two consist of depression IWs. All waves propagated from the right to the left

used to calculate the amplitudes of the leading internal solitons. The calculation results are 48 and 37 m, respectively. All the results are listed in Table 6.2. One can see that the soliton amplitudes are three times larger than that in the case of the Portuguese Continental Shelf. But, they are still within a reasonable range, because previous investigations showed that amplitudes of IWs in SCS reached as large as 100–140 m [Hsu and Liu, 2000; Liu *et al.*, 1998; Ramp *et al.*, 2004].

#### 6.4.3.3. *Test 3: RADARSAT-1 SAR image 2001 of SCS*

RADARSAT-1 SAR image 2001 of SCS refers to the RADARSAT-1 SAR image of the northeastern SCS taken on April 24, 2001 as shown in Figure 6.9(a). From GLPs shown in Figure 6.9(c), we measure the horizontal distances between positive and negative peaks $d_{p-p}$ of the single soliton $A$ and the leading soliton of wave packet $C$

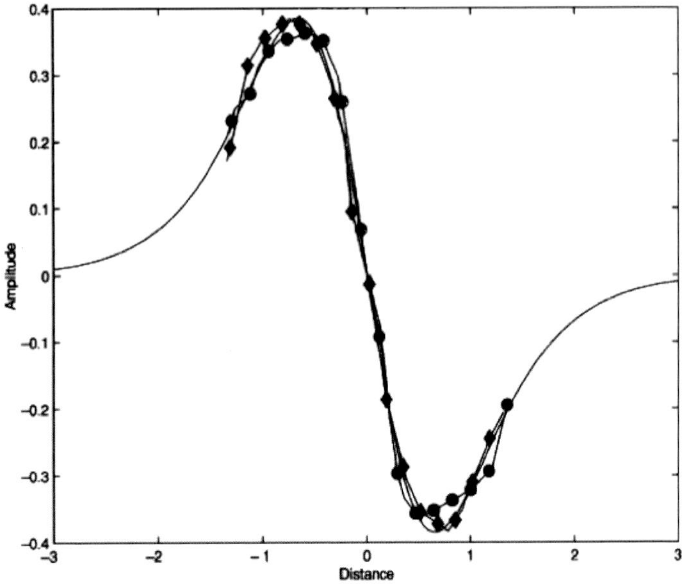


Fig. 6.13 Best fits of the normalized form of theoretical model to the normalized SAR image intensity of leading waves of IW packets IW22 (solid dots) and IW23 (diamonds) in Figure 6.12 [Zheng *et al.*, 2001b]

Table 6.2 Characteristic half widths and amplitudes of ocean IWs on the northern Continental Shelf of SCS in May 4, 1998*

| **No.** | $\mathbf{\Delta_1}$, **km** | **r** | $\eta_{01}$, **m** | $d_{p\text{-}p}$, **km** | $\mathbf{\Delta_2}$, **km** | $\eta_{02}$, **m** |
|---|---|---|---|---|---|---|
| IW22 | 1.0 | 0.997 | 24 | 0.90 | 0.70 | 48 |
| IW23 | 1.0 | 0.998 | 24 | 1.10 | 0.80 | 37 |
| Mean | | | $24 \pm 0$ | | | $42 \pm 6$ |

*Parameter definitions are the same as that in Table 6.1, but for $h_1 = 150$ m and $h_2 = 200$ m

are 0.7 and 0.8 km, respectively. Using the peak-to-peak method (6.4.12), we obtain their characteristic half widths are 0.53 and 0.61 km, respectively. Furthermore, from the interpretation map in Figure 6.9(b), we estimate that the water depth is 350 m, thicknesses of the upper and lower layers $h_1$ and $h_2$ at A are 100 and 250 m, respectively. From (6.2.11), (6.2.12) and (6.2.15), we obtain the soliton amplitude of $A$ is 20 m. For the leading soliton of wave packet

C, we estimate that the water depth is 160 m, thicknesses of the upper and lower layers $h_1$ and $h_2$ are 75 and 85 m, respectively. Thus, we obtain the soliton amplitude of $C$ is 15 m. In comparison to amplitudes of 20 and 25 m of leading solitons in IW packets $P_2$ and $P_3$ measured by the mooring buoy at $S$, the SAR-derived amplitudes are close to buoy-measured ones, but reasonably smaller because of deeper thermocline and water depths.

The results of above three tests indicate that the theories and methods presented in this section give reasonable estimates of characteristic half widths and amplitudes of internal solitons with an acceptable accuracy compared to field measurements. Therefore, they are valid and may serve as a practical technique for interpreting ocean IW radar (including SAR and real aperture radar) images.

### 6.4.4. *Discussion*

#### 6.4.4.1. *Wind background*

Wind friction is a major external forcing for generating sea surface wavelets, which serve as backscatters of radar waves. On an ocean radar image, therefore, wind field signatures usually accompany ocean phenomena. Sometimes the wind field signatures appear as an image background if the wind field is spatially uniform or its scale of variation is much larger than that of the ocean features of interest. In certain cases, the wind field signatures may mix with ocean features, if the wind field is not spatially uniform or modulated by other processes with a variation scale close to that of the ocean phenomena of interest. This section will quantitatively estimate the relative weight of wind field signatures in the case of a uniform wind field using Eqs. (1.4.10)–(1.4.12) and scale analysis. The scale of the wind term in (1.4.10) depends on $m$, $u_*$ and $c$. While coefficient $m$ is 0.04 [Plant, 1980], and the typical value of $c$ is 30 cm s$^{-1}$. Thus, we have

$$O\left[m\left(\frac{u_*}{c}\right)^2\right] = O(4 \times 10^{-5} u_*^2), \tag{6.4.13}$$

where $u_*$ is in cm s$^{-1}$. From (1.4.12), we have the scale of the current modulation term

$$O\left(S_{\alpha\beta}\frac{\partial U_\beta}{\partial x_\alpha}\omega^{-1}\right) = O\left(\frac{C\eta_0}{h_1\Delta}\omega^{-1}\right). \tag{6.4.14}$$

On the basis of the data observed on the Portuguese Continental Shelf [da Silva *et al.*, 1998], we have the following scales for the parameters in (6.4.14): $C = O(0.5\,\mathrm{m\,s^{-1}})$, $\eta_0 = O(10\,\mathrm{m})$, $h_1 = O(30\,\mathrm{m})$ and $\Delta = O(150\,\mathrm{m})$. For the gravity–capillary band, the representative frequency is 5 Hz [Long and Huang, 1976]. Thus, we have

$$\left(S_{\alpha\beta}\frac{\partial U_\beta}{\partial x_\alpha}\omega^{-1}\right) = O(4\times 10^{-5}). \tag{6.4.15}$$

From (6.4.13), we obtain that for $u_* = 3.2$ cm s$^{-1}$, which corresponds to 1 m s$^{-1}$ of the sea surface wind $U_{10}$, the wind term is one order of magnitude higher than the current modulation term. For $u_* = 10$ cm s$^{-1}$, which corresponds to 3 m s$^{-1}$ of $U_{10}$, the wind term is two orders higher than the current modulation term. In other words, the ocean IW signatures are detectable on a radar image only for low wind conditions. This is consistent with observations and statistical results obtained by da Silva *et al.* [see Figures 4 and 5 in da Silva *et al.*, 1998]. All 453 solitons they observed were detected on SAR images taken at wind speeds lower than 5 m s$^{-1}$.

#### 6.4.4.2. *Dynamical range of radar linearity*

The theories and methods given in this section are developed based on the assumption that radar receiver is a perfectly linear system, *i.e.*, the intensity of radar return signals recorded on a radar image depends linearly on the roughness of the target surface, which is defined quantitatively by a radar backscatter cross section per unit area. In practice, however, the radar receiver system is designed to be linear only within a certain dynamical range. Beyond the range, either the system output will be saturated if the radar echo signal is too high, or the system output will cut off if the radar echo signal is too low. In other words, within the linearly dynamic range, the ocean internal soliton should appear as a double sign signature on the radar image. Beyond this range, the soliton probably displays a truncated or a single sign signature. The theories and methods in this section are directly applicable only to the linear case. For the other two cases, the radar image signal needs first to be linearized using the dynamic transfer curve of the radar receiving system in order to use the methods. However, this linearization process may further induce uncertainty and measurement errors.

#### 6.4.4.3. *Error analysis*

The measurement errors of the methods depend upon the SAR image resolution. For the first test, the SAR pixel size is 12.5 m. The average characteristic half width is 150 m. Therefore, the relative error of the characteristic half width is 8%, and the relative error of the soliton amplitude is 16%. For the second and third tests, the SAR image resolution is 100 m [Hsu and Liu, 2000]. The average characteristic half width is 1 km. The relative error of the characteristic half width is 10%. Therefore, the relative error for the soliton amplitude is 20%.

For three tests, the curve fitting method always gives underestimated results. In other words, the method gives overestimated characteristic half widths. We believe that there are two major reasons causing this systematic error. One is a technical reason, as discussed in the above section. The radar return signal may be truncated due to a limited dynamic range of the radar receiver, so that the amplitude of the IW image curve becomes shorter and the width becomes relatively broader. In this case, the characteristic half width obtained by curve fitting will be overestimated. However, the peak-to-peak method would not be seriously affected and may generate results close to the observations, as shown in Table 6.1. The other reason is a natural one. On the continental shelf, wave–current interaction, wave–wave interaction, dissipation and fission of a single soliton into several small solitons are frequently observed [Zheng *et al.*, 2001a]. These processes may cause the IW shape to be biased from a typical KdV soliton formulation, which is used to develop the methods. This is also a possible cause for the systematic error. In addition, in some cases the ocean IWs are quite close to linear waves, so that the theories and methods in this section are invalid.

# Chapter 7

# Ocean Internal Waves: II. Statistical Analysis

## 7.1. Introduction

Sections 6.3 and 6.4 have detailed a list of important data and dynamic parameters of IWs, which can be directly extracted from satellite SAR images or indirectly calculated using the data extracted from the images. Thus, it is feasible to do the statistical analysis for these data and parameters using SAR images. This chapter will take SCS as an example to explain theories and methods developed for the statistical analysis as well as the major results.

Satellite images and field observations have revealed that SCS, in particular the northern SCS (NSCS) from around 19° to 22°N and from 114° to 122°E, is an ocean area where energetic internal waves occur frequently [Buijsman *et al.*, 2010a; Cai *et al.*, 2002; Hsu and Liu, 2000; Lien *et al.*, 2005; Liu *et al.*, 1998, 2004; Meng, 2002; Orr and Mignerey, 2003; Zhang *et al.*, 2011; Zhao, 2004; Zhao *et al.*, 2004; Zheng *et al.*, 2007]. The wave amplitudes may reach as high as 140–160 m [Duda *et al.*, 2004; Ramp *et al.*, 2004], and the wave crest lines as long as 250 km [Zheng *et al.*, 2007].

The ocean images taken by ERS-1/2 SAR, RADARSAT-1/2 SARs and ENVISAT ASAR will serve as a baseline for statistical analysis. Figure 7.1 shows an example of satellite SAR images, which is a RADARSAT ScanSAR image taken on April 26, 1998. The image covers an area of 528 by 478 km centered at 21°09′N, 116°18′E. The

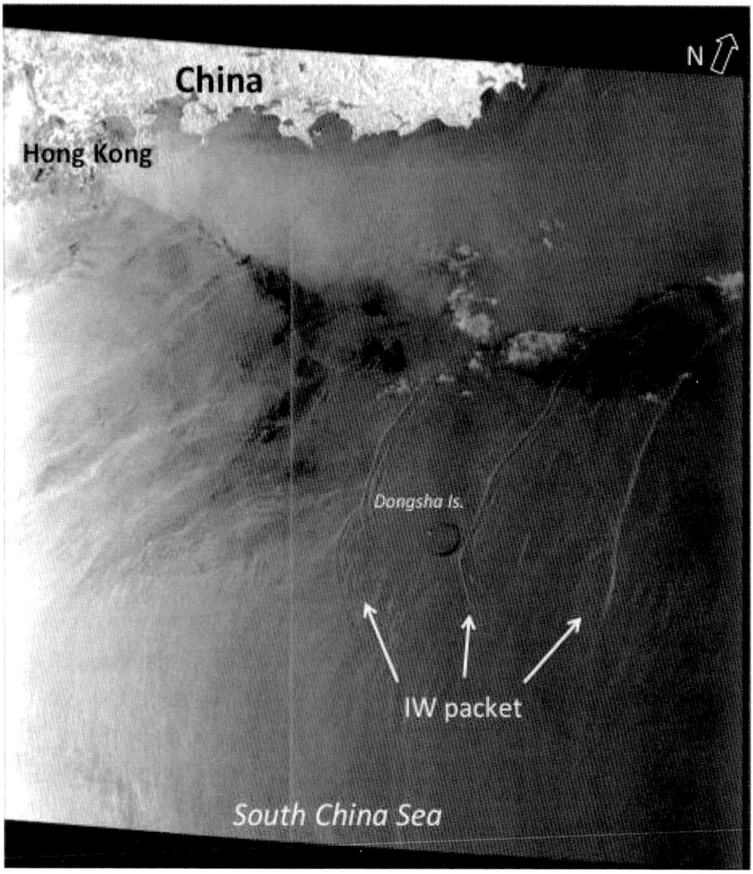


Fig. 7.1 A RADARSAT ScanSAR image of NSCS taken on April 26, 1998 (downloaded from http://www.ifm.uni-hamburg.de). The dark, horseshoe-like feature at the center is the Dongsha Coral Reef. Grouped bright lines are IW packets

bright area on the top shows the south China coastal region near Hong Kong. The horseshoe-like feature at the center is the Dongsha Island (small bright spot) and the Dongsha Coral Reef (dark ring). Grouped bright lines distributed in the central portion are imagery of westward-propagating IW packets.

## 7.2. Geographic Distribution

Wang *et al.* [2012] collected more than 2500 of ERS-2 SAR, ENVISAT ASAR, Aqua/Terra MODIS and HJ-1A/1B (Chinese ocean satellites) images for statistical analyses of geographic distribution and propagation direction of IWs in SCS from 2005 to 2010. First, the satellite images are geometrically corrected by re-projection. Then the corrected images are imported into geographic information system software, and the leading crests of internal wave packets are extracted to represent the wave packets. The propagation directions are determined by the curvature direction or the normal of leading crest, *i.e.*, the angle of the normal with respect to the true north is defined as the propagation direction of IWs as shown in Figure 7.2.

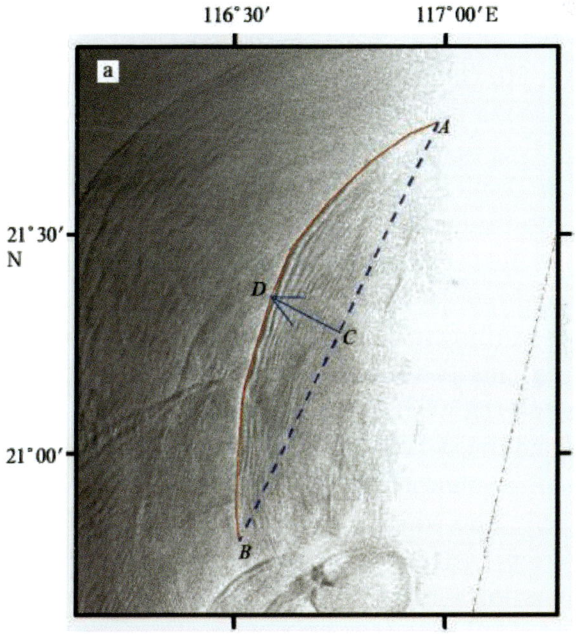


Fig. 7.2 Method for determination of IW propagation direction. Arc AB is taken from the leading crest of an IW packet. C is a center of chord AB. CD is a normal of arc AB [Wang *et al.*, 2012]

From more than 3500 IW packets extracted from the satellite images, a geographic distribution map of IWs in SCS is obtained as shown in Figure 7.3(a), in which red arcs represent IW packets. One can see that the majority of IW packets are distributed on the continental slopes of the northern and the western SCS, and the minority on the shallow continental shelves of the southern SCS, the Gulf of Tonkin and NSCS. For the SCS deep basin, the IW packets are imaged only in its northeastern corner west of the Luzon Strait. These distribution features imply that the majority of IWs originate from the west of the Luzon Strait, and the fission of solitary IWs propagating along the shoaling topography and thermocline is a major mechanism for multiplying of IW numbers (see Section 4.2.2.2 for details). Figures 7.3(b) and (c) show the geographic distributions of propagation directions and propagation paths of IW packets, which are derived from satellite images using the methods shown in Figure 7.2. One can see that the majority of IWs propagate toward the coast, and the minority backward near Dongsha Island and the mouth of Mekong River as well as in the Gulf of Tonkin.

Using the results shown in Figures 7.3(a)–(c), the distribution features of sub-regions can be further examined [Wang *et al.*, 2012]. For example, for the northeastern SCS, over 2200 IW packets are extracted from the satellite images. According to the bottom topography and the distribution of IWs, the northeastern SCS can be further divided into two sub-regions bordered by the Dongsha Island: the eastern and the western sea areas. In the eastern sea area of the Dongsha Island, most IWs propagate westward with angles between 220° and 330° as shown in Figure 7.4(a). The average propagation angle is 269.6°. The white hollow arrows in Figure 7.4(b) represent the propagation path and direction of internal waves. While there are exceptions for a few IW packets propagate eastward with the angles between 50° and 150° east of the Dongsha Island as shown in Figure 7.4(b), implying the reflection or refraction of westward propagating IWs by the sharply changed bottom topography near the Dongsha Island.

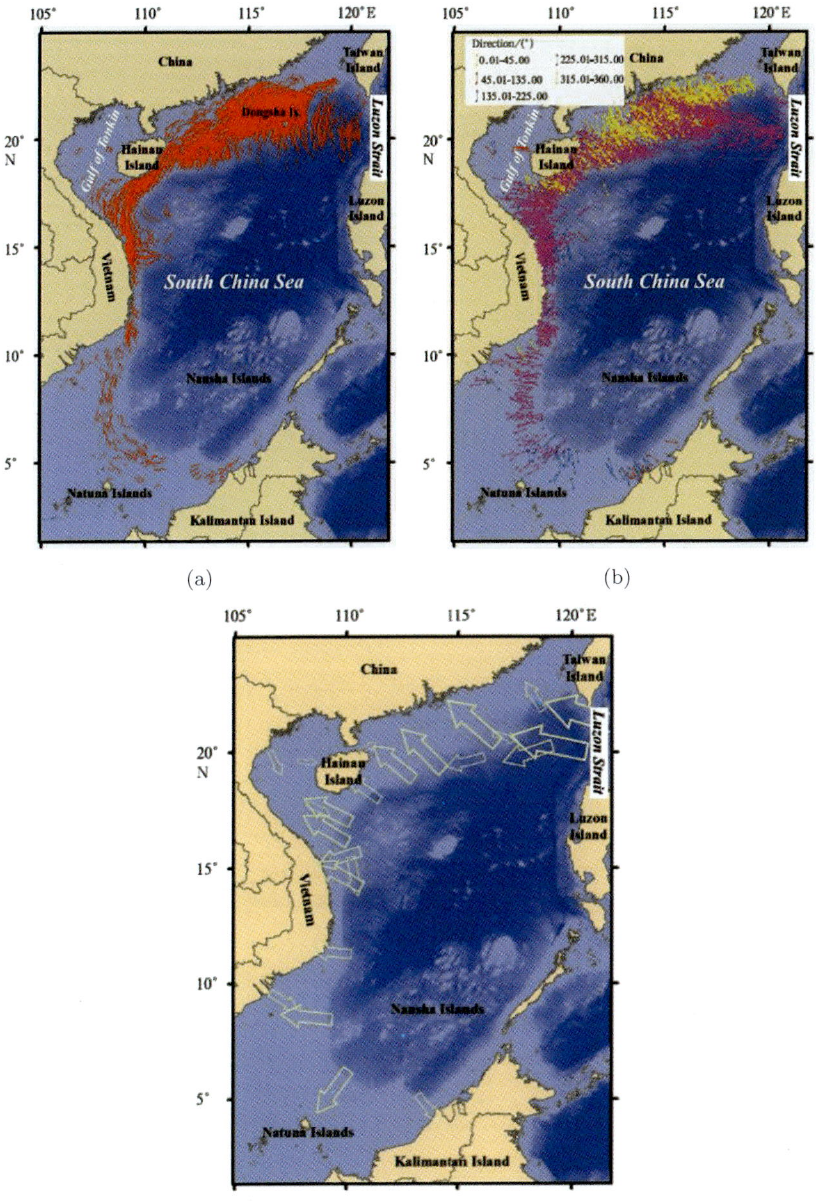


(a) (b) (c)

Fig. 7.3 Geographic distribution map (a), propagation direction map (b) and propagation paths (c) of IW packets in SCS Sizes of arrows in (c) represent relative quantities of the IW packets [Wang *et al.*, 2012]

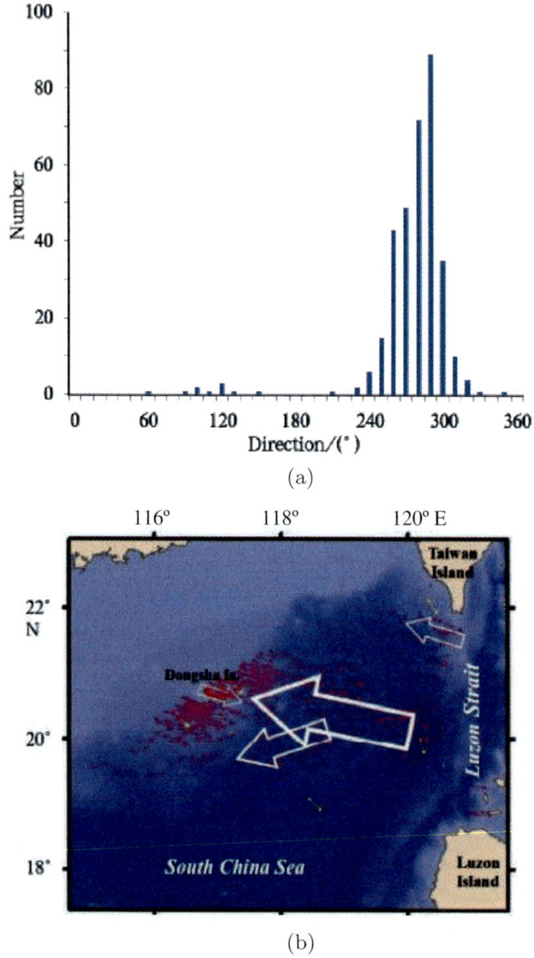


Fig. 7.4 Propagation direction histogram (a) and geographic distribution map (b) of IW packets in the eastern sea area of the Dongsha Island [Wang *et al.*, 2012]

## 7.3. Yearly Distribution of SAR-Observed IW Occurrence Frequency

It is of interest to see if there is any interannual variability in the yearly IW occurrence frequency in NSCS. Seven years of SAR images of IWs from 1995 to 2001 are used for this statistical analysis.

For a calendar year, the SAR-observed IW occurrence frequency is defined as

$$p_i = q_i \left( \frac{N_i}{\sum_j N_j} \right), \tag{7.3.1}$$

where $N_i$ ($N_j$) is the number of days, on which IWs are imaged by SAR and $i(j)$ represents the $i$th ($j$th) year. $q_i = D_{1997}/D_i$ is a yearly de-weighted factor, where $D_{1997}$ is a number of SAR working days over the study area in 1997, and $D_i$ is the SAR working days in the $i$th year (71 (1995), 68 (1996), 34 (1997), 57 (1998), 65 (1999), 77 (2000) and 57 (2001)), after Meng, [2002]. The de-weighted factor is required because the SAR working days over the study area during the statistical period were not evenly distributed, but varied in different years. The number of SAR working days in 1997 is a minimum, so the yearly de-weighted factor for 1997 is taken as 1. The factors for other years are deduced accordingly. The statistical results are shown in Figure 7.5(a). One can see a remarkable interannual variability of SAR-observed IW occurrence frequency. In 1995, 1998 and 2000, the frequencies are 2–4 times higher than that in other years.

On the other hand, the SAR working days used in the statistical calculation are inhomogeneously distributed in a year; and the SAR-observed IW occurrence frequencies have the seasonal variability as discussed in the next section. These two points might cause a statistical bias in the interannual variability. In order to further confirm the existence of the interannual variability, the IW occurrence frequency (IW occurrence days divided by SAR working days) in an occurrence peak phase from April to July each year is calculated. The results are shown in Figure 7.5(b). One can see that the interannual variability is still there. In 1995, 1998 and 2000, the frequencies are remarkably higher than other years. This interannual variability implies that there are long-term and large-scale processes playing roles in modifying IW occurrence frequency and SAR imaging in NSCS.

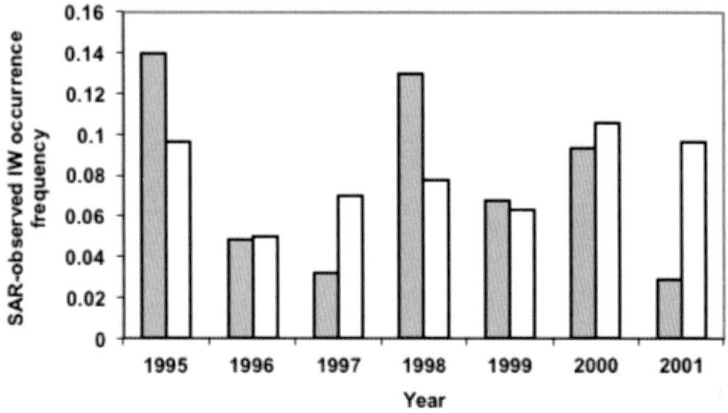


(a)

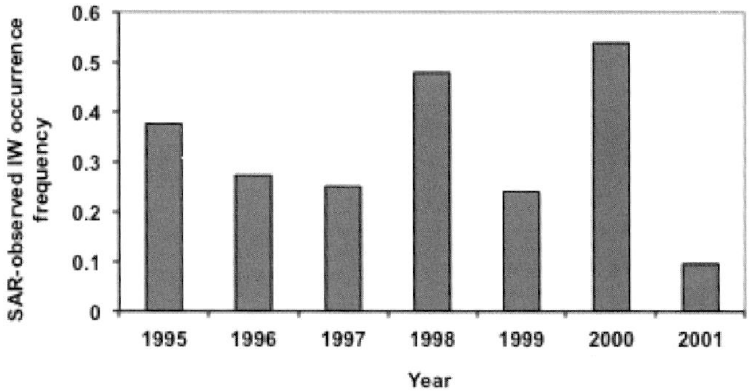


(b)

Fig. 7.5 (a) Yearly distribution of SAR-observed IW occurrence frequencies in NSCS calculated using the full year data (dark bars) and in Meng [2002] (open bars) [Zheng *et al.*, 2007]. (b) Same as (a) but for the data from April to July only

## 7.4. Monthly SAR-Observed IW Occurrence Frequency

The monthly SAR-observed IW occurrence frequency is defined as

$$p_{mi} = q_{mi}\left(\frac{n_i}{\sum_{j=1}^{12} n_j}\right), \tag{7.4.1}$$

where $n_i$ ($n_j$) is a number of total days in the $i$th ($j$th) month of seven years, at which IWs were imaged, and $q_{mi} = d_8/d_i$ is a monthly de-weighted factor. Here, $d_8$ is a minimum number of SAR working days

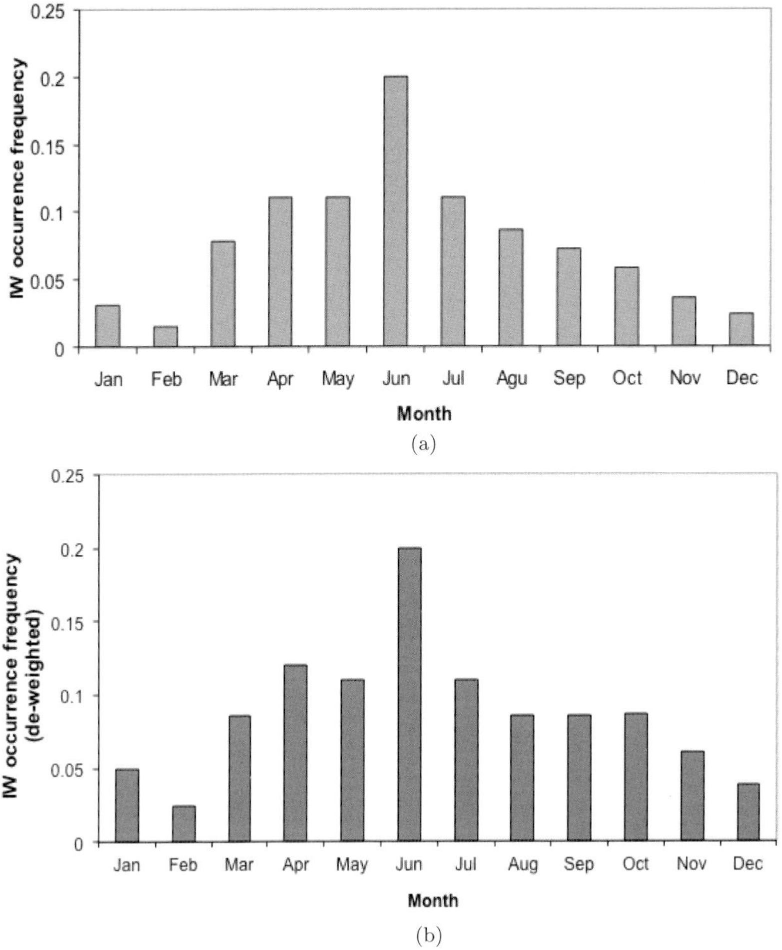


Fig. 7.6 (a) Monthly distribution of SAR-observed IW occurrence frequencies in NSCS and (b) same as (a) but de-weighted by low sea states [Zheng *et al.*, 2007]

over the study area in August, and $d_i$ is the SAR working days in the $i$th month. The calculation results are shown in Figure 7.6(a). Considering the dependence of SAR imaging on the sea states (see the next section for details), the calculation results are further de-weighted by the low sea state frequencies (for wave height less than 1.9 m). The method is to multiply a de-weighted factor of $f_m/f_i$, in which $f_m$ is the maximum frequency of low sea states, and $f_i$ is the frequency in $i$th month. The results are shown in Figure 7.6(b). One can see that both statistical results give similar and remarkable seasonal variability. The high occurrence frequencies are distributed from April to July and reach a peak in June with a maximum frequency of 20%. The low occurrence frequencies are distributed in winter from December to February of the next year with a minimum frequency of 1.5% in February. Other months appear to be transition periods. In the next two sections, the external and internal causes leading to this uneven distribution will be examined.

## 7.5. Statistics of Boundary Conditions for IW SAR Imaging

As mentioned above, the IW occurrence frequencies shown in Figures 7.5 and 7.6 are derived from SAR images. According to SAR imaging theories, however, only low wind and low sea state conditions are favorable for IW imaging [Zheng *et al.*, 2001b]. The statistical distributions of these two conditions may serve as dynamic filters to modify the SAR-observed IW occurrence frequency. But, the two conditions do not constitute direct forcing to generate IWs.

### 7.5.1. *Wind condition statistics*

Winds over the study area are dominated by the East Asia monsoon. In summer (JJA), the southwesterly winds with an average wind speed of $6\,\mathrm{m\,s^{-1}}$ prevail. In winter (DJF), the wind direction totally reverses, northeasterly winds with an average speed of $9\,\mathrm{m\,s^{-1}}$ prevail over the whole region [Metzger and Hurlburt, 1996; Pohlmann, 1987]. Statistical distributions of monthly mean wind frequencies are derived from long-term historical data sets of field observations [Editorial Group of Marine Climate Atlas of Institute of Geography, Chinese Academy of Sciences and Publishing & Compilation Group of Institute of Marine Scientific and Technological Information, National Bureau of Oceanography, China, 1982]. Statistical cells of $2^\circ$ latitude by $2^\circ$ longitude were used in the data sets. The cell centered at $22^\circ$N $118^\circ$E is chosen to represent the study area. The results are shown in Figure 7.7. One can see the seasonal variability

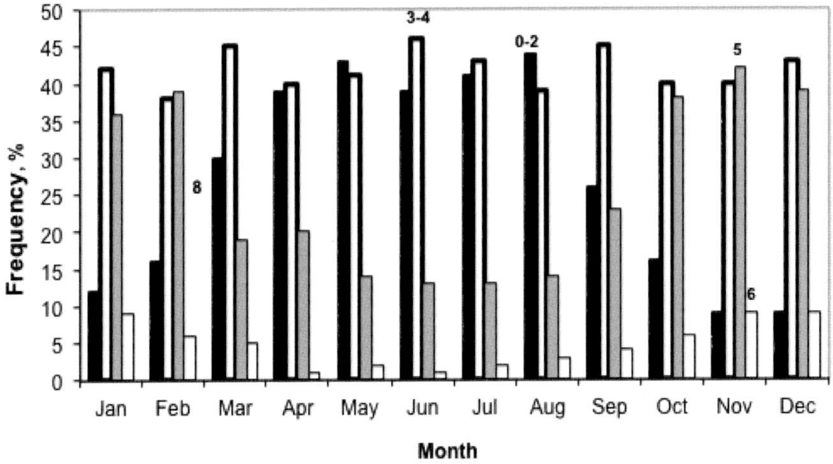


Fig. 7.7 Statistical distributions of monthly mean wind frequencies at statistical cell centered at 22°N 118°E in NSCS. The numerals marked at peak bars represent wind scale ranges. Statistical errors are less than 1% [Zheng *et al.*, 2007]

of wind scales. High frequencies of low wind conditions (scales 0–1 and 2–3, equivalent to the wind speed lower than $5\,\mathrm{m\,s^{-1}}$) are distributed from April to September and reach a peak phase from May to July with a maximum frequency of 58% in July. The low frequencies are distributed from November to March of the next year. The frequency distributions of high wind conditions (scales 4–5, 6–7 and 8, equivalent to the wind speed higher than $5\,\mathrm{m\,s^{-1}}$) are exactly out of phase with low wind conditions. High frequencies are distributed from October to next March and reach a peak phase from November to next January with a maximum frequency of 87% in December.

### 7.5.2. *Sea state statistics*

Statistical distributions of monthly mean sea state frequencies are derived from the long-term historical data sets of field observations in the same literature as the wind data and at the same statistical cell. The statistical distribution of monthly sea state frequencies is shown in Figure 7.8. One can see that high frequencies of wind wave scales 0–2 (for the wave height less than 0.7 m) are distributed from April to

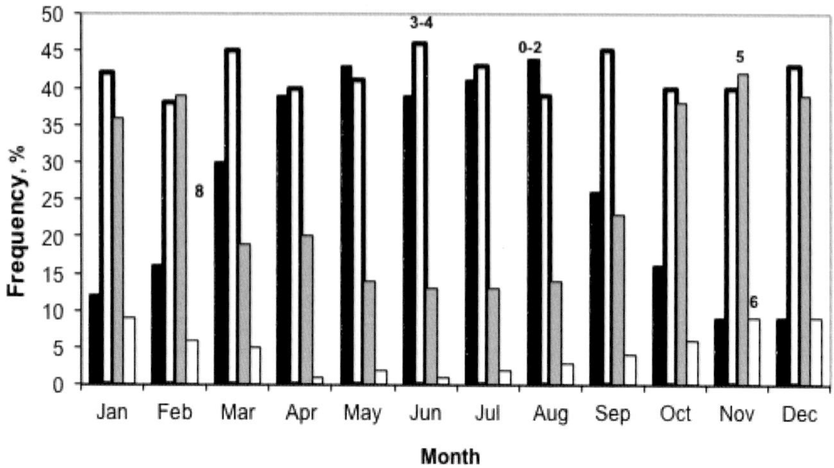


Fig. 7.8 Statistical distribution of monthly mean sea state frequency at the statistical cell centered at 22°N 118°E. Numerals marked at peak bars are wave scale ranges. Statistical errors are less than 1% [Zheng *et al.*, 2007]

August and reach a peak value of 44% in August. The low frequencies are distributed from October to next February with frequencies lower than 16%. March and September appear as transition phases. The frequencies of scales 3–4 (for the wave height from 0.8 to 1.9 m) have an almost even distribution with a mean of 41.8% and a standard deviation of 2.5%. The distributions of scales 5 (for the wave height from 2.0 to 3.4 m) and 6 (for the wave height from 3.5 to 6.0 m) are in opposite phase with scales 0–2.

In order to estimate the dependence of IW SAR imaging on the sea surface boundary conditions, we calculate the correlation coefficient and confidence level of the SAR-observed IW occurrence frequency shown in Figure 7.6(a) with the sea surface wind scale shown in Figure 7.7 and the sea state scale shown in Figure 7.8. The results are listed in Table 7.1. One can see that for the wind scale, the highest correlation coefficient (0.766) and the highest confidence level (>99%) occur at wind scale 2–3, not at the lowest wind scale 0–1. The reason may be that at the lowest wind scale, there are not enough small waves generated on the sea surface, which serve as major backscatters of radar pulses. For the sea state, the highest correlation

Table 7.1 Correlation coefficient and confidence level of the SAR-observed occurrence frequency with monthly mean sea surface wind and the monthly mean sea state

| | | | | | |
|---|---|---|---|---|---|
| Wind scale | 0–2 | 3–4 | 4–5 | 6–7 | 8 |
| Wind speed, m s$^{-1}$ | 0–1 | 2–5 | 6–10 | 11–17 | ≥18 |
| Corr. Coefficient | 0.489 | 0.766 | –0.610 | –0.766 | –0.598 |
| Confidence level | >80% | >99% | >95% | >99% | >95% |
| Wave scale | 0–2 | 3–4 | 5 | 6 | |
| Wave height, m | 0–0.7 | 0.8–1.9 | 2.0–3.4 | 3.5–6.0 | |
| Corr. Coefficient | 0.792 | 0.478 | –0.823 | –0.844 | |
| Confidence level | >99% | >80% | >99% | >99% | |

coefficient (0.792) and the highest confidence level (>99%) occur at the lowest sea state scale. The negative correlation coefficients mean that the two processes vary out of phase (phase difference of $\pi$), *i.e.*, the higher the wind speed and sea state frequencies, the lower the SAR-observed IW occurrence frequencies.

## 7.6. Statistics of Structure of IW Field and Packets

Zheng *et al.* [1993b] did statistics for IWs on the continental shelf of the Middle Atlantic Bight. The data they used are the space shuttle photographs taken by the US astronauts with an Aero Linhof Technika 45 camera during mission STS-40 in June 1991. The spatial resolution of the photographs is 23 m at nadir, which is close to that of satellite SARs. In the sun glint area, IWs show up clearly in the photographs. Figure 7.9 shows an example of the space shuttle photographs and interpretation map with isobaths. One can see that the total of 34 IW packets are distributed over an area from 38° to 40°N latitude and 72° to 75°W longitude, where the water depth near the continental shelf break is from 50 to 150 m. The crest lines of the internal waves are nearly parallel to local isobaths with the longest continuous length of 150 km.

### 7.6.1. *Two-dimensional structure of the IW field*

By examining Figure 7.8(b), one can see that the IW field has a three-level structure, *i.e.*, packet groups, packets and solitons.

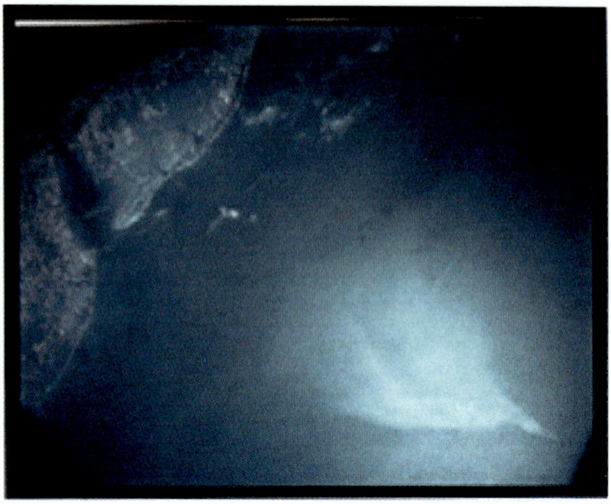

(a)

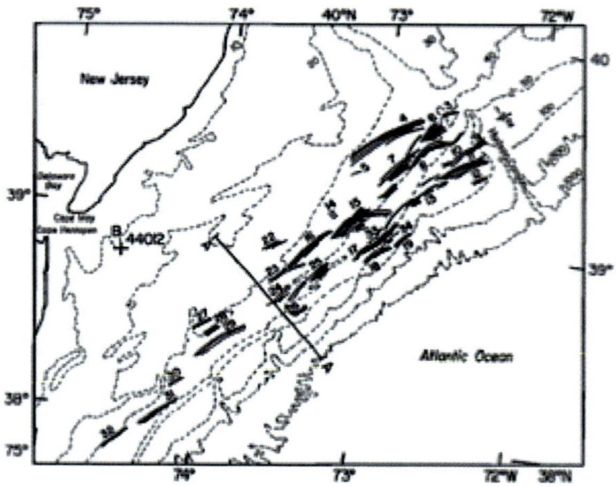


(b)

Fig. 7.9 (a) Example of space shuttle photographs taken by the astronauts on June 8, 1991 during mission STS-40 of the space shuttle Columbia. The land features on top are in New Jersey, Delaware, Maryland and Virginia, USA and the lower part is the Middle Atlantic Bight. (b) IW interpretation map. The groups of solid lines coded with numbers are taken from imageries of IW packets. Isobaths are in fathoms (1 fm = 1.8 m). B is the location of NOAA data buoy 44012. Segment AA is transect for CTD data measurement (not shown) [Zheng *et al.*, 1993b]

A packet group consists of several packets. Each packet contains a different number of solitons. In a group, packets containing a smaller number of solitons are on the shoreward side of a packet containing about 10 solitons. The average separate distance in the shoreward direction between two packet groups, *i.e.*, the envelope or packet group wavelength is $(17.5 \pm 2.6)$ km. In a packet, solitons with the longest crest line and the largest wavelength are located in the front of the packet. The average wavelength of the packet is $(7.9 \pm 1.6)$ km. The average wavelength of 80% of the total solitons is $(0.6 \pm 0.1)$ km.

### 7.6.2. *Statistics of number of solitons in a packet*

Knowing the number of solitons in an IW packet is of great importance for studying the fission law of solitons and for tracing back to the generation source [Djordjevic and Redekopp, 1978]. This work can be performed with space shuttle photographs and SAR images. Distribution histogram of the number of solitons in a packet derived from Figures 7.9(a) and (b) are shown in Figure 7.10(a). One can see that the IW packets can be divided into three groups according to the number of solitons. The first group, being in the majority, contains 82% of the total. The average number of solitons is 5 ranging from 2 to 7. The second group contains 15% of the total. The average number is 11 ranging from 10 to 12. The third group has just one packet of 17 solitons.

### 7.6.3. *Statistics of IW wavelength*

The IW wavelength is a basic parameter for the dynamic analysis of the IW field, which is defined as a horizontal width between the frontal edges of two consecutive solitons in the space shuttle photographs. A total of 50 wavelengths are measured. Distribution histogram of the wavelengths is shown in Figure 7.10(b). One can see that the wavelengths are distributed into three groups. The first group, being in the majority, contains 80% of the total. The average wavelength is $(0.6 \pm 0.1)$ km ranging from 0.3 to 0.8 km. The second group contains 14% of the total. The average wavelength is $(1.2 \pm 0.1)$ km ranging from 1.1 to 1.4 km. The third group contains

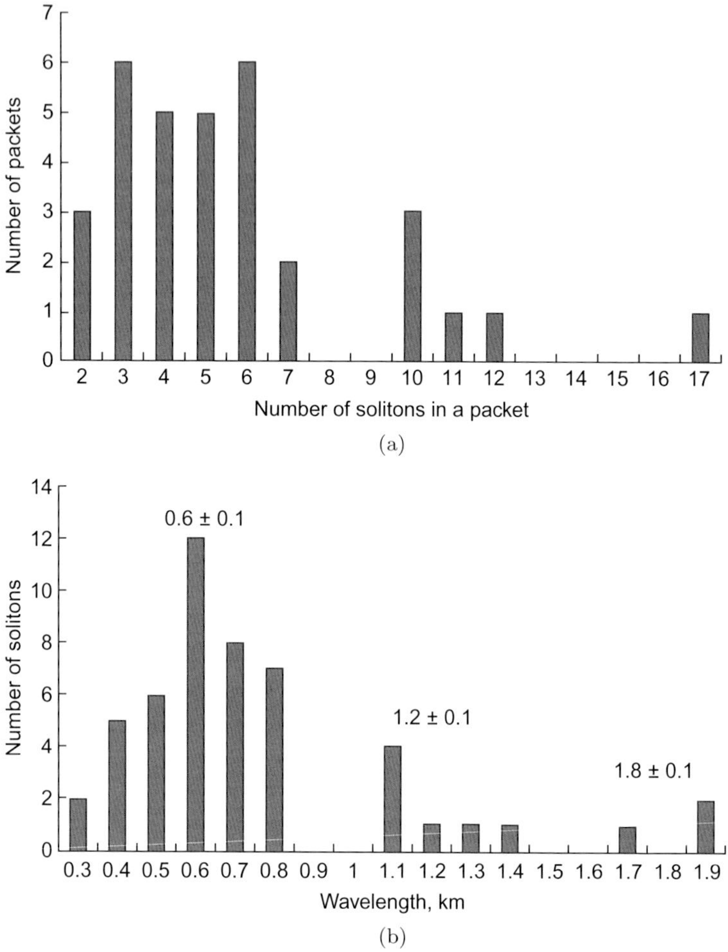


Fig. 7.10 (a) Histogram of the number of solitons in a packet in the Atlantic Bight and (b) same as (a) but for IW wavelengths [Zheng *et al.*, 1993]

6% of the total. The average is $(1.8 \pm 0.1)$ km ranging from 1.7 to 1.9 km. One can see that the distribution of wavelengths is discontinuous, and the average wavelengths of the second and the third groups are just double and triple the average of the first group, respectively. This implies that the fission law for the solitary waves propagating along a shoaling continental slope (see Section 6.2.2.2 for details) plays an important role in formation of wavelength structure of the IW fields in the Middle Atlantic Bight.

## 7.7. Generation Sources of IWs in NSCS

As mentioned in Section 7.1, NSCS is characterized by a populated distribution of strong internal waves, which occur all the year around [Hsu and Liu, 2000; Liu *et al.*, 1998; Zheng *et al.*, 2001b, 2007]. According to their generation sources, these waves are categorized into two categories, the locally generated IWs and the transbasin IWs. The locally generated IWs are distributed on the continental shelf west of 118°E, while the transbasin IWs are distributed in the whole basin [Zheng *et al.*, 2007]. In the deep basin east of 118°E, the transbasin IWs appear as solitary waves with amplitudes as large as 160 m [Duda *et al.*, 2004], while on the western continental shelf, they fission into multi-wave packets of smaller waves, and co-exist with the locally generated IWs. The previous investigators suggested that the transbasin IWs originate from the Luzon Strait, which is lying between Taiwan Island and Luzon Island and occupies a water area around from 18.5°N to 22.0°N and from 120°E to 122°E. Its north–south width is 386 km and the west–east length is about 200 km. This means that the Luzon Strait is a large-scale water body compared with the IW characteristic wavelength scale of 1 km [Zheng *et al.*, 2007]. The bottom topography is also quite complex (see Figure 7.4). Thus, it is of significance further to find out the accurate location(s) of generation source(s) for analyzing the propagation and evolution of IWs in SCS.

### 7.7.1. *Statistics of origination locations*

The generation sites of IWs in the Luzon Strait are debatable. Bole *et al.* [1994] inferred the generation site as a 4-km wide channel between Batan Island and Sabtang Island, the southernmost two islands in the Batanes Islands (also called Batan Islands, see Figure 7.11). Ramp *et al.* [2004] also suggested the Batanes Islands to be a generation site of all the largest solitons, and used the site as a coordinate origin for calculating the propagation parameters

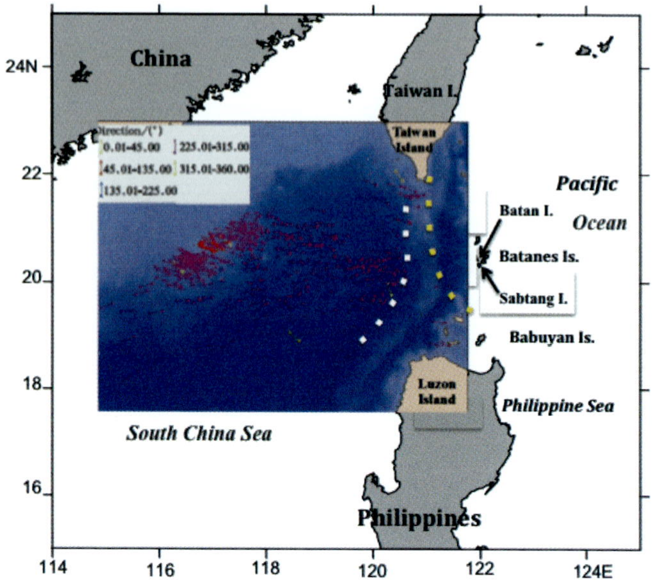


Fig. 7.11 Composite map of geographic distribution of IW packets with map of NSCS. White dashed line shows the boundary of IW origination locations, and the yellow dashed line shows the boundary of the Kuroshio western wing

of IWs in the NSCS. Yuan *et al.* [2006] suggested the Kuroshio west wing to be an IW disturbance source based on the dynamic analysis.

In fact, Figures 7.3(a) and (b) have shown the statistical results of origination locations of IWs west of the Luzon Strait, which are derived from more than 3500 IW packets extracted from the satellite images [Wang *et al.*, 2012]. Figure 7.11 shows a composite map of geographic distribution of IW packets east of the Dongsha Island with map of NSCS. The white dashed line shows the boundary of IW origination locations west of the Luzon Strait, and the yellow dashed line shows the boundary of the Kuroshio western wing derived from trajectories of Argos satellite-tracked drifters deployed during the period of World Ocean Circulation Experiment (WOCE) from 1979 to 2003 [Yuan *et al.*, 2006]. One can see that the closest distance from the boundary of IW packet origination to the Kuroshio is 50 km and that to the Batanes Is. is about 150 km. These distances imply large-scale gaps for IWs with characteristic wavelength scale of 1 km [Zheng *et al.*, 2007]. Within the gaps, there are no any IW signatures observed on 2500 images taken by various satellites for six years from 2005 to 2010. This implies clearly that neither Batanes Is. nor the Kuroshio are generation sites for IWs.

On the basis of the above observations and statistical analysis, the following hypotheses are proposed for IW generation sites. The solitary IWs are generated in a certain area defined as a source region. The source region, working like an oscillator, has the potential to provide the necessary and sufficient conditions for wave generation: (1) initial disturbance formation and (2) wave amplitude and crest line growth. The source region also dissipates the disturbance energy, so that only fully developed waves are able to radiate out of the source region [Zheng *et al.*, 2008b].

### 7.7.2. *Initial disturbance formation: Internal tides*

As shown in Figure 7.11, the Luzon Strait is a 386 km wide opening connecting SCS to the western Pacific (WP). Bottom topography of the strait features two north-to-south running submarine ridges, the Heng-Chun Ridge to the west and the Lan-Yu Ridge to the east. According to Farmer *et al.* [2009], the two ridges are mostly critical to both the diurnal and semidiurnal internal tidal rays. The TPXO7.2 tide model gives that the cross-strait barotropic tidal currents of K1, M2, O1 and S2 are in respective ratios of 1:0.8:0.7:0.3 [Egbert and Erofeeva, 2002]. Field measurements at 20.6°N, 121.8°E (depth 930 m) revealed that the barotropic tidal currents may reach $1\,\mathrm{m\,s^{-1}}$ [Alford *et al.*, 2010]. The combination of steep submarine ridges and strong barotropic tidal currents leads the Luzon Strait to be one of the most energetic barotropic to baroclinic conversion sites in the global oceans [Arbic *et al.*, 2012; Egbert and Ray, 2000, 2003; Niwa and Hibiya, 2004]. Field measurements near one deep channel on the eastern ridge observed isopycnal displacements as high as 200–300 m high [Pinkel *et al.*, 2012]. Field measurements using gliders, moored and shipboard instruments revealed the internal tide radiation from the Luzon Strait [Johnston *et al.*, 2013; Rainville *et al.*, 2013; Rudnick *et al.*, 2013].

Numerical modeling also reveals that the Luzon Strait is a generation site of energetic internal tides. Most of the internal tidal energy is transported away by low-mode internal tides, while high-mode internal tides are prone to dissipate locally [Klymak *et al.*, 2006; St. Laurent *et al.*, 2003]. Niwa and Hibiya [2004] estimated that the conversion rate for M2 is 14 GW (1 GW = $10^9$ W), of which ≈50% radiates away as internal tides (4.2 GW into the SCS and 3.2 GW into the WP). Jan *et al.* [2008] estimated that the conversion rates for the semidiurnal (M2 + S2) and diurnal (O1 + K1) constituents are about 11 and 19 GW, respectively, and ≈40% of the converted tidal energy is transported away by internal tides. Jan *et al.* [2008] also estimated that the spring-neap averaged energy fluxes into SCS and WP are about 7 and 6 GW, respectively. Alford *et al.* [2011] estimated that the conversation rate is about 24.1 GW (≈14.6 GW radiated away as internal tides) for the four principal tidal constituents, based on numerical simulations [Simmons, 2008]. Alford *et al.* [2011] reported that the total westward and eastward energy fluxes are about 7.9 and 6.1 GW, respectively [Zhao, 2014].

Zhao *et al.* [2014] derived the internal tide radiation patterns originating from the Luzon Strait from the sea surface height (SSH) data of multiple satellite altimeters. Figure 7.12 shows the radiation patterns of M2 internal tides. One can see that in SCS, the westward M2 internal tides bifurcate into two beams with different propagation directions of ∼165° versus ∼220° from due east. The northwestward beam is stronger than the southwestward beam. Due to these bifurcated strong and weak M2 beams, in the northeastern SCS west of the Luzon Strait, the semidiurnal constituent dominates in the northern region, and the diurnal dominates in the southern region. This south–north asymmetry was recently revealed using field moorings [Xu *et al.*, 2014].

It is believed that the solitary IWs near in the northeastern SCS deep basin originate from nonlinear steepening of the internal tides from the Luzon Strait [Farmer *et al.*, 2011; Lien *et al.*, 2005; Zhao *et al.*, 2004]. Figures 7.13(a)–(d) show the spatial distribution of the IWs and internal tides, which may serve as new evidence for this

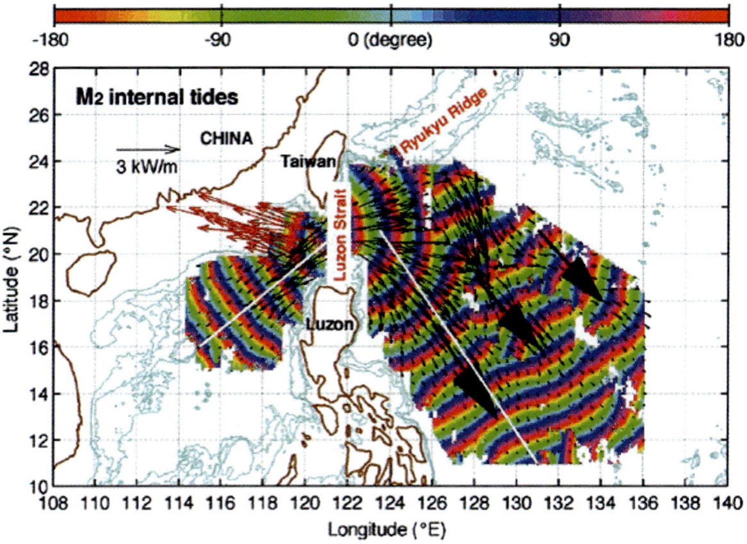


Fig. 7.12 Radiation patterns of the mode-1 M2 internal tides from the Luzon Strait. There are two beams in SCS. The red arrows show the energy flux of the northwestward beam. The bold arrows in WP represent three distinct beams formed by the internal tides from the Luzon Strait and the Ryukyu Ridge. White lines indicate the central paths of two M2 beams. Color codes represent the Greenwich phase [Zhao, 2014]

concept. The green curves indicate 116 IW packets observed in SAR images [Zhao *et al.*, 2004]. The red and blue arrows represent the southwestward and northeastward energy fluxes, respectively. From Figure 7.13(a), one can see that IWs and the northwestward M2 beam are spatially coincident, and propagate westward with the central directions about 165° from due east. In contrast, there are no IWs associated with the weak southwestward beam (blue arrows). Figures 7.13(b)–(d) show the phase, energy flux and propagation direction along the two beams [Zhao, 2014]. Internal tides K1 and O1 mainly propagate southwestward, inconsistent with the IW occurrence in NSCS. This feature supports the previous suggestion that the diurnal internal tides play a secondary role by modifying the IW generation [Buijsman *et al.*, 2010b; Helfrich and Grimshaw, 2008; Li and Farmer, 2011; Zhang *et al.*, 2011].

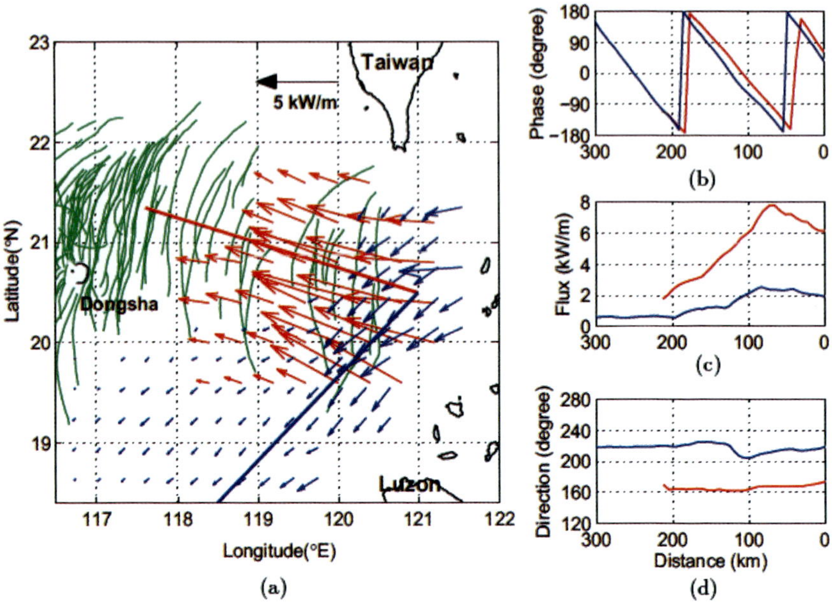


Fig. 7.13 (a) Mode-1 M2 internal tides and IWs from the Luzon Strait. The red and blue arrows represent the energy fluxes of the northwestward and southwestward internal tides. Green curves represent IW packets observed from the SAR images. (b–d) Greenwich phase, energy flux and propagation direction of the M2 internal tides along the red and blue lines in Figure 7.13(a) [Zhao, 2014]

### 7.7.3. *Dynamic condition for IW growth: Shoaling thermocline*

As analyzed in Section 6.2.2.2, the shoaling thermocline provides a dynamic condition for growth and fission of a single soliton, which is the form of initial disturbance of a solitary IW packet. The soliton amplitude growth ratio (SAGR) can be derived from the solvability of PKdV equation (6.2.6), which demands [Newell, 1985]

$$\frac{\partial \eta}{\partial \tau} = -\frac{2}{3}\Gamma \eta, \tag{7.7.1}$$

where $\tau$ is a rescaled distance coordinate

$$\tau = \frac{1}{6}\int^{x} D^{1/2} dX, \tag{6.2.17}$$

and

$$\Gamma = \frac{9}{4}\frac{D_T}{D}. \tag{6.2.22}$$

Transforming (7.7.1) into the $x-y$ coordinates defined in Figure 6.5 yields

$$\frac{d\eta}{dx} = \left[\left(-\frac{3}{2D(x)}\frac{dD(x)}{dx}\right)\right]\eta. \tag{7.7.2}$$

The solution of (7.7.2) is

$$\eta = \eta_0 D(x)^{-3/2}, \tag{7.7.3}$$

where $\eta_0$ is the amplitude of initial solitary disturbance at $x_0$. Based on (7.7.3), a SAGR is parameterized as

$$\beta(x) = \frac{\eta}{\eta_0} \tag{7.7.4}$$

or

$$\beta(x) = D(x)^{-3/2}. \tag{7.7.5}$$

To simplify the analysis, Eq. (6.2.6) does not include a dissipation term. In a real ocean, however, the dissipation induced by the eddy viscosity and bottom friction [Liu *et al.*, 2004] and mixing in the upper ocean [Zhang *et al.*, 2012, 2014] always exist. In the studied case particularly, where the length of soliton growth zone or thermocline shoaling area is as long as 200 km as shown in Figure 6.7, the dissipation effect cannot be ignored. Therefore, the SAGR becomes a necessary condition to overcome the effect of dissipation. In other words, not all the initial disturbances may be finally radiated out of the source region, depending on the balance between the growth and dissipation. This may be verified by the low monthly SAR-observed IW occurrence frequency shown in Figure 7.6. The maximum monthly occurrence frequency in a year is only 20% in June, although the internal tidal disturbance occurs daily and the favorable sea state frequency is as high as 40%. Thus, the IW occurrence frequency $p$ in NSCS (beyond the source region) should be proportional to the SAGR at the source region boundary, *i.e.*,

$$p = a\beta(x_{\mathrm{up}}) - b, \tag{7.7.6}$$

where $x_{\mathrm{up}}$ represents the location of source region boundary, $a$ and $b$ are constants. Non-zero constant $b$ implies that the IW occurrence needs a threshold value of SAGR.

### 7.7.4. *Field-measured shoaling thermocline*

The ocean water temperature profile data from January 1980 through December 1999 [Boyer and Levitus, 1994] are used to determine the thermocline depth distribution in the Luzon Strait. The depth of isotherm of 15°C is defined as the thermocline depth [Zheng *et al.*, 2007]. The data within a band of (21° ± 1°) N from 115.5° to 124.5°E are used to generate monthly mean thermocline depths with a resolution of 2° latitude by 0.5° longitude, which are used to represent ocean stratification in the Luzon Strait. The results are shown in Figure 7.14. One can see that the thermocline depth decreases

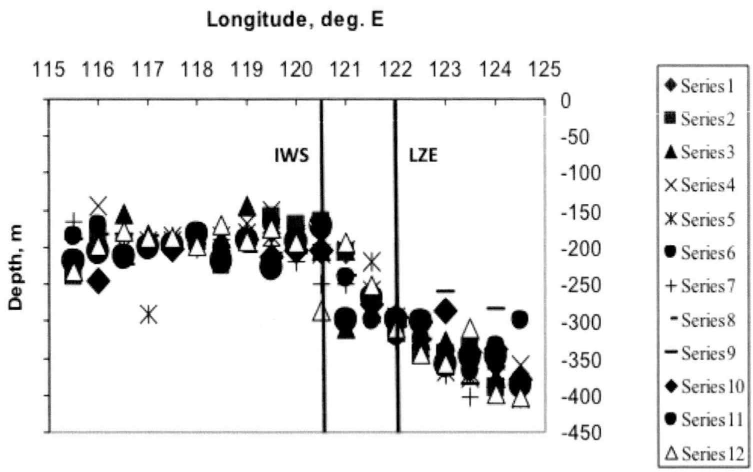


Fig. 7.14 Zonal distribution of thermocline depth in the Luzon Strait Series 1 represents the monthly mean thermocline depth in January, Series 2 in February and so on. Vertical lines IWS and LZE represent boundary of IW origination locations and eastern boundary of Luzon Strait, respectively [Zheng *et al.*, 2007]

westward from a mean of $(380 \pm 30)$ m at 124.5° E to $(178 \pm 21)$ m at 119°E. According to Eq. (7.7.4), if an IW disturbance is formed at 122°E, where the mean thermocline depth is $(310 \pm 10)$ m, the amplitude will grow to 1.9 times of the initial value after the wave propagates westward for 1.5° longitude to $(200 \pm 30)$ m at 120.5°E. This means that the zonal distribution of thermocline depth in the Luzon Strait enables to provide the dynamic conditions for the amplitude growth to the westward propagating IW type disturbance.

Using the data taken from Figure 7.14 and comparing to the physical model shown in Figure 6.5, the SAGR is calculated by

$$\mathrm{SAGR} = \left(\frac{h_0}{d}\right)^{3/2}, \tag{7.7.7}$$

where $h_0$ and $d$ are the thermocline depths on the deep side and shallow side of the shoaling thermocline, respectively. The calculation results are shown in Figure 7.15, in which the results derived from

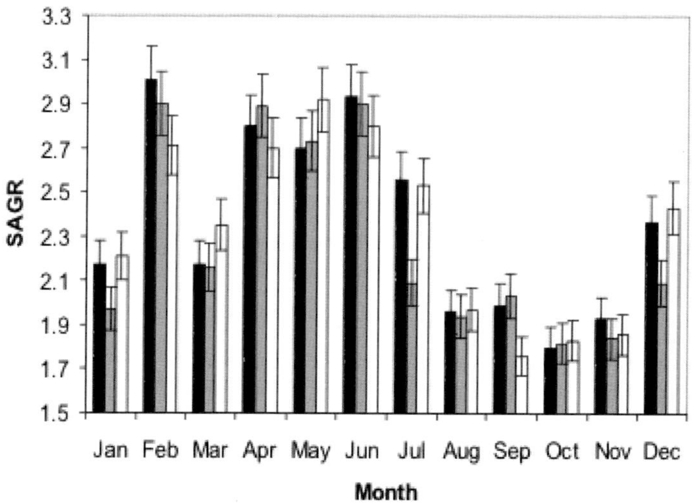


Fig. 7.15 Monthly mean SAGR in the Luzon Strait derived from the temperature profile data from 1980 to 1993. Dark bars are derived from the thermocline depth defined by 15°C isotherm, light dark and open bars from 14°C and 16°C isotherms, respectively. Error bars show a maximum error of 10% [Zheng *et al.*, 2007]

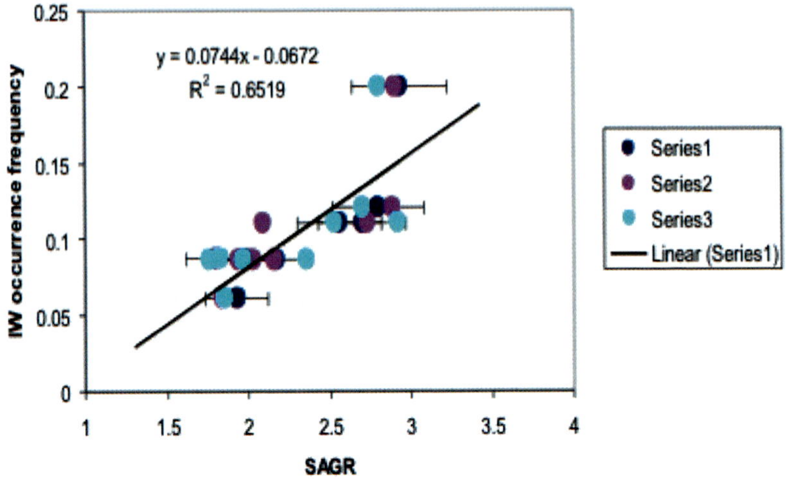


Fig. 7.16 Monthly IW occurrence frequency versus the monthly mean SAGR in NSCS (no winter months DJF included). The straight line is derived from a linear regression of the data series 1 (15°C thermocline). Data points derived from 14°C (series 2), and 16°C (series 3) are also plotted for comparison. Error bars represent an estimated maximum error of 10% [Zheng *et al.*, 2007]

thermocline defined by 14°C and 16°C isotherms are also shown for comparison.

The monthly SAR-observed IW occurrence frequency (shown in Figure 7.6(b)) versus the monthly mean SAGR (except winter months) derived from upper layer depths defined by 14°C, 15°C and 16°C isotherms is plotted in Figure 7.16. One can see that as a first-order approximation, the theoretical model (linear relation) gives a trend of data points with a correlation coefficient ($R^2$) of 0.6519 (15°C data). The regression equation is

$$P_c = 0.0744\,\mathrm{SAGR}(\beta) - 0.0677. \qquad (7.7.8)$$

The threshold value of SAGR($\beta$) is 0.90, which is defined as an SAGR value for the IW occurrence frequency equal to zero.

## 7.8. Discussion

### 7.8.1. *Interannual variability of SAR-observed IW occurrence frequency*

Figures 7.5(a) and (b) show the interannual variability of SAR-observed IW occurrence frequency in NSCS. The yearly distribution of SAR-observed IW occurrence frequencies from 1995 to 2001 indicates that the occurrence frequencies in 1995, 1998 and 2000 are 2–4 times higher than other years. The dynamic analysis of IW generation indicates that the IW generation requires the following necessary and sufficient conditions: initial disturbance formation and wave amplitude growth. Sections 7.7.2 and 7.7.3 provide evidences for that the internal tides constitute a major source to generate initial IW disturbance at the Luzon Strait, and thermocline shoaling provides the internal forcing for IW growth and fission in NSCS, respectively. Therefore, they should have links with the interannual variability of IW occurrence frequency. In fact, internal tides do not have remarkable interannual variability, but thermocline shoaling does. On the other hand, the local circulation including the Kuroshio intrusion may have impact on thermocline distribution [Wang *et al.*, 2006]. Thus, the interannual variability of the SCS circulation and the Kuroshio intrusion are possible causes to the interannual variability of SAR-observed IW occurrence frequency in NSCS [Park and Farmer, 2013].

### 7.8.2. *Eastward propagating initial disturbances*

The bottom topography in the Luzon Strait as shown in Figure 7.11 is near symmetric about the north–south submarine ridges. Radiation patterns of the mode-1 M2 internal tides from the Luzon Strait are also near symmetric about the central line of the strait as shown in Figure 7.12. Thus, the internal tides would have the same probability to induce westward and eastward propagating initial disturbances. The analysis in Section 7.6 indicates that the westward propagating initial disturbances indeed obtain a chance to amplify and to radiate out of the source region. However, the eastward propagating initial disturbances run into a deepening thermocline. According to (7.7.5), the deepening thermocline cannot provide the dynamic conditions for solitary IW growth. Hence, the eastward propagating initial disturbances may not be able to overcome the dissipation effect, and therefore may not radiate out of the source region. This cxplains the asymmetrically latitudinal distribution of IW packets in NSCS shown in Figure 7.4.

According to Eq. (7.7.4) and the thermocline depth distribution shown in Figure 7.14, for an IW disturbance formed at 122°E and propagating eastward for 2° longitude to 124°E, its amplitude will decline to 0.65 time of the initial value, which is smaller than 0.90 of the threshold value of SAGR($\beta$). Thus, the eastward propagating IW type disturbance would be damped, implying that the internal tides are the necessary conditions for initial disturbance formation, but not the sufficient conditions for IW generation. Due to dissipation of the initial disturbance, the eastward propagating initial disturbances have no chance to amplify, therefore they are hardly observed on the east side of submarine ridge in the eastern Luzon Strait.

# Chapter 8

# Ocean Bottom Topography: I. Longitudinal Flow Passing Over Parallel Topographic Corrugations

## 8.1. Introduction

Seawater is a high conductivity dielectric medium. Based on the principles of electromagnetic wave propagation, microwave pulses that radars use to detect the targets on the earth surface are capable of penetrating only into a skin layer of seawater with a maximum depth on the order of centimeters. This implies that the radar is unable "to see" the ocean bottom topographic features and underwater objects directly. But previous investigators have confirmed since launch of SEASAT satellite in 1978, that the ocean bottom topographic features, such as tidal flats, submerged sand waves, banks and shoals, indeed show up on SAR images [Alpers *et al.*, 2004; Alpers and Hennings, 1984; Apel, 1987; Donato *et al.*, 1997; Fu and Holt, 1982; Hennings 1990; Hsu *et al.*, 1997; Li *et al.*, 2009, 2010; Lodge, 1983; Shuchman *et al.*, 1985].

Owing to high spatial resolution, SAR images are able to reveal many details of the ocean bottom topographic features, which cannot be found in existing navigation charts. Thus, previous investigators suggested use SAR images for ocean bottom topographic mapping [Alpers *et al.*, 2004; Hsu *et al.*, 1997; Vogelzang, 1997]. However, this resource cannot practically be used for operational or engineering mapping, unless dynamic links between sea bottom topographic features and SAR imagery features, including imaging mechanisms and quantitative relationships between the two, are clarified, and accurate inverse methods are established. Hence, previous investigators have dedicated continuous efforts to answer this 'what-is-what' question. Alpers and Hennings [1984] suggested a radar imaging mechanism of ocean bottom topography, which is a topography-via-tidal-current-modulating surface wave model. Later, their model was further developed to include the effects of advection [Hennings, 1990; Vogelzang, 1989] and the effects of the sand wave shape [van der Kooij *et al.*, 1995]. Vogelzang [1997] used the continuity equations to describe modulation of sand waves to the tidal current. Shuchman *et al.* [1985] developed a hydrodynamic and electromagnetic numerical model to investigate SAR imaging conditions of ocean bottom topography.

In order to interpret the satellite images associated with ocean bottom topography, we should solve the following three questions. (1) How does ocean bottom topography generate disturbance signals in the water body above? (2) How do disturbance signals propagate from the bottom layer to the upper ocean? (3) How do upper ocean disturbance signals generate the satellite sensor signals? The answers to the three questions constitute a physical basis for quantitative retrieval of ocean bottom topographic information from satellite images and identification of other processes, such as atmospheric disturbances, rainfall cells, ocean internal waves, fronts, coastal upwelling and river discharge plumes. Previous investigators have developed SAR imaging theories of ocean bottom topography under assumptions of traverse flow and un-stratified ocean [Alpers and Hennings, 1984; Romeiser and Alpers, 1997; Shuchman *et al.*, 1985; Valenzula *et al.*, 1983, 1985; van der Kooij *et al.*, 1995; Vogelzang, 1989, 1997; Vogelzang *et al.*, 1992]. For a stratified ocean, Zheng *et al.* [2006b] derived dynamic solutions for vertical propagation of disturbance signals induced by underwater periodic topography from the ocean bottom to the surface. Zheng *et al.* [2012a] found that the seamount-induced vortex trains at a depth of 600 m in the Gulf Stream region appeared on SAR images.

Parallel topographic corrugations are frequently seen in ocean bottom features on the continental shelves. They appear in the forms of sand bars, tidal creeks and underwater sand ridges, which are often distributed in the areas with rich sediment deposit and strong tidal currents. Well-developed submerged sand ridges have been reported in entrances of shallow bays such as the Bohai Sea [Liu *et al.*, 1998] and the Delaware Bay [Zheng *et al.*, 1993a], entrances of shallow straits such as English Channel [Lodge, 1983] and Taiwan Strait [Zheng *et al.*, 2006b], modern estuaries such as Yangtze River Estuary, and paleo-estuaries such as paleo-Yangtze River and the paleo-Yellow River Estuaries [Liu *et al.*, 2007]. The width of topographic corrugation and separate distance are in the order of 5–10 km, and their lengths may reach as long as 100 km. The relative height of the individual sand ridge may reach 10–30 m. This chapter will analyze the secondary circulation induced by the longitudinal flow passing over parallel ocean bottom topographic corrugations in a shallow, un-stratified ocean. Thus, the effects of the horizontal velocity shear inevitably resulting from the sidewall friction, and the vertical velocity shear induced by the surface and the bottom Ekman layers are considered in the dynamic analysis.

## 8.2. Physics Model

Figures 8.1(a) and (b) sketch a physics model for the dynamical analysis of the longitudinal flow passing over parallel bottom topographic corrugations in a shallow, un-stratified ocean [Zheng *et al.*, 2012b]. One can see that the problem of secondary circulation induced by the topographic corrugations is simulated by a flow in a long canal with a rectangular cross section of two flat sidewalls and a free surface. The sidewalls have height $D$ and the bottom has width $2b$. The origin of the Cartesian coordinate system is set at the bottom. Vertical axis $z$ is positive upward. The horizontal axis $y$ is perpendicular to the central line of the canal and positive leftward. Horizontal axis $x$ is parallel to the canal and positive downstream. The three-dimensional scales of the canal, length $L_1$, width $L_2$ $(= 2b)$ and height $L_3$ $(= D)$ satisfy $L_1 \gg L_2 \gg L_3$. The mean flow $(\bar{u}, \bar{v}, \bar{w})$ is driven by a

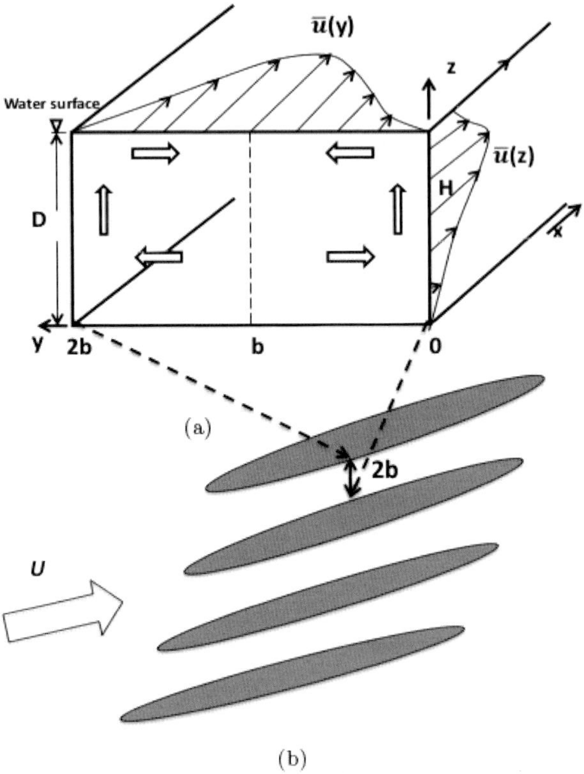


Fig. 8.1 Physics model for a secondary circulation (large arrows) induced by a shear flow (a) in parallel ocean bottom topographic corrugation (b)

pressure gradient externally imposed by a large-scale process, such as the tidal waves or the ocean circulation, thus, it is considered as a stable process. Due to the confinement of sidewalls, the mean flow is one-dimensional and parallel to the $x$-axis, *i.e.*, $(\bar{v}, \bar{w}) = 0$, and has horizontal and vertical velocity shears. The horizontal shear can be described by a parabolic profile as a plane Poiseuille flow (Kundu, 1990)

$$\bar{u}(y) = -\frac{y}{\mu}\frac{dP_0}{dx}\left(b - \frac{y}{2}\right), \tag{8.2.1}$$

where $\mu$ is the dynamic viscosity, $dP_0/dx$ is the externally imposed pressure gradient. On the other hand, considering the existence of the surface and the bottom Ekman layers, the vertical shear is supposed to have a sinusoidal profile with an apex at $H$

$$\bar{u}(z) = \sin\frac{\pi z}{2H'} \quad 0 \le z \le D. \tag{8.2.2}$$

Thus, we have

$$\bar{u}(y,z) = -\frac{y}{\mu}\frac{dP}{dx}\left(b - \frac{y}{2}\right)\sin\frac{\pi z}{2H'} \quad 0 \le y \le 2b, \quad 0 \le z \le D. \tag{8.2.3}$$

## 8.3. Dynamic Analysis

### 8.3.1. *Derivation of disturbance governing equations*

Consider the governing equations for a flow consisting of mean flow and disturbance

$$\frac{\partial U}{\partial t} + U\frac{\partial U}{\partial x} + V\frac{\partial U}{\partial y} + W\frac{\partial U}{\partial z} - fV = -\frac{1}{\rho}\frac{\partial P}{\partial x} + A\nabla^2 U + F_x, \tag{8.3.1}$$

$$\frac{\partial V}{\partial t} + U\frac{\partial V}{\partial x} + V\frac{\partial V}{\partial y} + W\frac{\partial V}{\partial z} + fU = -\frac{1}{\rho}\frac{\partial P}{\partial y} + A\nabla^2 V + F_y, \tag{8.3.2}$$

$$\frac{\partial W}{\partial t} + U\frac{\partial W}{\partial x} + V\frac{\partial W}{\partial y} + W\frac{\partial W}{\partial z} = -\frac{1}{\rho}\frac{\partial P}{\partial z} - g + A\nabla^2 W + F_z, \tag{8.3.3}$$

$$\frac{\partial U}{\partial x} + \frac{\partial V}{\partial y} + \frac{\partial W}{\partial z} = 0, \tag{8.3.4}$$

where $f$ is the Coriolis parameter, $P$ the pressure, $\rho$ the water density, $A$ the kinetic viscosity, $F_x$, $F_y$ and $F_z$ the components of external forcing, $g$ the gravitational acceleration and

$$U = \bar{u} + u, \; V = \bar{v} + v, \; W = \bar{w} + w, \; P = \bar{p} + p. \tag{8.3.5}$$

The boundary conditions are

$$U = V = W = 0, \quad \text{for } z = 0 \tag{8.3.6}$$

and

$$U = V = 0, \quad \text{for } y = 0, \quad \text{and} \quad y = 2b. \tag{8.3.7}$$

Substituting (8.3.5) into Eqs. (8.2.3), (8.3.1)–(8.3.4) yields the disturbance governing equations

$$\frac{\partial u}{\partial t} + \bar{u}\frac{\partial u}{\partial x} + v\frac{\partial \bar{u}}{\partial y} + w\frac{\partial \bar{u}}{\partial z} - fv = -\frac{1}{\rho}\frac{\partial p}{\partial x} + A\nabla^2 u, \tag{8.3.8}$$

$$\frac{\partial v}{\partial t} + \bar{u}\frac{\partial v}{\partial x} + fu = -\frac{1}{\rho}\frac{\partial p}{\partial y} + A\nabla^2 v, \tag{8.3.9}$$

$$\frac{\partial w}{\partial t} + \bar{u}\frac{\partial w}{\partial x} = -\frac{1}{\rho}\frac{\partial p}{\partial z} + A\nabla^2 w, \tag{8.3.10}$$

$$\frac{\partial u}{\partial x} + \frac{\partial v}{\partial y} + \frac{\partial w}{\partial z} = 0. \tag{8.3.11}$$

The boundary conditions are

$$u = v = w = 0, \quad \text{for } z = 0 \tag{8.3.12}$$

and

$$u = v = 0, \quad \text{for } y = 0 \quad \text{and} \quad 2b. \tag{8.3.13}$$

In order to examine the role of velocity shear in generating the secondary circulation, the following approximations are applied: (1) ignoring the viscous terms, (2) assuming the $x$-coordinate scale of mean flow $\bar{u}$, $\bar{L}$, is much larger than that of disturbance $L$, thus, we have $\partial/\partial t \ll \bar{u}\partial/\partial x$, (3) the $x$-coordinate scale of disturbance is much larger than that of the $y$-coordinate scale, thus, we have $\partial/\partial x \ll \partial/\partial y$, and (4) in Eqs. (8.3.8)–(8.3.10), the velocity shear terms are much larger than other terms. Thus, we have simplified disturbance equations

$$\mathrm{v}\frac{\partial \bar{u}}{\partial \mathrm{y}} + \mathrm{w}\frac{\partial \bar{u}}{\partial \mathrm{z}} = 0, \tag{8.3.14}$$

$$\bar{u}\frac{\partial v}{\partial x} = 0, \tag{8.3.15}$$

$$\bar{u}\frac{\partial w}{\partial x} = 0, \tag{8.3.16}$$

$$\frac{\partial v}{\partial y} + \frac{\partial w}{\partial z} = 0. \tag{8.3.17}$$

### 8.3.2. *Solutions of disturbance governing equations*

From (8.3.15) and (8.3.16), we have $\frac{\partial v}{\partial x} = \frac{\partial w}{\partial x} = 0$, *i.e.*, $v$ and $w$ are independent of $x$. From (8.3.14) and (8.3.17), we drive the equation of $w$

$$\frac{\partial w}{\partial y}F - \frac{\partial w}{\partial z} = -w\frac{\partial F}{\partial y}, \tag{8.3.18}$$

where $F$ is defined as

$$F \equiv \frac{\partial \bar{u}}{\partial z} \Big/ \frac{\partial \bar{u}}{\partial y}. \tag{8.3.19}$$

The solutions of (8.3.18) satisfy equations

$$\frac{dy}{F} = \frac{dz}{-1} = \frac{dw}{-w\partial F/\partial y}. \tag{8.3.20}$$

From (8.3.20) we have

$$dy = -F dz, \tag{8.3.21}$$

$$\frac{dw}{dz} = w\frac{\partial F}{\partial y}, \tag{8.3.22}$$

$$\frac{dw}{dy} = -w\left(\frac{\partial F}{\partial y}\Big/ F\right). \tag{8.3.23}$$

From (8.3.19) we have

$$F = \frac{\pi}{4H}\frac{y(2b-y)}{b-y}\cot\frac{\pi z}{2H}. \tag{8.3.24}$$

Substituting (8.3.24) into (8.3.21) and integrating the two sides of the equation yield

$$\int \frac{b-y}{y(2b-y)}dy = -\frac{\pi}{4H}\int \cot\frac{\pi z}{2H}dz + C. \tag{8.3.25}$$

From Eq. (8.3.25) we obtain

$$y\,(2b-y)\sin\frac{\pi z}{2H} = C.$$

In this case, solving (8.3.22) and (8.3.23) yields two independent solutions

$$w_1 = c_1 e^{\int \frac{\partial F}{\partial y}dz} \tag{8.3.26}$$

and

$$w_2 = c_2 e^{-\int \frac{\partial F}{\partial y}/F dy}. \tag{8.3.27}$$

Thus, a general solution of (8.3.18) is

$$w(y,z) = \frac{1}{1-\Lambda}\left(c_1 e^{\int \frac{\partial F}{\partial y}dz} - c_2 e^{-\int \frac{\partial F}{\partial y}/F dy}\right), \tag{8.3.28}$$

where $\Lambda, c_1$ and $c_2$ are arbitrary constants. Taking $\Lambda = c_2 = 0$, we obtain

$$w = c_1 e^{\int \frac{\partial F}{\partial y}dz}. \tag{8.3.29}$$

Substituting (8.3.24) into (8.3.29) yields

$$w(y,z) = w_0\left(\sin\frac{\pi z}{2H}\right)^{\left[1+\frac{y(2b-y)}{2(b-y)^2}\right]}, \tag{8.3.30}$$

where $w_0$ is a constant to be determined. From (8.3.14) we have

$$v = -w\left(\frac{\partial \bar{u}}{\partial z}\Big/\frac{\partial \bar{u}}{\partial y}\right). \tag{8.3.31}$$

Then we derive

$$v(y,z) = -v_0 \left[\frac{y(2b-y)}{2(b-y)}\right] \left(\sin\frac{\pi z}{2H}\right)^{\frac{y(2b-y)}{2(b-y)^2}} \cos\frac{\pi z}{2H}, \qquad (8.3.32)$$

where $v_0 = \pi(2H)^{-1}w_0$.

Solutions (8.3.30) and (8.3.32) are shown graphically in Figures 8.2(a)–(c). One can see that secondary circulation is two-dimensional motion in the $y-z$ plane. For plotting the solutions, we take $D = 30\,\text{m}$, $2b = 10\,\text{km}$, $w_0 = 0.01\,\text{m}\,\text{s}^{-1}$, and the vertical velocity as 10 times of its values.

### 8.3.3. *Analysis of secondary circulation solutions*

From Figures 8.2(a)–(c), one can see the following dynamic features of secondary circulation. (1) Secondary circulation consists of a pair of current vortexes with opposite polarities distributed symmetrically on two sides of the central line of the cannel, a cyclonic vortex on the right and an anti-cyclonic vortex on the left. The mean flow shear drives upwelling along two sidewalls, and the stronger, the closer to sidewalls. (2) In the case of presence of the surface Ekman layer and the direction of Ekman current component opposite to the mean flow (for $H < D$), two vortexes converge at the central line of the canal in the upper layer as shown in Figure 8.2(a). Thus, there is a surface current convergent zone along the central line of the canal. (3) In the case of absence of the surface Ekman layer (for $H = D$), there are no convergent or divergent zones in the sea surface, but below the surface, two vortexes diverge at the central line of canal in all the layers as shown in Figure 8.2(b). (4) In the case of presence of the surface Ekman layer, but the direction of Ekman velocity component identical with the mean flow (for $H > D$), the same as (3), two vortexes diverge at the central line of the canal in all the layers as shown in Figures 2(c). Thus, there is a surface current divergent zone along the central line of the canal.

Hereafter, the physics model shown in Figure 8.1(a) will be used to represent grouped submerged sand ridges. The dynamic conditions for formation of the convergent (divergent) zone are the presence of a longitudinal, shear mean flow in the two adjacent submerged sand

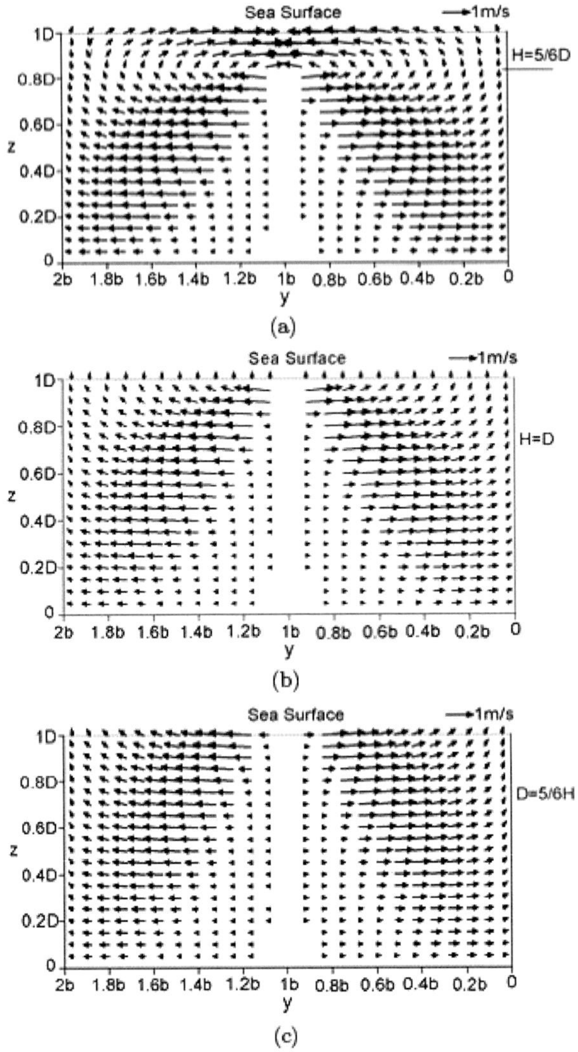


Fig. 8.2 Analytical solutions of the secondary circulation induced by a shear flow in a long, rectangular canal. (a) There is an upper Ekman layer, in which the Ekman current has a negative component in the mean flow direction. (b) No upper Ekman layer. (c) There is an upper Ekman layer, in which the Ekman current has a positive component in the mean flow direction [Zheng *et al.*, 2012b]

ridges and the surface Ekman layer with the direction of Ekman current component opposite to (identical with) the mean flow. In the other words, $V_E \cdot V < 0$ ($V_E \cdot V > 0$), in which $\mathbf{V_E}$ is an Ekman velocity vector and $\mathbf{V}$ is a mean flow velocity vector, is a necessary condition for formation of the sea surface convergent (divergent) zone.

## 8.4. Application of Theoretical Solutions

The theoretical solutions derived from Section 8.3 reveal that under favorable conditions ($V_E \cdot V < 0$), the mean flow such as the tidal flow is capable of generating the surface convergent zone along the central line of the canal between two adjacent submerged sand ridges. Thus, floating materials, such as bubbles and foams, pollutants, macro-algal and floating sea ice, would be concentrated in the convergent zone. Satellite radar and visible band sensors are capable of imaging the phenomena. In this section, the solutions are used to interpret satellite images of the Liaodong Shoal centered at 39°25′N 120°45′E, which is near the entrance of the Bohai Sea, China.

### 8.4.1. *Hydrodynamic conditions of the Liaodong Shoal*

Figure 8.3(a) shows a true-color image of the Bohai Sea taken by a Moderate Resolution Imaging Spectroradiometer (MODIS) on board the US National Aeronautics and Space Administration (NASA) Aqua satellite on February 6, 2010. The image was pre-processed and geometrically corrected. The spatial resolution of the image is about 1 km. From the bathymetric map shown in Figure 8.3(b), one can see that the Liaodong Shoal is a shallow water area with the water depth ranging from 20 to 30 m. A group of six finger-like, submerged sand ridges are distributed from east to west coded by numbers 1 to 6. Among them, sand ridges 2–5 are orientated nearly north–southwards and parallel to each other with an average spacing of about 10 km. The heights of sand ridges are ranging from 10 to 20 m, the average width is

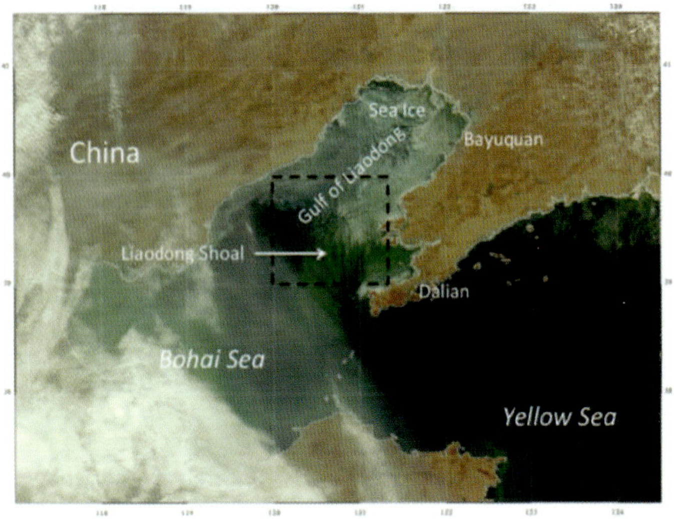


(a)

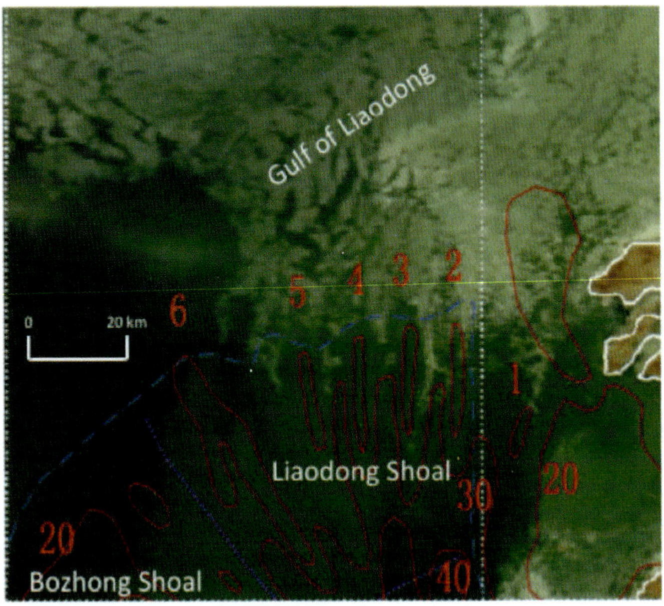


(b)

Fig. 8.3 (a) True-color image of the Bohai Sea taken by MODIS onboard the US NASA Aqua satellite at 5:00 UTC on February 6, 2010 [Zheng *et al.*, 2012b]. (b) Zoomed image of the Liaodong Shoal area (black dashed square in (a)). Red lines are isobaths. Numbers 20–40 on isobaths are in m. Submerged sand ridges are coded by 1–6 from east to west [Zheng *et al.*, 2012b]

3 km, and the average length is about 20 km [Liu *et al.*, 1998]. In the Bohai Sea, the tidal motion is a major dynamic process, and the semidiurnal tide (M2) is a dominant tidal constituent. In the Liaodong Shoal region, the major axis of M2 current ellipse with the amplitude of $0.75\,\mathrm{m\,s^{-1}}$ is parallel to the submerged sand ridge orientation [Fang and Yang, 1985]. This implies that the tidal flows are longitudinal to underwater topographic corrugation most of the time. Thus, the physical model shown in Figure 8.1 is a good approximation for the case. In addition, the Bohai Sea is located in the East Asia monsoon region, thus, seasonal wind-driven currents constitute the important components of the local circulation system.

### 8.4.2. *Floating sea ice signatures in MODIS image*

In winter, the Gulf of Liaodong freezes starting from mid November to early December and ending in late February to mid March of the next year. The ice thickness is ranges from 10 to 50 cm. The maximum thickness may reach 1 m. With the exception of very shallow water areas close to the coastline, in open water, sea ice exists in the form of floating ice. The average drifting speed is $0.4$–$0.5\,\mathrm{m\,s^{-1}}$ with a maximum speed of $1\ \mathrm{m\ s^{-1}}$ [Bai *et al.*, 1999]. The Liaodong Shoal is just located at the south margin of ice region, thus, floating sea ice or pack ice often appears there. From Figure 8.3b, one can see that floating sea ice image appears as bright, finger-like patterns in the submerged sand ridge region. All the floating sea ice fingers are sandwiched by two adjacent sand ridges. The lengths of the floating sea ice fingers range from 10 to 30 km. The widths are much smaller than the separate distance of the two adjacent sand ridges. This implies that pack ice flows southward along the central lines of troughs or canals formed by two adjacent submerged sand ridges. If we treat ice packs as sea surface current drifters, the floating ice finger patterns just show the trajectories of the sea surface current. Thus, this finger-like pattern reveals the following important hydrodynamic information: (1) the local ocean current flows southward and the strongest currents occur along troughs or canals formed by two adjacent submerged sand ridges, and (2) along each central line of trough or canal, there is a surface current convergent zone, where floating sea ice is gathered together. These two points agree well with the solutions of secondary circulation derived in Section 8.3.

In order to conform the suitability of the solutions for interpreting the floating sea ice satellite image, it is necessary to examine if the hydrodynamic conditions at the imaging time satisfy the conditions for formation of a surface current convergent zone along the central line of the canal between two adjacent submerged sand ridges.

As mentioned above, the M2 tide is a dominant tidal constituent and a major dynamical process in the Liaodong Shoal region. The tide table data at the nearest station, Bayuquan (40° 39′00″N 122°13′00″E) of China, is used to extrapolate the tidal phase at the Liaodong Shoal. On February 6, 2010, high (low) tide times are 0326 (1025), and 1553 (2145) CST (China Standard Time) at Bayuquan. Referring to co-tidal charts and tidal current ellipses of the M2 tide [Fang and Yang, 1985; Guo and Yanagi, 1998], high (low) tide times at the central Liaodong Shoal (39° 25'N 120° 45'E) are estimated as about 3.5 h earlier than that at Bayuquan, *i.e.*, 0000 (0655) and 1223 (1815) CST. At the imaging time 1330 CST on February 6, 2010, the tide phase was 1 h 7 min after the high tide, *i.e.*, an ebb tide period. The tidal current was southward and parallel to submerged sand ridges. From Figures 8.3(a) and (b), one can see that floating ice was concentrated in the canals of the sand ridge region.

The weather conditions before the imaging time are shown in Figure 8.4, which was a surface weather chart of East Asia at 00UTC (08CST) on February 6, 2010 downloaded from website of Korea Meteorological Administration (KMA). The weather chart shows a clear sky and easterly wind of 5 m s$^{-1}$ over the Liaodong Shoal region. Thus, the Ekman velocity is northwestward in the surface layer and satisfies the condition $V_E \cdot V < 0$. The scale of the Ekman velocity is estimated as 15 cm s$^{-1}$. The depth of the Ekman layer can be estimated by [Pedlosky, 1987]

$$\delta_E = \left(2A_v f^{-1}\right)^{\frac{1}{2}}, \tag{8.4.1}$$

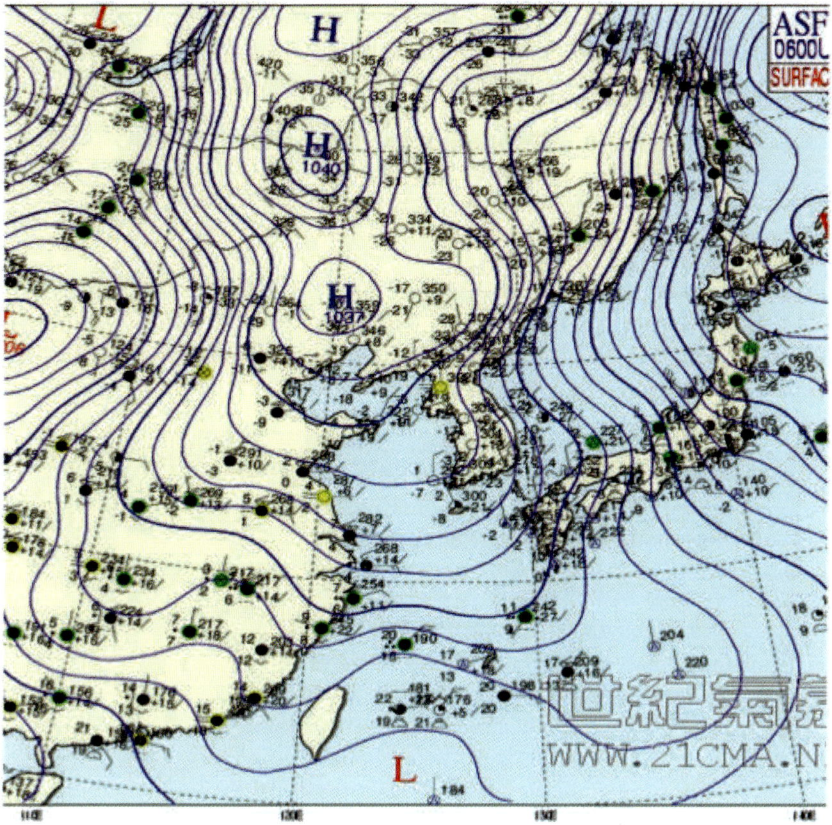


Fig. 8.4 Surface weather chart of East Asia at 00UTC (008CST) on February 6, 2010 (downloaded from http://web.kma.go.kr)

where $A_v$ is the vertical viscosity, and $f$ is the Coriolis parameter. The Liaodong Shoal is a shallow water region with strong tidal currents and complex sea bottom topography. Thus, it is reasonable to choose $A_v$ as $5 \times 10^2\,\mathrm{cm}^2\,\mathrm{s}^{-1}$, a value close to its top limit of $10^3\,\mathrm{cm}^2\,\mathrm{s}^{-1}$. In this case, from (8.4.1) the depth of the Ekman layer is derived as 30 m. In other words, it is possible for the Ekman layer to reach the ocean bottom. The above analysis indicates that the hydrodynamic conditions at the imaging time satisfy the conditions for formation of the surface convergent zone along the central line of the canal formed by two adjacent submerged sand ridges.

### 8.4.3. *Floating sea ice signatures in RADARSAT-2 SAR image*

Figure 8.5 shows a RADARSAT-2 SAR image of the Liaodong Shoal region taken at 0656 CST on March 6, 2010. The image was taken by ScanSAR Wide mode with a nominal swath width of 500 km, a spatial resolution of $100 \times 100$ m, and HH polarizations. In the ocean region of the image, the dark area is seawater, and the bright areas are floating sea ice, because under the conditions of low sea states and middle incidence angle (30°–60°), radar backscattering section of sea ice is larger than open seawater. From the SAR image, one can see

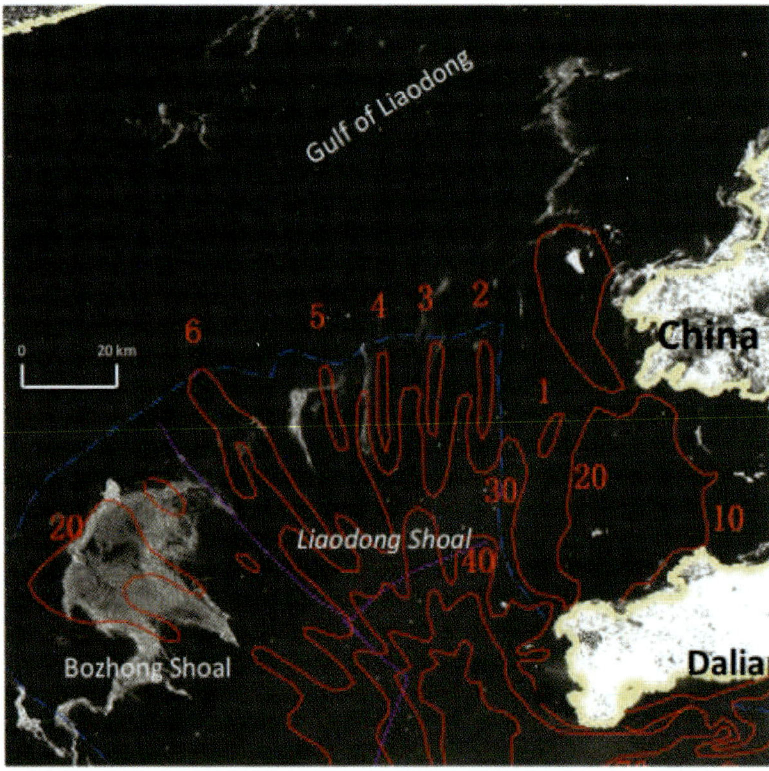


Fig. 8.5 RADARSAT-2 SAR image of the Liaodong Shoal region taken at 06:55:42 CST on March 3, 2010. Bright patterns in the water area (in dark) are imagery of floating sea ice. Bottom topography is the same as Figure 8.3 [Zheng *et al.*, 2012b]

that floating sea ice image appears as bright band patterns in the submerged sand ridge region of the Liaodong Shoal. It is the same as MODIS images shown in Figures 8.3(a) and (b) that floating sea ice bands are sandwiched by two adjacent sand ridges (cases between sand ridges 4–5, and 5–6 are particularly clear). The lengths of the floating sea ice finger range from 20 to 30 km. This implies that ice packs are concentrated in the central canals formed by two adjacent submerged sand ridges.

Using the same method as in Section 8.4.2, we estimate that the tidal phase of the Liaodong Shoal at the imaging time, 0656 CST on March 3, 2010, is between high tide (0034 CST) and low tide (0720 CST), *i.e.*, an ebb tide. Thus, floating sea ice was carried to the canals by the southward tidal current of the ebb tide from the Gulf of Liaodong in the north. The weather chart (not shown) shows a sustaining easterly wind of about $4\,\mathrm{m\,s^{-1}}$ before the imaging time. The wind driven Ekman current in the upper ocean layer was favorable for formation of convergent zones in the sand ridge canals.

### 8.4.4. *Sea surface roughness signatures in RADARSAT-1 SAR image*

Figure 8.6a shows a portion of a RADARSAT-1 SAR image of the Liaodong Shoal area taken at 0600 CST on March 23, 2003. The full size image with a swath of 500 km and a spatial resolution of 100 m was taken with ScanSAR Wide B imaging mode and HH polarizations. One can see that bright, finger-like image patterns coded by S2–S7 are distributed over the Liaodong Shoal region [Li *et al.* 2009]. Compared submerged sand ridges to bright SAR imagery patterns shown in Figure 8.6(b), one can see the following features. (1) Submerged sand ridges and bright, finger-like SAR imagery patterns are of one-to-one correspondence and have the same orientation. (2) Lengths of bright SAR imagery fingers are 2–3 times longer than the sand ridges. In particular, fingers S2, S3 and S4 extend 20–40 km northward, far beyond the sand ridge region. (3) In the sand ridge region, the locations of submerged sand ridges and bright SAR imagery fingers do not overlay each other. All the bright

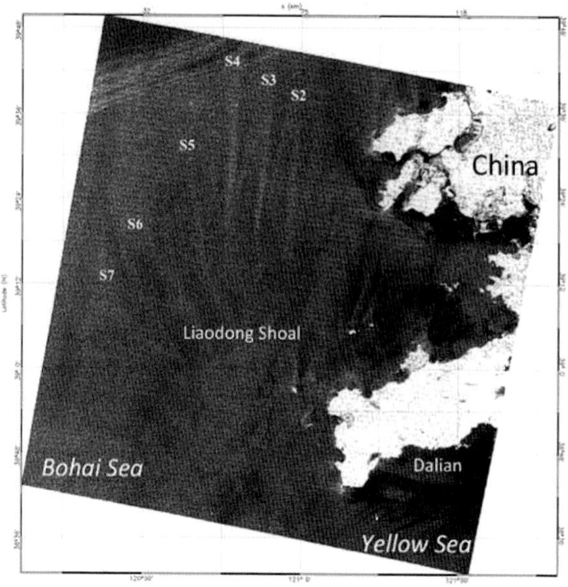


(a)

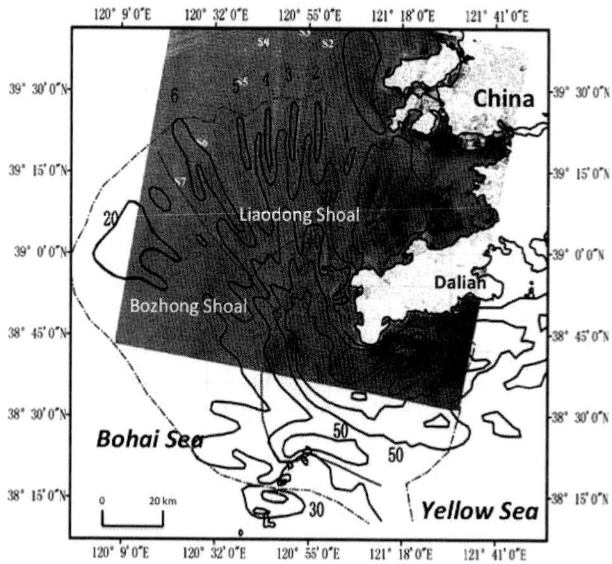


(b)

Fig. 8.6 (a) A RADARSAT-1 SAR image of the Liaodong Shoal region taken at 06:00:03 CST on March 23, 2003 and (b) bathymetry-overlaid image of (a). Bottom topography is the same as Figure 8.3 [Zheng *et al.*, 2012b]

SAR image fingers are sandwiched by two adjacent sand ridges. In the other words, bright SAR image fingers correspond to the canals formed by two sand ridges, not to crests of submerged sand ridges.

According to radar imaging theories, the radar signal intensity from the ocean surface depends on the power spectral density of small waves with the wavelength satisfying the Bragg resonance condition [Plant, 1990]. On the other hand, the power spectral density of the Bragg waves is depending on the sea surface wind stress and modulated by the ocean current [Yuan, 1997; Zheng *et al.*, 2001b]. In the case shown in Figures 8.6(a) and (b), bright SAR image fingers coincide with the cannels formed by two adjacent sand ridges. This implies that there are surface current convergent zones along the canals, as predicted by the solutions of secondary circulation derived in Section 8.3.

The imaging time of Figure 8.6(a) was 0600 CST on March 23, 2003. Using the same method as in Section 8.4.2, we estimate that the tidal phase of the Liaodong Shoal is between high tide (0300 CST) and low tide (0917 CST), *i.e.*, an ebb tide. The tidal currents formed the southward shear-flows in the canals of sand ridges. The weather chart (not shown) shows a southerly wind of about $4\,\mathrm{m\,s^{-1}}$ before and at the imaging time. The wind driven Ekman current in the upper ocean layer and the tidal current provided favorable conditions for formation of surface current convergent zones in the canals between two adjacent sand ridges.

Statistics of SAR imaging conditions of submerged sand ridges of the Liaodong Shoal using total 27 images taken by RADARSAT-1/2 and ENVISAT C band SARs and Italian Cosmo-Skymed satellite X band SAR from January 14 to March 10, 2010 is used to further confirm the theoretical solutions. The statistical results are listed in Table 8.1. One can see that nine images contain the submerged sand ridge imagery, occupying 33% of the total. All the nine cases occurred under the hydrodynamic conditions favorable for formation of surface current convergent zones in the canals of submerged sand ridge region and low wind condition (wind speed $<6\,\mathrm{m\,s^{-1}}$). Other 18 images contain none of the submerged sand ridge imagery, occupying 67% of the total. The hydrodynamic conditions for the 18 images

Table 8.1 Statistics of SAR imaging of submerged sand ridges in the Liaodong Shoal from January 14 to March 10, 2010

| | Convergent Zone | | Divergent Zones | | |
|---|---|---|---|---|---|
| **Sand Ridge Imagery** | **Low W. <6 m s$^{-1}$** | **High W. >6.5 m s$^{-1}$** | **Low W. <6 m s$^{-1}$** | **High W. >6.5 m s$^{-1}$** | **No Ekman Layer** |
| Presence | 9 | 0 | 0 | 0 | 0 |
| Absence | 0 | 3 | 7 | 7 | 1 |

include 14 cases favorable for formation of surface current divergent zones, no matter low wind or high wind, one case of no Ekman layer (zero wind speed) and three cases favorable for formation of surface current convergent zones but at high wind conditions (wind speed >6.5 m s$^{-1}$). These statistical results further reveal that besides the surface hydrodynamic conditions, the wind conditions also play a decisive role in SAR imaging of the submerged sand ridges. In other words, presence of the surface current convergent zones and low wind condition constitute sufficient and necessary conditions for SAR imaging of the cannels in the submerged sand ridge region.

## 8.5. Summary

This chapter analyzes the secondary circulation induced by the longitudinal flow passing over the parallel ocean bottom topographic corrugation. Considering the sidewall friction, the surface and the bottom Ekman layers, it is supposed that the flow has the horizontal and the vertical velocity shears. For this shear-flow, the disturbance equations are solved and analytical solutions of secondary circulation are obtained. The theoretical results are applied to interpretation of satellite images in the submerged sand ridge region of the Liaodong Shoal. The major conclusions are summarized as follows.

A flow with a parabolic horizontal velocity shear and a sinusoidal vertical velocity shear induces a two-dimensional secondary circulation in the cross section perpendicular to the canal walls. The secondary circulation consists of a pair of vortexes with opposite signs distributed symmetrically on two sides of the central line of the canal, a cyclonic vortex on the right and an anti-cyclonic vortex on the left.

In the case of presence of the surface Ekman layer and the direction of Ekman current opposite to the mean flow, *i.e.*, $\partial\bar{u}/\partial z < 0$, the two vortexes converge at the central line of the canal in the upper layer. In other words there is a surface current convergent zone along the central line of the canal. In the case of absence of the surface Ekman layer, *i.e.*, $\partial\bar{u}/\partial z = 0$, there is no the convergent or divergent zone to be formed in the sea surface, but divergent zone below the surface. In the case of presence of the surface Ekman layer and the direction of Ekman current component identical with the mean flow, *i.e.*, $\partial\bar{u}/\partial z > 0$, there is a surface current divergent zone along the central line of the canal.

Two cases of satellite images of floating sea ice in the Bohai Sea show that floating sea ice follows the ebb tide current entering the canals of submerged sand ridges and forming finger-like floating sea ice bands. This verifies that there are the surface current convergent zones along the central lines of the canals. One case of SAR imagery of the same area shows that long, finger-like, bright patterns are distributed along the central lines of the canals, verifying that there are the surface current convergent zones with a high density of sea surface roughness or radar scatters. In other words, long, finger-like, bright patterns on SAR images correspond to the locations of the canals (or tidal channels) formed by two adjacent sand ridges, not to the location submerged sand ridges themselves. This is a very important inference for interpretation of SAR images of submerged sand ridges in shallow water region.

All the above analyses, results and conclusions are supposed to stand only for the very shallow and un-stratified water region, where the signals generated by ocean bottom topographic disturbance may vertically propagate to the sea surface directly. The next chapter will analyze hydrodynamic conditions for ocean bottom topographic disturbance signals to propagate vertically to the ocean surface in the case of traverse flow passing over periodical bottom topography in the deep and stratified ocean.

Chapter 9

# Ocean Bottom Topography: II. Traverse Flow Passing Over Periodical Topography

## 9.1. Introduction

Most ocean radar imaging models available to date concentrate on derivation of solutions for modulation of variable current to the surface waves [Alpers *et al.*, 2004]. While the underlying ocean is treated as homogeneous, namely, the ocean is un-stratified and has no-shear. In most cases, however, the real ocean is vertically stratified even in very shallow water. Donato *et al.* [1997] carried out intensive air-borne radar imaging flights over ocean bottom sand waves on the continental shelf east of Cape Hatteras, where the water is only 37 m deep, but strongly stratified. Comparing the radar imagery with simultaneously measured accurate bathymetry, they found that the major features of sand waves on the radar images, such as the maximum signal intensity and the peak value locations, disagreed with theoretical predictions derived from the previous models. Thus, they suggested that any radar imaging theory must account for both influences of stratification and topography. Zheng *et al.* [2006b, 2012a] revealed that ocean bottom topographic features located underwater 100 m at the south Taiwan Strait and underwater 600–800 m in the Gulf Stream region still showed up on the SAR images. For such ocean areas, it is unimaginable that the ocean is still treated as homogeneous. In fact, conductivity-temperature-depth (CTD) profiler measurements shown later clearly demonstrate that the vertical stratification becomes an unavoidable hydrodynamic feature. Therefore, new hydrodynamic models need to be developed, which must be able to reveal how the lower layer disturbances generated by the bottom topography propagate via all layers to the ocean surface, and finally modulate the sea surface waves, which serve as backscatters of the radar pulses.

Chapter 8 has examined the SAR imaging problem for the longitudinal flow passing over parallel ocean bottom topographic corrugations in a shallow, un-stratified ocean. This chapter specifies another SAR imaging problem, which is for the traverse flow passing over periodic topography in the stratified ocean. The SAR imaging hydrodynamic model for this problem includes both stratification and topography. As an application example, the model and solutions are used to interpret satellite SAR images of periodic bottom topography (this term is more general than 'sand waves') in the Taiwan Strait. The results reveal the SAR imaging mechanisms for underwater features, and provide theoretical basis and quantitative relations for interpretation of the SAR images [Zheng *et al.*, 2006b].

## 9.2. Study Area and Hydrodynamic Conditions

### 9.2.1. *Taiwan Banks*

The study area is the Taiwan Banks located at the south outlet of the Taiwan Strait, which covers a water area of 21.5°–24°N latitude and 117°–120°E longitude as shown in Figure 9.1. One can see that about 80% of the total area is shallower than 100 m, the rest 20% on the southeast corner of the strait is the shelf break with depths steeply increasing from 100 to 2000 m within a distance less than 50 km. The Taiwan Banks located 22.5° to 23.5°N and 117.5° to 119°E are a large shoal-like threshold lying across the south outlet of Taiwan Strait. It connects the SCS continental shelf and the coastal shoals of China mainland on the western side, and separates from Taiwan Island by the Penghu Channel on the eastern side. The total area is about 8000 $km^2$. The average water depth is 30 m. The submarine

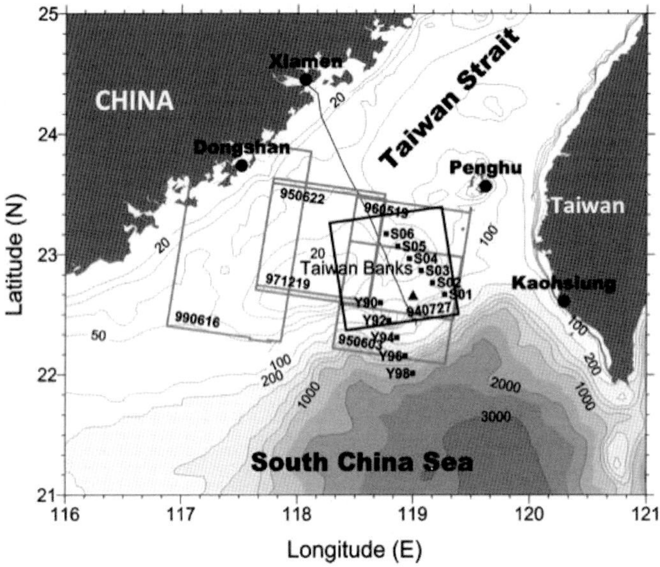


Fig. 9.1 Chart of study area Coded boxes show SAR image coverage. Triangle in Box 940727 represents the location where 3-day moored tidal measurements were taken. The thin line extending from Xiamen to Boxes 940727 and 950603 represents the ship track for bathymetric measurements using an onboard Acoustic Doppler Current Profiler (ADCP). CTD stations along two transects for cruises of August–September 1994 are also marked. Isobaths are in m [Zheng *et al.*, 2006b]

morphology of Taiwan Banks is featured by two-scale subsystems of periodic topography. The large-scale subsystem is composed of tidal sand ridges with relative heights of 10–20 m extending as long as 10 km. The small-scale subsystem is composed of a vast distribution of hundreds of sand waves. The orientation of sand waves is near perpendicular to tidal sand ridges. Field measurements reveal that the heights of sand waves reach 3 to 15 m, and the widths 350–500 m. The sand waves are distributed in groups, in which the sand waves run near parallel to each other with separate distances much smaller than their widths [Cai *et al.*, 1992]. Using LANDSAT-5 Thematic Mapper (TM) images, Li *et al.* [2001b, 2003] estimated that the sand waves roughly take a west–east orientation with lengths from 0.5 to 5 km, and the wavelengths range from 0.35 to 1.5 km.

These sand waves show up clearly on the SAR images as shown in Figure 9.5.

### 9.2.2. *Circulation systems*

Circulation around Taiwan Banks consists of two major subsystems: the South China Sea Warm Current subsystem, which flows northeastward along the shelf break south of Taiwan Banks [Guan and Fang, 2006; Li *et al.*, 2000a] and enters the eastern strait through the Penhu Channel flowing northward year-round [Chuang, 1986]; and a seasonally varying coastal current subsystem on the west strait modulated by the annual cycle of monsoon wind forcing, which drives a narrow southward jet (so-called Zhejiang–Fujian Coastal Current) along the coasts in winter [Zheng and Klemas, 1982] and coastal upwelling with weak horizontal currents and strong upwelling fronts in the channel west to Taiwan Banks [Li and Li, 1989; Li *et al.*, 2000a]. In any case, these circulation systems have little influence on the study area, which is located at the center of the strait.

### 9.2.3. *Tides*

The tidal currents play a key role in hydrodynamics of the study area. Tides in the Taiwan Strait are dominated by semidiurnal tide M2 with the amplitude as high as 2 m in the middle strait. As M2 tide propagates from the Pacific into the shallow East China Sea, its amplitude is amplified. A part of this amplified tidal wave propagates southward along the China coast and enters the Taiwan Strait with further amplification. The southward propagating wave and its reflection by the abruptly deepened topography south of the Taiwan Strait lead to the coexistence of a progressive wave in the west and a partial standing wave in the east portion of the strait. The southern branch of M2 tide from the Pacific also enters the Taiwan Strait passing through the Luzon Strait. In comparison to the northern branch, however, its amplitude is relatively small [Yin, 1984].

Tidal currents are weak in the deep water and become stronger over the shallow shelf [Jan *et al.*, 2004]. Major axes of tidal ellipses are essentially aligned along the strait. In the study area, the northward flood tidal velocity is about $1.8\,\mathrm{m\,s^{-1}}$ and the southward ebb tidal velocity is about $1.3\,\mathrm{m\,s^{-1}}$ [Hsu *et al.*, 1997].

### 9.2.4. *Vertical stratification*

Hydrographic data were collected by a joint cruise program, which was carried out by R/V Ocean Researcher 3 from Taipei and R/V Yan Ping from Xiamen. The observations were carried out from August 28 to September 10, 1994 [Li *et al.*, 1998a]. There were two transects, S01–S06 and Y90–Y98 as marked in Figure 9.1, just across the Taiwan Banks. The location, water depth and time of CTD casting at each station are listed in Table 9.1. The CTD data from these stations were obtained on September 3 and 7, 1994, 36 and 40 days later than the date SAR image 940727 was taken. The data are used to represent approximately the ocean stratification at the SAR imaging time. The vertical profiles of density stratification derived from the CTD data at each station and the Brunt–Väisälä frequency, which is defined as

$$\mathbf{N}^2 = -\mathbf{g}\frac{\partial \ln \boldsymbol{\rho}}{\partial \mathbf{z}}, \tag{9.2.1}$$

where $g$ is the gravitational acceleration, and $\rho(\mathbf{z})$ is the density, corresponding to each density profile are shown in Figures 9.2(a) and (b), respectively.

One can see that 8 of 11 stations show strong vertical stratification, and rest three stations (S04–S06) show weak continuous stratification. These field measurements and derived parameters will serve as a basis for physics ocean model development and dynamic analysis later on.

### 9.2.5. *Sea surface winds*

Winds over the Taiwan Strait are dominated by the East Asia monsoon, prevailing northeasterly winds in winter and southwesterly winds in summer. The northeast monsoon lasts for nine months a year, beginning in mid-September, reaching a peak phase from

Table 9.1 CTD stations within the study area

| Station | Location | Depth, m | Casting Time, UTC |
|---|---|---|---|
| S01 | 22° 40.00′N 119° 16.05′E | 81 | 1159 September 3, 1994 |
| S02 | 22° 46.03′N 119° 10.03′E | 63 | 1057 September 3, 1994 |
| S03 | 22° 52.02′N 119° 04.11′E | 37 | 1003 September 3, 1994 |
| S04 | 22° 58.04′N 118° 58.09′E | 25 | 0906 September 3, 1994 |
| S05 | 23° 04.09′N 118° 52.01′E | 26 | 0801 September 3, 1994 |
| S06 | 23° 10.13′N 118° 45.98′E | 31 | 0651 September 3, 1994 |
| Y90 | 22° 36.0′N 118° 43.0′E | 56 | 0905 September 7, 1994 |
| Y92 | 22° 27.2′N 118° 47.6′E | 92 | 1023 September 7, 1994 |
| Y94 | 22° 18.6′N 118° 51.7′E | 140 | 1135 September 7, 1994 |
| Y96 | 22° 09.3′N 118° 56.0′E | 450 | 1250 September 7, 1994 |
| Y98 | 22° 00.5′N 118° 59.8′E | 1709 | 1420 September 7, 1994 |

October to January, and weakening continuously thereafter till May next year. The southwest monsoon lasts for only three months from June to August. The northeast monsoon is much stronger than the southwest monsoon. Analyses of satellite scatterometer winds from 1997 to 2001 indicate that the maximum monthly mean wind speeds in the central strait are $8\,\mathrm{m\,s^{-1}}$ in winter, and $3\,\mathrm{m\,s^{-1}}$ in summer, respectively [Kuo and Ho, 2004]. This implies that summer is a low sea state season favorable for SAR imaging of ocean dynamic processes. Later on, one can see that 83% of SAR images containing the ocean bottom topographic features collected for this chapter were taken in summer season.

There were not simultaneous wind measurements available at the SAR imaging time. Thus, the US National Oceanic and Atmospheric Administration (NOAA) National Centers for Environmental Prediction (NCEP) reanalysis wind field is used for estimating the sea state at the SAR imaging time. The case for image 940727 is plotted in Figure 9.3. One can see that a reasonable estimate of the wind in the area covered by 940727 at the imaging time of 940727 was a southeasterly wind of $3\,\mathrm{m\,s^{-1}}$, implying a low sea state that is a favorable condition for SAR imaging.

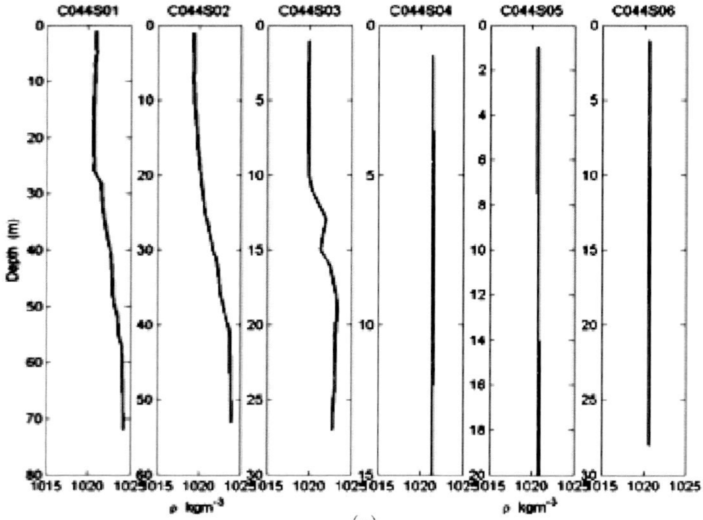


(a)

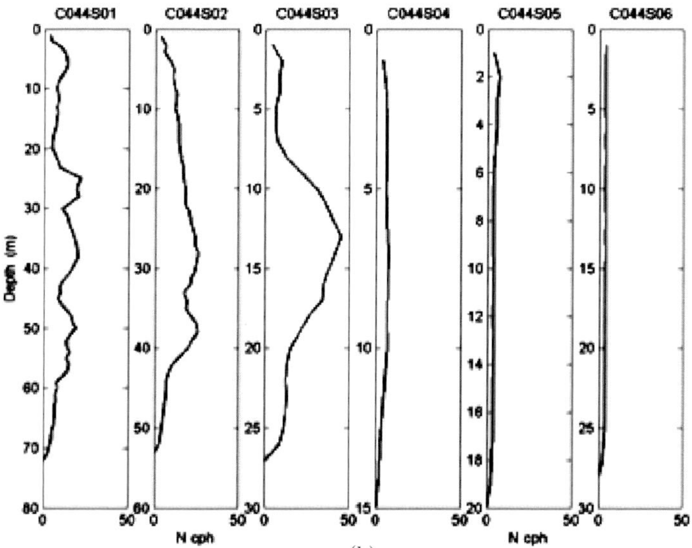


(b)

Fig. 9.2 (a) Vertical profiles of density stratification derived from the CTD data measured along transect S01–S06 by R/V Ocean Researcher 3 on September 3, 1994. (b) Vertical distributions of the Brunt–Väisälä frequencies corresponding to each density profile in (a). (c) and (d) Same as (a) and (b), but along transect Y90–Y98 by R/V Yan Ping on September 7, 1994 [Zheng *et al.*, 2006b]

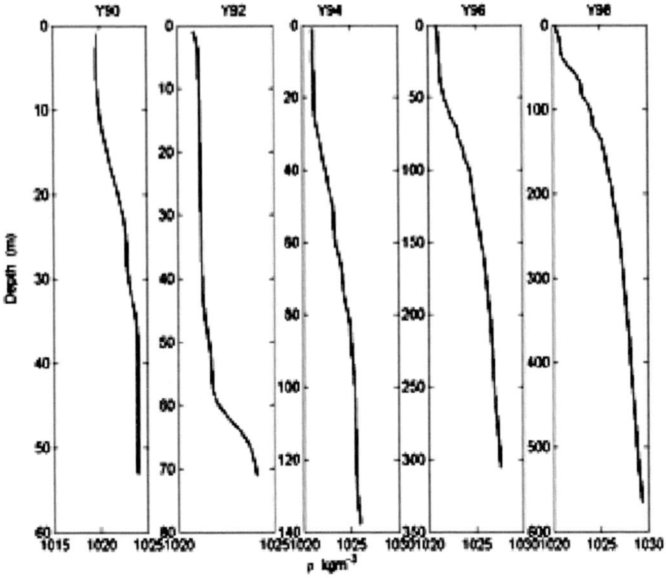


(c)

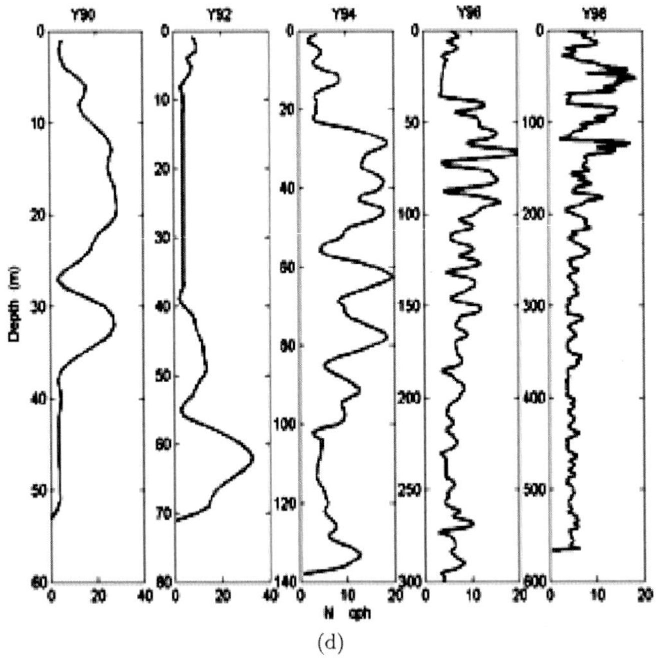


(d)

Fig. 9.2 (*Continued*)

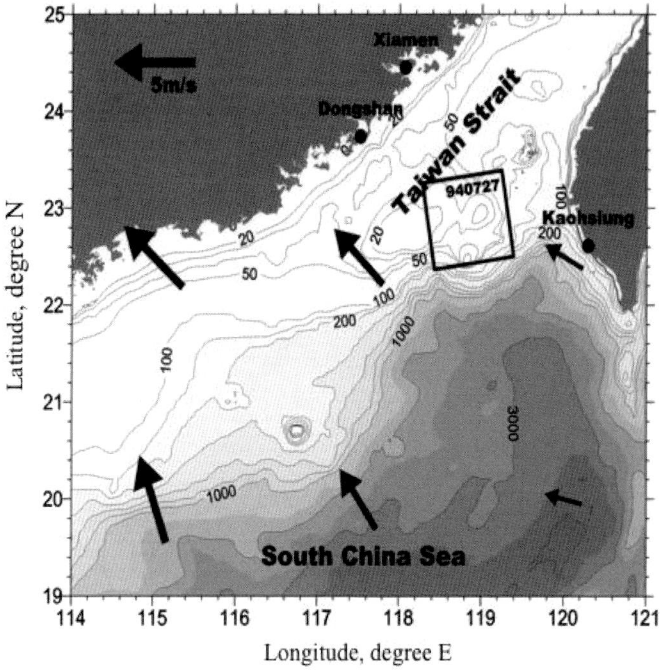


Fig. 9.3 NCEP wind field at the SAR imaging time of image 940727 shown in Figure 9.5. Isobaths are in m [Zheng *et al.*, 2006b]

### 9.2.6. *ADCP water depth profiles*

An ADCP onboard R/V Yan Ping measured bathymetry along the ship track on April 21–22, 2004, which is shown as the thin line from Xiamen to Taiwan Banks in Figure 9.1, and a bright line shown on image 940727 as an inset of Figure 9.4. The ADCP sampling time interval was 4 s that is equivalent to a horizontal distance resolution of about 12 m for an average navigation speed of 6 Kn. This horizontal resolution assures the topographic features of interest with a horizontal length scale of 0.5 km would not be filtered out. The measured bathymetric transect is shown in Figure 9.4. In comparison, the SAR image brightness along the ship track shown on image 940727 is also shown. One can see that both curves contain multi-scale signals, which will be examined later.

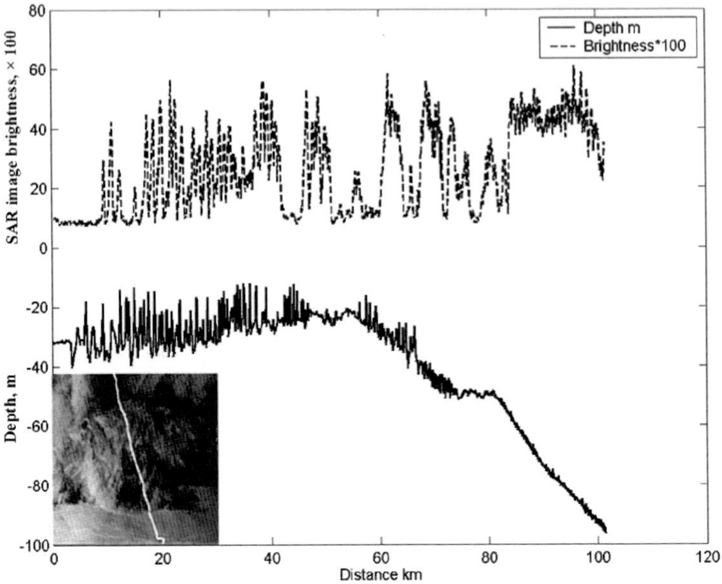


Fig. 9.4 Bathymetric transect measured by a shipboard ADCP (lower) and corresponding image brightness in arbitrary unit (upper) on SAR image 940727 (inset) [Zheng *et al.*, 2006b]

## 9.3. SAR Images

### 9.3.1. *Information of SAR images*

Six ERS-1/2 SAR images with coverage within the study area are collected for the analysis. Four images (940727, 960619, 971219 and 990616) are downloaded from website www.ifm.uni-hamburg.de/ers-sar. Other two (950603 and 950622) are provided by National Taiwan Ocean University SAR data base (courtesy of Drs. C.-R. Ho and M.-K. Hsu). The brief information of the images is listed in Table 9.2. The ground coverage of each image has been projected in Figure 9.1. The six images covered a time span of six years from 1994 to 1999. Ground coverage of each scene or image is 100 by 100 km (except 990616, a mosaic of two adjacent scenes). The pixel size of raw data is 12.5 m, while the ground resolution of all images used for this study has been reduced to 100 m (1000 by 1000 pixels per scene).

Table 9.2 Brief information of SAR images collected for this study

| Code | Satellite | Date and Time (UTC) | Orbit # | Frame # | Center Position (Corrected) |
|---|---|---|---|---|---|
| 940727 | ERS-1 | July 27, 1994 14:31 | 15850 | 441 | 23°02′N 118°32′E |
| 950603 | ERS-1 | June 3, 1995 2:37 | 20303 | 3159 | 22°36′N 118°52′E |
| 950622 | ERS-1 | June 22, 1995 2:40 | 20575 | 3159 | 23°05′N 118°12′E |
| 960519 | ERS-2 | May 19, 1996 2:37 | 05640 | 3159 | 22°58′N 118°50′E |
| 971219 | ERS-2 | December 19, 1997 2:40 | 13928 | 3141 | 23°07′N 118°12′E |
| 990616 | ERS-2 | June 16, 1999 2:43 | 21715 | 3123-3141 | 23°08′N 117°30′E |

Among the six images, image 940727 shown in Figure 9.5 serves as a main target, because of right ground coverage and more imagery information than others. Hundreds of grouped, wave-like, bright–dark patterns on the central image show the surface manifestations of bottom topographic features of the south Taiwan Banks. CTD data show that the depth for bottom topographic features to show up on the SAR image reaches around 100 m.

### 9.3.2. *Tidal phase and velocity at SAR imaging time*

The tidal current is considered as a major forcing to generate the bottom topographic disturbance signals on the SAR image. Thus, it is important to clarify the correspondence of the SAR imaging time to the tidal phase. As described in Section 9.2.3, the amplitude of tidal currents in the study area is about 1.5 $\mathrm{m\,s^{-1}}$, and the major axis of the prevailing M2 tide is generally parallel to the NNE-SSW orientation of the strait. Namely, the observed periodic topography is generally perpendicular to the tidal current axis. In order to determine a statistical phase relation between the tidal phase and SAR imaging time, the tide gauge data recorded at two nearby stations, Xiamen (24° 27.0′N 118° 04.0′E) at the west coast of the Taiwan Strait, and Kaohsiung (22° 36.6′N 120° 17.5′E) at the east coast, are used to extrapolate the tidal phase at SAR imaging time. The results as shown in Figure 9.6 indicate that 5 of 6 cases (except 971219 with respect to Kaohsiung) fall within a ±1.5 h span with respect to the local ebb tide. This implies that a certain

Fig. 9.5 ERS-1 SAR image 940727 of the Taiwan Banks taken at 14:31 UTC on July 27, 1994. The image is centered at 22°29′N 118°54′E, located at the south outlet of the Taiwan Strait. Grouped, wave-like, bright-dark patterns show the surface manifestations of periodic bottom topography [Credit: ESA]

level of tidal current velocity is needed to sustain the existence of hydrodynamic disturbances of topography, although their generated time cannot be determined directly by SAR observations. Based on 3-day moored measurements at the sand wave area marked by a triangle in Figure 9.1 [Chen *et al.*, 1999] and the results by previous investigators [Hsu *et al.*, 1997; Jan *et al.*, 2004; Yin, 1984], a reasonable estimate for the tidal velocity is about $1.5\,\mathrm{m\,s^{-1}}$.

## 9.4. Hydrodynamic Analysis

### 9.4.1. *Physics model*

Figure 9.7 shows a two-dimensional, three-layer model ocean with a Cartesian coordinate system. The origin is located at the ocean bottom. The $z$-axis is positive upward. The horizontal $x$-axis is

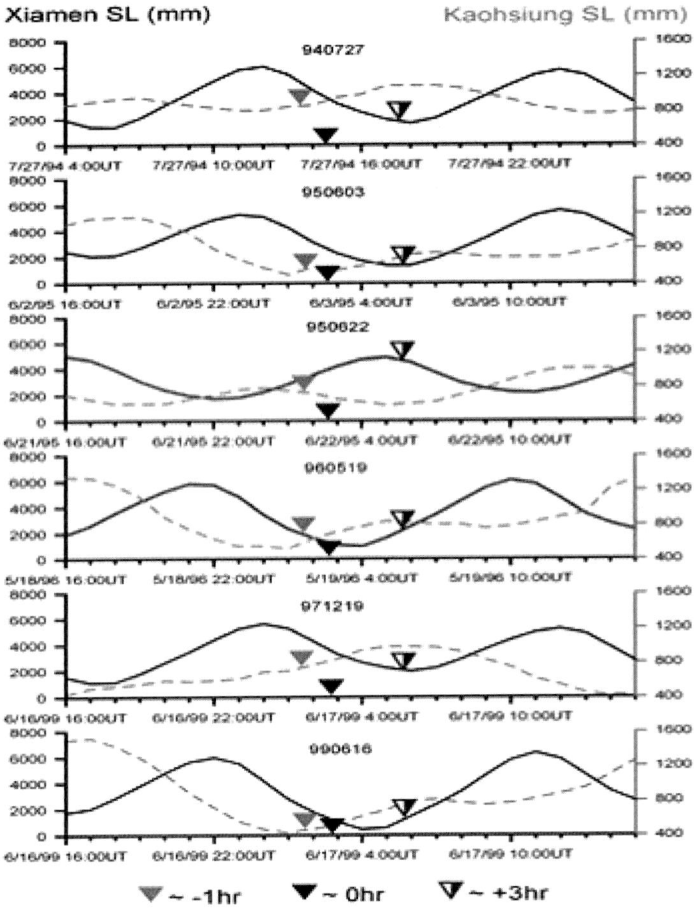


Fig. 9.6 SAR imaging times (dark reverse triangles) and tides at two nearby tide gauges, Xiamen (24° 27.0′N 118° 04.0′E, dark line) on the west coast of the Taiwan Strait, and Kaohsiung (22° 36.6′N 120° 17.5′E, dashed line) on the east coast. Split reverse triangles and smaller reverse triangles show the corresponding tidal phases with respect to Xiamen (3-h lag) and Kaohsiung (1-h lead), respectively [Zheng *et al.*, 2006b]

perpendicular to the axis of periodic bottom topography and positive downstream. In other words, the ocean current is a traverse flow with respect to the topography. Assuming that the ocean with a depth of $D$ can be divided into three layers. The lower layer has a depth of $H$, the middle layer, $2\Delta H$, and the upper layer, $H_3$ $[=D-(H+2\Delta H)]$.

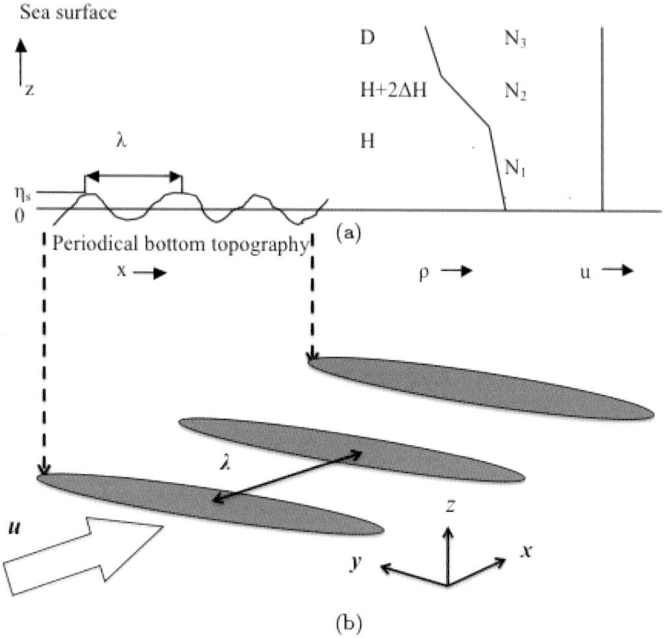


Fig. 9.7 (a) Schematic illustration of a two-dimensional, three-layer model ocean and (b) traverse flow passing over periodic bottom topography

The Brunt–Väisälä frequencies in the three layers are constants $N_1$, $N_2$ and $N_3$, respectively. As the lower rigid boundary of the ocean, the periodic bottom topography constitutes a small-amplitude feature, sinusoidal in $x$, *i.e.*, $\eta = (\eta_s/\pi)e^{ikx}$, where $\eta_s$ ($\ll$H) is the amplitude, $k(= 2\pi/\lambda$, and $\lambda$ is the wavelength) is the wavenumber. Only the real part is meant in applications. For convenience, we assume that there is not vertical shear in the horizontal velocity in each layer.

### 9.4.2. *Wave equations and solutions*

The methods used for the analysis of the atmospheric gravity wave are applied to the studied case [Gossard and Hooke, 1975]. The fluid is considered to be incompressible. The wave is assumed two-dimensional, and propagates in the $x-z$ plane. Thus, the momentum equations are

$$\frac{\partial u}{\partial t} = -\frac{1}{\rho_0}\frac{\partial p}{\partial x}, \tag{9.4.1}$$

$$\frac{\partial w}{\partial t} = -\frac{1}{\rho_0}\frac{\partial p}{\partial z} - g\frac{\rho}{\rho_0}, \tag{9.4.2}$$

$$\frac{\partial \rho}{\partial t} + w\frac{\partial \rho_0}{\partial z} = 0, \quad \frac{\partial u}{\partial x} + \frac{\partial w}{\partial z} = 0. \tag{9.4.3}$$

Here, $u$ and $w$ are the horizontal and vertical velocity components, respectively, $p$ is the pressure, and $\rho(= \rho_0 + \varepsilon\rho_1 + \varepsilon^2\rho_2 + \ldots)$ is the density. From (9.4.2) and (9.4.3), we have

$$\frac{\partial^2 w}{\partial t^2} + \frac{1}{\rho_0}\frac{\partial}{\partial t}\left(\frac{\partial p}{\partial z}\right) - \frac{g}{\rho_0}\frac{\partial \rho_0}{\partial z}w = 0. \tag{9.4.4}$$

From (9.4.1) and (9.4.3), we have

$$\frac{1}{\rho_0}\frac{\partial^2 p}{\partial x^2} = \frac{\partial}{\partial t}\left(\frac{\partial w}{\partial z}\right). \tag{9.4.5}$$

Suppose (9.4.4) and (9.4.5) have solutions of the form

$$w, p = (w_z, p_z)\exp i(kx - \sigma t), \tag{9.4.6}$$

where $\sigma(= 2\pi/T$, and $T$ is the period of horizontal component) is the angular frequency. Further eliminating $p$ in (9.4.4) yields the wave equation in $w_z$ as

$$\frac{\partial^2 w_z}{\partial z^2} + \frac{1}{\rho_0}\frac{\partial \rho_0}{\partial z}\frac{\partial w_z}{\partial z} + \frac{k^2}{\sigma^2}(N^2 - \sigma^2)w_z = 0. \tag{9.4.7}$$

Here, $N$ is the Brunt–Väisälä frequency. Suppose $\rho_0 = \rho_s e^{-\alpha z}$, in which $\rho_s$ is the density at a reference depth, and $\alpha$ is a constant, and using the variable transformation of the form

$$W_z = \left(\frac{\rho_0}{\rho_s}\right)^{1/2} w_z. \qquad (9.4.8)$$

Eq. (9.4.6) becomes

$$\frac{\partial^2 W_z}{\partial z^2} + \left[\frac{k^2}{\sigma^2}\left(N^2 - \sigma^2\right) + \left(\frac{1}{2\rho_0}\frac{\partial \rho_0}{\partial z}\right)^2\right] W_z = 0. \qquad (9.4.9)$$

If the coefficient of the second term is a constant

$$n^2 = \frac{k^2}{\sigma^2}\left(N^2 - \sigma^2\right) + \left(\frac{1}{2\rho_0}\frac{\partial \rho_0}{\partial z}\right)^2, \qquad (9.4.10)$$

Eq. (9.4.9) has plane-wave solutions in the form of

$$W_z = A \exp i(kx + nz - \sigma t). \qquad (9.4.11)$$

Here, $n$ is the vertical wavenumber corresponding to $k$ as the horizontal wavenumber. Equation (9.4.10) is called the dispersion relation or characteristic equation.

For a three-layer model ocean as shown in Figure 9.7, solutions satisfying the kinematical boundary conditions at the interface between two layers, *i.e.*,

$$\frac{\partial W_i}{\partial z} = \frac{\partial W_{i+1}}{\partial z}, \quad i = 1, 2 \qquad (9.4.12)$$

are as follows [Gossard and Munk, 1954]

$$W_1 = W_H \frac{\sinh \gamma_1 z + M \cosh \gamma_1 z}{\sinh \gamma_1 H + M \cosh \gamma_1 H} \exp[i(kx - \sigma t)], \qquad (9.4.13)$$

$$W_2 = W_H \frac{\sinh n_2(z - H) + M_0 \cosh n_2(z - H)}{M_0} \exp[i(kx - \sigma t)], \qquad (9.4.14)$$

$$W_3 = W_H \frac{\sin n_2\,(2\Delta H) + M_0 \cos n_2\,(2\Delta H)}{M_0} \cdot \exp\left[-\gamma_3\{z - (H + \Delta H)\} + i\,(kx - \sigma t)\right], \qquad (9.4.15)$$

where $W_H$ is the wave amplitude at the depth $H$, and $\gamma_1, \gamma_3 = -i(n_1, n_3)$. Applying the dynamic boundary condition (9.4.12) at $z = H + 2\Delta H$ yields

$$M_0 = \frac{n_2 \cos 2n_2\Delta H + \gamma_3 \sin 2n_2\Delta H}{-\gamma_3 \cos 2n_2\Delta H + n_2 \sin 2n_2\Delta H}. \tag{9.4.16}$$

Similarly at $z = H$, we obtain

$$M = \frac{(n_2/M_0)\sinh \gamma_1 H - \gamma_1 \cosh \gamma_1 H}{\gamma_1 \sinh \gamma_1 H - (n_2/M_0)\cosh \gamma_1 H}. \tag{9.4.17}$$

### 9.4.3. *Stationary wave solutions*

As shown in Figure 9.7(a), the lower boundary of the model ocean is a small-amplitude sinusoidal bottom topographic feature. If this boundary is moved with velocity $C_0$ in the negative $x$-direction beneath an ocean half-space, from the continuity equation (9.4.3) we derive the boundary perturbation of vertical velocity as

$$W(0) = (W_s/\pi)\exp[ik(x + C_0 t)]. \tag{9.4.18}$$

This is equivalent to a traverse flow with a speed $C_0$ flowing in the positive $x$-direction across the topography and establishing a system of stationary lee waves [Gossard and Hooke, 1975]. Substituting (9.4.18) into (9.4.13) and letting $H = 0$ yield a stationary solution in the form of

$$W_1(x, z, t) = (W_s/\pi M)(\sinh \gamma_1 z + M \cosh \gamma_1 z)\, e^{ikx}, \tag{9.4.19}$$

if a condition

$$\sigma = kC_0 \tag{9.4.20}$$

is satisfied. We call this condition the lower layer resonance condition (LLRC). Similarly, from (9.4.14) and (9.4.15), we have

$$W_2(x, z) = (W_s/\pi M_0)[\sin n_2(z - H) + M_0 \cos n_2(z - H)]e^{ikx}, \tag{9.4.21}$$

and

$$W_3(x,z) = (W_s/\pi M_0)[\sin n_2(2\Delta H) + M_0 \cos n_2(2\Delta H)] \cdot \exp\left[-\gamma_3 \left\{z - (H + 2\Delta H)\right\} + ikx\right]. \tag{9.4.22}$$

Besides the dynamic boundary condition, this case requires a special boundary condition, *i.e.*, $W_3$ is evanescent, *i.e.*, trending toward a small quantity at $z = D$ (sea surface). This requires $\gamma_3$ in (9.4.15) to be real. In other words, $n_3$ must be imaginary. Writing $n_3 = in_p$, then $\gamma_3 = n_p$, in which

$$n_3^2 = \frac{k^2}{\sigma^2}(N_3^2 - \sigma^2) + \left(\frac{1}{2\rho_0}\frac{\partial \rho_0}{\partial z}\right)^2. \tag{9.4.23}$$

The condition for $\gamma_3$ being real requires

$$N_3 < \sigma \tag{9.4.24}$$

and

$$\left|\frac{k^2}{\sigma^2}\left(N_3^2 - \sigma^2\right)\right| > \left(\frac{1}{2\rho_0}\frac{\partial \rho_0}{\partial z}\right)^2. \tag{9.4.25}$$

On the other hand, $n_2$ is real, implying $N_2 > \sigma$. Therefore, the bottom topographic lee waves propagating vertically in the three-layer model ocean must satisfy a frequency relation

$$N_3 < \sigma < N_2. \tag{9.4.26}$$

We further define a coupling factor

$$M_3 = \frac{\sin n_2(2\Delta H) + M_0 \cos n_2(2\Delta H)}{M_0}. \tag{9.4.27}$$

Thus,

$$W_3(x,z) = (W_s/\pi)M_3 \exp[-\gamma_3\{z - (H + 2\Delta H)\} + ikx].$$

One can see that $M_3$ can be used as an index for the efficiency of wave momentum transfer from the middle layer to the upper layer.

In the case for $\gamma_3 = n_p$, we have

$$M_3 = \frac{n_2}{n_2 \cos n_2 (2\Delta H) + n_p \sin n_2 (2\Delta H)}. \tag{9.4.28}$$

From Eq. (9.4.28), one can see that the coupling factor $M_3$ depends on the stratification parameters of the middle and upper layers. For convenience, (9.4.28) may further be simplified as

$$M_3 = \frac{1}{\cos\Theta + \mu \sin\Theta}, \tag{9.4.29}$$

where $\Theta = n_2(2\Delta H)$, implying a phase thickness of the middle layer, and $\mu = n_p/n_2$, implying a relative decay rate in the upper layer, respectively. For further understanding the physical meaning, we re-write $\Theta$ as

$$\Theta = \left[\frac{k^2}{\sigma^2}\left(N_2^2 - \sigma^2\right) + \left(\frac{1}{2\rho_0}\frac{\partial \rho_0}{\partial z}\right)^2\right]^{1/2} (2\Delta H). \tag{9.4.30}$$

Assuming the second term in the square bracket is negligibly small in comparison to the first term, we have

$$\Theta = \left[\left(\frac{N_2}{\sigma}\right)^2 - 1\right]^{1/2} \left[2\pi\left(\frac{2\Delta H}{\lambda}\right)\right]. \tag{9.4.31}$$

One can see that $\Theta$ is dependent on the frequency ratio and the length scale ratio between those of the middle layer and the topographic lee waves.

The schematic expressions of (9.4.29) are shown in Figure 9.8, in which $\mu$ serves as an adjustable parameter. One can see that the topographic lee wave propagation into the upper layer occurs only for very narrow ranges of $\Theta$.

Now we discuss two special cases.

(1) For $\mu \ll 1$, we have

$$M_3 \approx \sec\Theta. \tag{9.4.32}$$

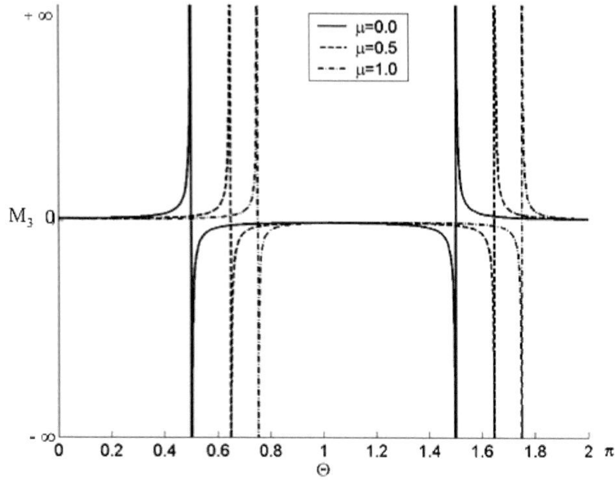


Fig. 9.8 Schematic expression of coupling factor $M_3$ (see the text for definition) [Zheng *et al.*, 2006b]

This implies

$$M_3 \to \infty, \quad \text{for } \Theta = \left(m + \frac{1}{2}\right)\pi, \quad m = 0, 1, 2, \ldots. \tag{9.4.33}$$

(2) For $\mu \sim 1$, we have

$$M_3 \approx \frac{\sqrt{2}}{2} \sec\left(\Theta - \frac{\pi}{4}\right). \tag{9.4.34}$$

This implies

$$M_3 \to \infty, \quad \text{for } \Theta = \left(m + \frac{3}{4}\right)\pi, \quad m = 0, 1, 2, \ldots. \tag{9.4.35}$$

We call (9.4.33) and (9.4.35) the upper layer resonance conditions (ULRCs).

From the continuity equation (9.4.3) and Eq. (9.4.22), we have the horizontal velocity component in the upper layer

$$u_3(x, z) = (n_p/ik)(W_s/\pi)M_3 \exp[-n_p\{z - (H + 2\Delta H)\} + ikx]. \tag{9.4.36}$$

At the sea surface

$$u_3(x, D) = (n_p/ik)(W_s/\pi)M_3 \exp[-n_p H_3 + ikx]. \tag{9.4.37}$$

where $H_3 = D - (H + 2\Delta H)$ is the upper layer thickness. This is the topographic lee wave solution at the sea surface. Compared to (9.4.19), one can see that the waves keep the horizontal patterns of the bottom topographic lee waves, but have a phase shift of $\pi/2$.

### 9.4.4. *SAR image of topographic lee waves*

In order to extract the data and information of ocean bottom topographic lee waves from SAR images, it is necessary to derive theoretical expression of topographic lee wave SAR image. In the other words, we must determine the quantitative relations between sea surface radar return signals and hydrodynamic parameters of the topographic lee waves. Chapter 1 of this book has given the systematic theories of ocean wave-radar modulation transfer function, which is used to derive the relations. On the other hand, the functional form of ocean current velocity at the sea surface induced by bottom topographic lee waves has been derived in (9.4.37). Based on (1.4.10) and (1.4.12), we obtain an analytical expression of SAR image of bottom topographic lee waves

$$\sigma_\circ|_{os} = -\left[16\pi k_0^4 \left|g_{ij}(\theta)\right|^2\right] \cdot \left[\frac{\cos^2\varphi}{2m_3\omega K^4}\right] \cdot \left[\frac{n_p W_s}{\pi} M_3 e^{-n_p H_3}\right] \cos kx. \tag{9.4.38}$$

One can see that the bottom topographic lee waves on the SAR image appear as the form of cosinoidal function with the same wavenumber of original topographic lee waves. The amplitude or intensity of radar signal depends on three factors, which are defined by radar wavenumber and incidence angle, sea surface gravity–capillary wave field and ocean vertical stratification as given in three square brackets in (9.4.38), respectively. Moreover, it is important to note that the minus sign of the amplitude does not certainly mean that the waves are antiphase with the original disturbance, because there is another determinant factor, coupling factor $M_3$, whose sign may change from case to case, depending on the vertical stratification in the study area. If coupling factor $M_3$ is negative (positive), the SAR imagery will be inphase (antiphase) with the topographic lee waves. All other factors have no influence on the sign, because they always take positive values. Furthermore, the radar signals would significantly be strengthened, if one of the upper layer resonance conditions (9.4.33) and (9.4.35) is met.

### 9.4.5. *Solutions for two and one layer oceans*

In the above derivation and analysis, a three-layer ocean model was considered, but there were no beforehand restraints for the thickness of any layers. Therefore, the solutions for a two layer ocean can be obtained by setting $H = 0$ in (9.4.21) and (9.4.22)

$$W_1\,(x, z) = (W_s/\pi M_0)[\sin n_1 z + M_0 \cos n_1 z]e^{ikx} \tag{9.4.39}$$

and

$$\begin{aligned} W_2(x, z) = {} & (W_s/\pi M_0)[\sin n_1(2\Delta H) + M_0 \cos n_1(2\Delta H)] \\ & \cdot \exp[-\gamma_2\{z - 2\Delta H\} + ikx]. \end{aligned} \tag{9.4.40}$$

Similar to the two-layer model ocean, we have a frequency relation $N_2 < \sigma < N_1$. Accordingly, the SAR image of bottom topographic lee waves become

$$\sigma_\circ|_{os} = -\left[16\pi k_0^4\,|g_{ij}(\theta)|^2\right] \cdot \left[\frac{\cos^2\varphi}{2m_3\omega K^4}\right] \cdot \left[\frac{n_p W_s}{\pi} M_2 e^{-n_p H_2}\right] \cos kx, \tag{9.4.41}$$

where $n_2 = in_p$, and subscript 2 represents the upper layer of the ocean instead of 3.

The solutions for a one-layer ocean can be obtained by setting $H = \Delta H = 0$ in (9.4.22)

$$W(x, z) = (W_s/\pi) \exp i(nz + kx). \tag{9.4.42}$$

The frequency relation is $N < \sigma$. The SAR image of bottom topographic waves becomes

$$\sigma_\circ\big|_{os} = -\left[16\pi k_0^4 \left|g_{ij}(\theta)\right|^2\right] \cdot \left[\frac{\cos^2\varphi}{2m_3\omega K^4}\right] \cdot \left[\frac{n_p W_s}{\pi} e^{-n_p D}\right] \cos kx. \tag{9.4.43}$$

## 9.5. Comparison with Observations

The hydrodynamic analysis above reveals the following dynamic features of ocean bottom topographic lee waves in a three-layer ocean. (1) In the lower layer, the ocean bottom topographic lee waves are generated by periodical bottom topography. The waves exist in the form of stationary waves, which satisfy a lower boundary resonance condition $\sigma = kC_0$, in which $k$ is a wavenumber of periodical bottom topography, and $C_0$ is a background current speed. (2) The bottom topographic lee waves may vertically propagate to the upper layer with an unchanged wavenumber $k$, if a frequency relation $N_3 < \sigma < N_2$ is satisfied. (3) The waves are significantly amplified if a upper layer resonance condition is satisfied. (4) The topographic lee waves on SAR images have the same wavelength of periodical bottom topography, and the imagery brightness peaks are either inphase (phase shift 0) or antiphase (phase shift $\pi$) with respect to the topographic lee waves, depending on signs of the coupling factor. In this section, we verify these major dynamical features using field observations and satellite remote sensing data.

### 9.5.1. *Stationary waves and wavelength*

Theories predict that the topographic lee waves are stationary waves, if LLRC (9.4.20), $\sigma = kC_0$, is satisfied. This prediction can be verified by multiple SAR images of topographic lee waves in Taiwan Banks within three years from 1994 to 1996 (940727, 950603, 950622 and 960619). Comparisons between these images reveal only a little change in the locations, patterns, and distribution ranges of the topographic lee waves. Li *et al.* [2001b] compared the SAR imagery of the topographic lee waves with LANDSAT TM visible band imagery taken at a different time and found a very good correspondence between the two. These facts indicate that the observed topographic lee waves are not only stationary waves, but are also in a quite stable condition.

The analytical expression of SAR image of bottom topographic lee waves (9.4.38) predicts that the SAR imagery of bottom topographic lee waves should appear as the form of sinusoidal function with the same wavenumber of original periodic bottom topography. The bathymetric profile measured by the on-board ADCP and collocated radar imagery brightness curve shown in Figure 9.4 show that fluctuations in the bottom topography and radar return are composed of multi-frequency/wavenumber components. Here, we do not intend to determine the one-by-one correspondence between the pulses in the two curves because of the limited resolution of the SAR image (100 m). But, the data may still be used to verify the theoretical solution (9.4.38). Instead of one by one comparison, a spectral or statistical analysis method is used for the verification. The Fourier spectra derived from high-pass filtered data are shown in Figure 9.9. One can see that the two spectra have almost the same patterns with an identical peak location at $1.2\,\mathrm{km}^{-1}$ corresponding to 0.83 km, which is a dominant or peak-power wavelength of topographic lee waves along the ADCP track. Thus, the theoretical prediction of an identical wavelength on the SAR images is verified.

### 9.5.2. *Hydrodynamic conditions for topographic lee waves*

In Section 9.4.3, we derived the frequency relation for the generation and vertical propagation of topographic lee waves in the stratified model ocean (9.4.26). In the studied case, the dominant or peak-power wavelength of topographic waves is 0.83 km, and the tidal speed is $1.5\,\mathrm{m\,s^{-1}}$. Thus, from the LLRC (9.4.20), we derive the angular frequency of topographic lee waves $\sigma = 6.5\,\mathrm{cph}$ (cycle per hour). Meanwhile, from Section 9.2.4 and Figure 9.2, we derive the average Brunt–Väisälä frequencies of each layer at the CTD stations as listed in Table 9.3. One can see that the frequency relation $N_3 < \sigma < N_2$ is satisfied for three-layer ocean at stations S01–S03 and Y90–Y98, and $N < \sigma$ for one-layer ocean at stations S04–S06.

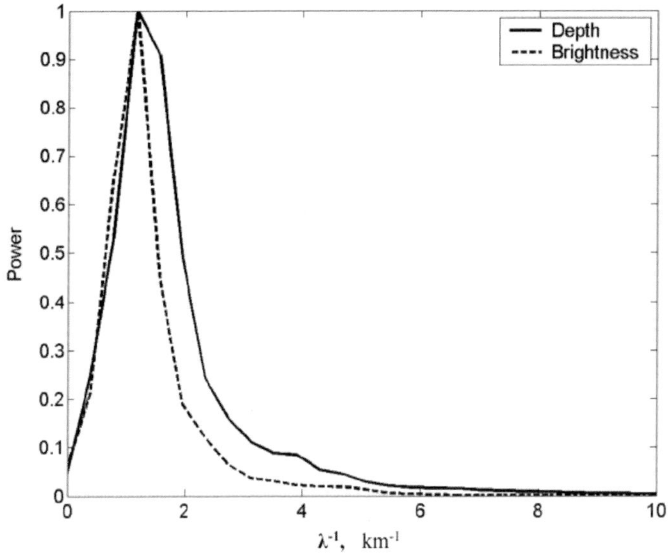


Fig. 9.9 Comparison of wavenumber ($\lambda^{-1}$) spectra of high-pass-filtered periodic ocean bottom topography (solid line) to that of corresponding SAR imagery brightness (dashed line) [Zheng *et al.*, 2006b]

Table 9.3 Stratification parameters at CTD stations

| Station | Depth m | CTD Depth, m | $H_3$ m | $2\Delta H$ m | $H_3$ m | $N_3$, cph | $N_2$, cph | $N_1$, cph |
|---|---|---|---|---|---|---|---|---|
| S01 | 81 | 73 | 20 | 40 | 21 | 6 | 15 | 3 |
| S02 | 63 | 54 | 5 | 40 | 9 | 6 | 20 | 3 |
| S03 | 37 | 27 | 7 | 13 | 17 | 6 | 35 | 3 |
| S04 | 25 | 15 | 25 | 0 | 0 | 5 | | |
| S05 | 26 | 20 | 26 | 0 | 0 | 4 | | |
| S06 | 31 | 28 | 31 | 0 | 0 | 3 | | |
| Y90 | 56 | 53 | 5 | 30 | 21 | 4 | 22 | 3 |
| Y92 | 92 | 71 | 40 | 30 | 22 | 4 | 22 | N/A |
| Y94 | 140 | 138 | 24 | 60 | 56 | 4 | 15 | 5 |
| Y96 | 450 | 300 | 40 | 80 | 330 | 5 | 15 | 5 |
| Y98 | 1709 | 570 | 30 | 160 | 1579 | 5 | 12 | 5 |

### 9.5.3. *Space phase of SAR imagery and radar signal enhancement*

From the solution of bottom topographic lee wave SAR imagery (9.4.38) for three-layer ocean, one can see that the space phase of SAR imagery with respect to the bottom topographic lee waves depends on the sign of coupling factor $M_3$ in the form of (9.4.29). The SAR imagery should be inphase (antiphase) with the topographic lee waves, if $M_3$ takes a minus (plus) sign. On the other hand, the sign of $M_3$ depends on $\Theta$ and $\mu$. Using (9.4.10), (9.4.23), (9.4.29) and field data listed in Table 9.3, we calculate these parameters at CTD stations. The results are listed in Table 9.4. One can see two important points. (1) In all the cases, $M_3$ takes the plus sign, implying that the SAR imagery is antiphase with the bottom topographic lee waves. (2) At station Y98, the radar return signals are enhanced by near satisfaction of the upper layer resonance conditions. Unfortunately, that station is beyond SAR image coverage.

On the other hand, examining imagery patterns on the SAR image, one can see that the wave-like patterns consist of two subsystems with different scales and orientations. One subsystem has a wavelength scale on the order of 5 km, lengths ranging from 20 to 40 km and a north–south (or top–bottom on the image) orientation. In comparison to the field observations, they are interpreted as imagery of periodic submarine sand ridges. The other subsystem riding on the large-scale system has a wavelength scale on the

Table 9.4 Vertical wavenumbers and coupling parameters between upper two layers at CTD stations

| **Station** | $n_2$, $k$ | $n_p$, $k$ | $\mu$ | $\Theta,\pi$ | $M_3$ |
|---|---|---|---|---|---|
| S01 | 2.1 | 0.38 | 0.18 | 0.20 | 1.1 |
| S02 | 2.9 | 0.38 | 0.13 | 0.28 | 1.4 |
| S03 | 5.3 | 0.38 | 0.072 | 0.17 | 1.1 |
| Y90 | 3.2 | 0.79 | 0.25 | 0.23 | 1.1 |
| Y92 | 3.2 | 0.79 | 0.25 | 0.23 | 1.1 |
| Y94 | 2.1 | 0.79 | 0.38 | 0.30 | 1.1 |
| Y96 | 2.1 | 0.64 | 0.30 | 0.40 | 1.7 |
| Y98 | 1.6 | 0.64 | 0.40 | 0.62 | 256 |

order of 1 km, and takes a west–east (or left–right on the image) orientation. They occupy two-thirds of the bright area. The spectra in Figure 9.9 show the peak-power wavelength as 0.83 km, which is comparable with previous results [Alpers *et al.*, 2004; Hsu *et al.*, 1997; Li *et al.*, 2001b]. The imagery features, including distribution patterns, scales, orientations, in particular narrow separations and wide widths of grouped sand waves, indicate that the bright imagery bands correspond to the troughs of bottom topographic lee waves. This agrees with theoretical prediction.

Moreover, the CTD data shown in Figures 9.2(a) and (b) indicate that the vertical stratification of the shallow water region, represented by stations S04–S06 and mainly distributed in the north study area, can be described well by a one-layer ocean model with weakly continuous stratification. The solution of one-layer ocean indicates that the imagery brightness peaks are also antiphase (phase shift $\pi$) with respect to the topographic lee waves. This phase shift of $\pi$ is close to the observations on the continental shelf east of Cape Hatteras [Donato *et al.*, 1997].

## 9.6. Summary

The goal of this chapter is to explore satellite SAR imaging mechanisms of periodic ocean bottom topography in a stratified ocean. The results are used for a quantitative interpretation of SAR images of bottom topographic lee waves. The Taiwan Banks area located at the south outlet of Taiwan Strait is selected as a study area, where the periodic bottom topography at depths up to 100 m show up on the SAR images with very high imagery contrast, and where simultaneous or quasi-simultaneous cruise data are available. On the basis of hydrographic conditions and the bottom features in the study area, a two-dimensional, three-layer ocean model with sinusoidal bottom topography is developed. From the solutions and inferences derived from the momentum equations of the ocean model and verified by observations, the major conclusions are summarized as follows.

(1) Under the conditions of the existence of a traverse flow and the smallness of the amplitude of periodic bottom topography compared

to the water depth, the bottom topography serves as a source of disturbance to induce stationary water waves in the lower layer. The waves satisfy the lower boundary resonance condition $\sigma = kC_0$, here, $\sigma$ is the angular frequency of the stationary water waves, $k$ is the wavenumber of the sinusoidal bottom topography, and $C_0$ is the traverse flow speed.

(2) The bottom topographic lee waves are the internal waves. The waves may propagate vertically to the upper layer with the same wavenumber, under the condition of that the frequency relations $N_j < \sigma < N_i$ are satisfied for three- and two-layer oceans, here, $N_i$ and $N_j$ ($i = 1, 2$ and $j = 2, 3$) are the Brunt–Väisälä frequencies of middle (lower) layer and upper layer, and $N < \sigma$ for a one-layer ocean, respectively. Physically, this relation requires the water density to increase sharply in the middle (lower) layer for the three- (two-) layer ocean. CTD measurements at all stations within the study area indicate that the relation stands in the SAR imaging season.

(3) The amplitude of topographic lee waves in the upper layer depends on coupling factor $M_3 = n_2[n_2 \cos n_2(2\Delta H) + n_p \sin n_2(2\Delta H)]^{-1}$, for the three-layer ocean. This implies that there is a resonance mechanism, *i.e.*, $M_3$ may reach a quite large number when the term in the square bracket takes a number much smaller than $n_2$. In this case, the radar return signals are significantly amplified.

(4) The physical expression of SAR image of topographic lee waves is derived on the basis of current-modulated small wave spectra developed by Yuan [1997]. The expression indicates that the topographic lee waves on SAR image should have the same wave patterns with the same wavelength of periodic bottom topography. The imagery brightness peaks are either inphase or antiphase with respect to the topography, depending on the sign of coupling factor $M_3$. In the case of SAR image 940727, $M_3$ is always a positive number, so that the bright imagery bands correspond to the troughs of periodic bottom topography.

# Chapter 10

# Ocean Bottom Topography: III. Stratified Flow Passing Over Isolated Topography

## 10.1. Introduction

Chapters 8 and 9 have documented SAR imaging mechanisms and interpretation methods in the cases of longitudinal flow passing over parallel topographic corrugations and traverse flow passing over periodical topography, respectively. This chapter will further analyze the case of the third type of ocean bottom topography: the isolated seamount.

The seamounts of interest are located in the Gulf Stream region off the southeastern coast of the US [Zheng *et al.*, 2012a]. Figure 10.1 shows a sea surface temperature (SST) image of the study area taken by the National Oceanic and Atmospheric Administration (NOAA) Advanced Very High Resolution Radiometer (AVHRR) onboard NOAA-18 satellite at 07:21:05 UTC on February 28, 2006. One can see that the Gulf Stream as shown as a high SST water belt (in red) flows northeastward along the continental slope between Florida Peninsula and Cape Hatteras, North Carolina. A black box shows the coverage of the SEASAT SAR image of July 25, 1978 as shown in Figure 10.3(a), which is a segment of the Gulf Stream over the Charleston Rise region. Typically, the Gulf Stream is 100 km wide and 800–1200 m deep. The current velocity is fastest near the surface, with the maximum velocity of $2.5\,\mathrm{m\,s^{-1}}$ [Stommel, 1976]. On the continental shelf, the Gulf Stream flow may reach near the

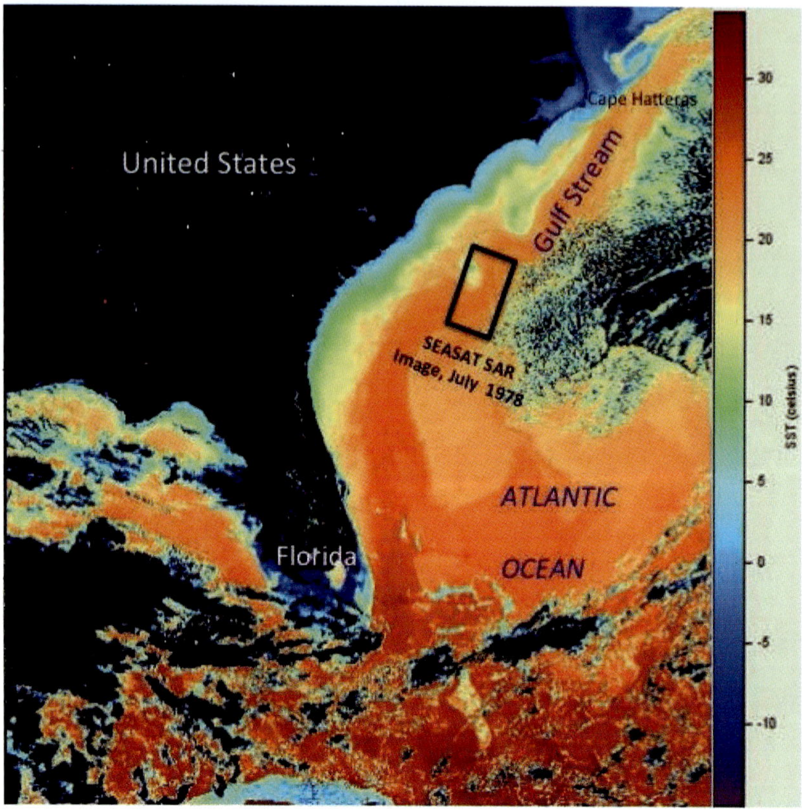


Fig. 10.1 AVHRR SST image of the Gulf Stream region off the US southeastern coast taken by the NOAA-18 satellite at 07:21:05 UTC on February 28, 2006. The black box represents the SEASAT SAR imaged area (Credit: NOAA NESDIS)

ocean bottom and still keep a strong current parallel to the isobaths [Hayes, 1981].

Figure 10.2 shows a resolution-reduced version of bathymetric map of the study area, covering the Charleston Rise region in the Atlantic from $31°00'$ to $32°00'$N and from $77°00'$ to $78°30'$W, which is used for book printing purpose. The water depth is between 500 and 800 m on the middle segment of continental slope. The ocean bottom topography is characterized by densely distributed seamounts named the Hoyt Hills. The original high-resolution bathymetric map shows that the heights of seamounts above the background sea floor are

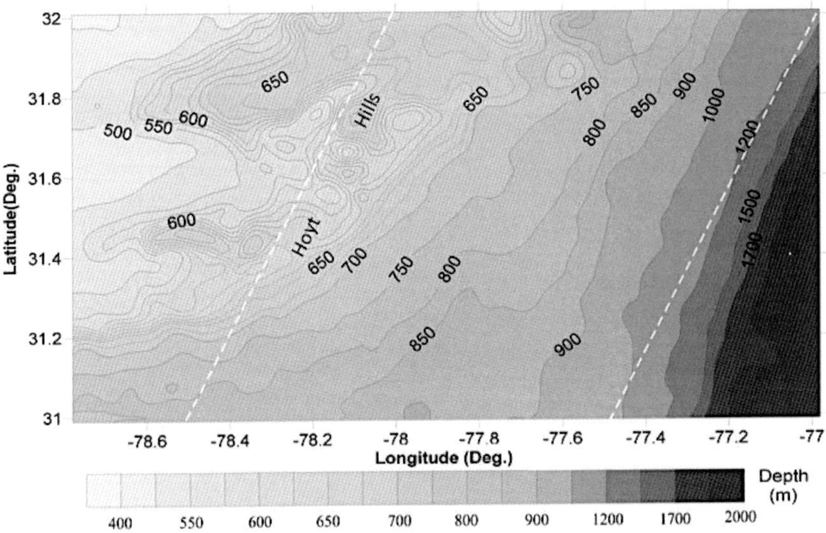


Fig. 10.2 Resolution-reduced bathymetric map of the study area. Dash lines represent boundaries of south portion of the SEASAT SAR image taken on July 25, 1978 (Figure 10.3(a)). (Original data are downloaded from website http://www.ngdc.noaa.gov/mgg/bathymetry/)

from 20 to 140 m. All seamounts are approximately elliptical. The mean major and minor axis lengths are $8.4 \pm 4.9$ km and $4.1 \pm 2.2$ km, respectively.

## 10.2. SAR Image Interpretation

The SEASAT SAR image of the Gulf Stream region taken on July 25, 1978 is shown in Figure 10.3(a). A map of seamounts extracted from the original high-resolution version of bathymetric map is overlaid on the SAR image. Compared with the Gulf Stream boundaries on infrared image shown in Figure 10.1, we mark the Gulf Stream boundaries and direction in Figure 10.3(a). From Figure 10.3(a), we find the following key points. (1) All SAR imagery streaks originate from seamounts, without any exception. (2) The streaks are not uniformly distributed, but grouped. In general, there are two bunches of streaks. One is originating from a seamount chain consisting of seamounts 1–4 and another from seamounts 8 and 9.

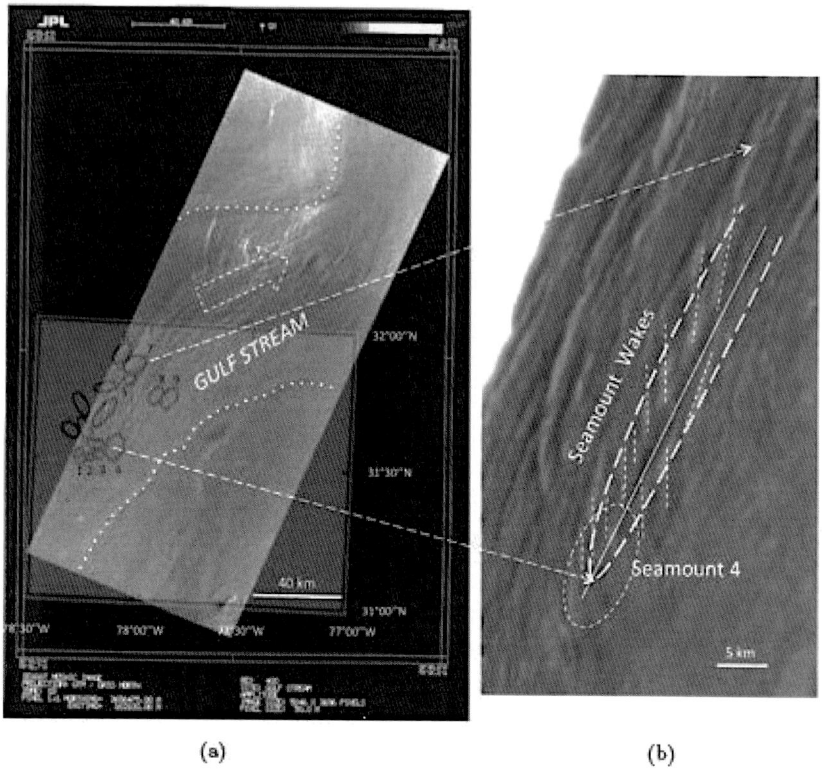


Fig. 10.3 (a) SEASAT SAR image of the Gulf Stream region on July 25, 1978. The swath width is 100 km, and the original spatial resolution is 25 by 25 m. Coded black ellipses represent locations, sizes and orientations of seamounts of the Hoyt Hills. Dotted lines and arrow represent the boundaries and direction of the Gulf Stream. (b) Details of seamount wakes or SOVTs on zoomed SAR image (Credit: NASA JPL)

(3) Starting from seamounts, the streaks extend downstream, and would be enhanced if passing over a large seamount, such as the cases of seamounts 5 and 7. This implies that dynamically, seamounts are generation sources for the disturbance signals represented by the SAR imagery streaks. (4) Inside the grouped streaks, there are secondary structures. Along the cross-stream direction, the width of individual streak varies. Along the downstream direction, the streaks are not smooth lines or bands, but contain the fine structure as

shown in Figure 10.3(b). This implies that there is similarity among the seamount streaks, the island ocean vortex trains (OVTs) [Zheng *et al.*, 2008a; Zheng and Zheng, 2014] and the mountain vortex trains in the atmosphere [Tsuchiya, 1969]. In the other words, the seamount streaks consist of seamount-induced OVTs (SOVTs) (see detailed analysis in Section 2.4.3). The SOVTs are concentrated in the central Gulf Stream flow with the flow direction as marked by a big arrow in Figure 10.3(a). The total length of the SOVTs reaches about 150 km. This implies that the generation and development of the SOVTs relay on the Gulf Stream.

In summary, the above imagery features of the streaks on the SEASAT SAR image indicate that they are imprints of seamount wakes on the sea surface, which are generated by interaction between seamounts and the Gulf Stream flow. In this process, seamounts serve as a disturbance source, and the Gulf Stream provides a dynamic source.

## 10.3. Dynamic Analysis

Previous investigators have developed SAR imaging theories [Alpers and Hennings, 1984; Romeiser and Alpers, 1997; Shuchman *et al.*, 1985; Valenzula *et al.*, 1983; Zheng *et al.*, 2006b, 2012b] for underwater linear topography as reviewed in chapters 8 and 9. This chapter deals with a new type of ocean bottom topographic feature, the isolated seamounts. The above interpretation results indicate that the SAR imagery streaks are imprints of the seamount wakes. In order to verify this interpretation, dynamically we should again answer three questions. (1) How does the seamount generate the disturbance signals in the above water body? (2) How do the disturbance signals propagate from the bottom layer vertically to the upper ocean? (3) How do the upper ocean disturbance signals generate the satellite sensor signals? For the third question, previous investigators have given the answers [Plant, 1990; Yuan, 1997; Zheng *et al.*, 2001b]. In this section, we look for the answers for the first two questions.

### 10.3.1. *Generation mechanisms of SOVTs*

The SAR image interpretation in Section 10.2 indicates that seamounts are generation sources for the disturbance signals represented by the SAR imagery streaks. This implies that the disturbance signals are generated in the ocean bottom layer. In this layer, the Brunt–Väisälä frequency $N$ ($N_1$) is about 4 cph as shown in Figure 10.4. The length scale of seamounts is $a = O(5\,\text{km})$. The current velocity of Gulf Stream at 500 m is $U = O(1\,\text{m}\,\text{s}^{-1})$ [Stommel, 1976]. We have $Na/U = 35 \gg 1$. Thus, the flow is approximately hydrostatic [Baines, 1995]. In this case, classic fluid dynamics has solved the problem on generation of wakes in the lee of a solid obstacle when the flow passes over [Kundu, 1990]. The form of the wakes is dependent on the Reynolds number, $Re$ ($=UD/\nu$ for a circular

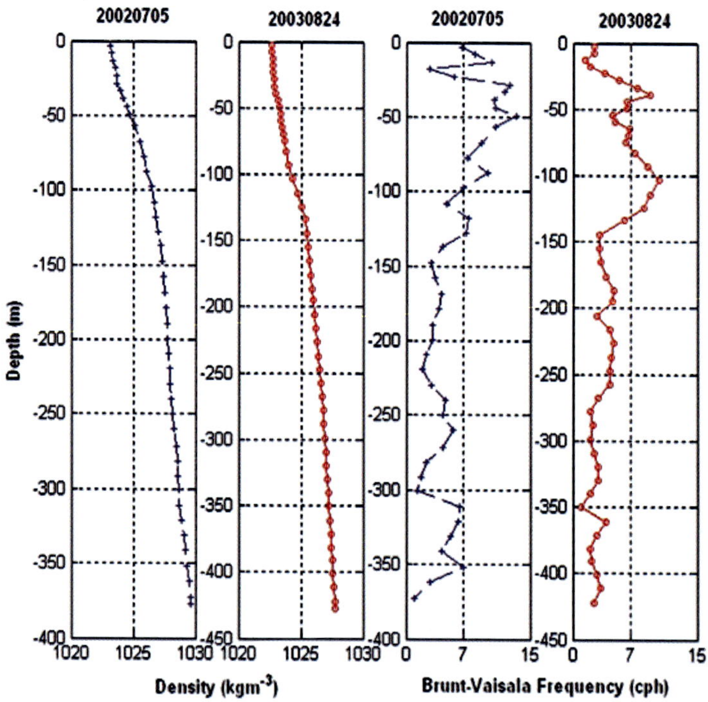


Fig. 10.4 Vertical profiles of seawater density derived from CTD data measured at NOAA Station 1770962 (31°52.8′N 79°10.8′W) on July 5, 2002 and August 24, 2003 and corresponding profiles of the Brunt–Väisälä frequency (downloaded from website http://www.nodc.noaa.gov/GTSPP/access_data/)

cylinder, here, $U$ is the upstream velocity, $D$ is the cylinder diameter and $\nu$ is the viscosity). The viscosity should be within a reasonable range from $1 \times 10^5$ to $5 \times 10^5\,\text{cm}^2\,\text{s}^{-1}$. The Reynolds number, *Re*, should be in a range of $100 < \text{Re} < 500$. Within this range, the wake should appear as an unstable and irregular vortex train.

According to Baines [1995], on the other hand, the behavior of the wakes not only depends on the Reynolds number, but also on another parameter, $Nh/U$, here, $N$ is the Brunt–Väisälä frequency of stratified flow, and $h$ is the seamount height. In our case, two vertical profiles of the Brunt–Väsälä frequency, which are calculated by the density profiles measured on July 5, 2002 and August 24, 2003 shown in Figure 10.4, give an average $N$ ($N_1$) as about 4 cph. Thus, we estimate that $N$ at the SEASAT SAR imaging time on July 25, 1978 was about 4 cph. The scale of $h$ is 100 m. In this case, $Nh/U$ is about 0.7. Under the conditions of $100 < \text{Re} < 500$ and $Nh/U \approx 0.7$, the wakes may appear in three possible forms: symmetric vortex shedding (SVS), asymmetric vortex shedding (AVS) and lee waves with attached vortices (WV) [Baines, 1995]. On the other hand, the details of SAR-imaged seamount wakes are shown in Figure 10.3(b), which contains a zoomed part of SEASAT SAR image.

As interpreted in Section 10.2, the seamount wake imagery is not uniformly distributed, but grouped. Inside the grouped imageries, there are secondary structures. Along the cross-stream direction, the width of individual imagery varies. Along the downstream direction, the imageries are not smooth lines or bands, but contain the fine, vortex-like structure. A typical example is a wave packet of 9 starting from the upstream slope of seamount 4 and extending downstream. The average wavelength, which is defined as the distance between two successive waves, is 4.9 km. The wave packet is distributed on the left hand of seamount, while on the right hand side distributed is a discontinuously linear pattern, which also extends downstream as far as the wave packet. This complex internal structure agrees with the predicted AVS, implying that the seamount wakes essentially comprise the SOVTs, but modified when they propagate from the ocean bottom to the surface. Other seamount wake imageries have similar structure to that of seamount 4, implying the same generation mechanism.

### 10.3.2. *Vertical propagation mechanism*

For a stratified ocean, Zheng *et al.* [2006b] derived dynamical solutions for vertical propagation of disturbance signals from the ocean bottom to the upper layer (also see Chapter 9 for details). For a three-layer ocean, the pass frequency band is $N_3 < \sigma < N_2$, here, $N_3$ and $N_2$ are the Brunt–Väisälä frequencies of the upper and middle layers, and $\sigma$ is the angular frequency of the disturbance signal. In our case, we choose the nearest NOAA Station 1770962 (31°52.8′N 79°10.8′W) to represent the study area. We use the conductivity-temperature-depth (CTD) data measured on July 5, 2002 and August 24, 2003 to estimate the vertical stratification in the same season at the SEASAT SAR imaging time, July 25, 1978 due to lack of simultaneous measurements. From the vertical profiles of the seawater density and the corresponding profiles of the Brunt–Väsälä frequency on the two days are shown in Figure 10.4, one can see that the ocean can be treated as a three-layer ocean, (1) bottom layer ($N_1$) for depth >150 m, (2) middle layer ($N_2$) from 20 to 150 m and (3) upper layer ($N_3$) from the surface to 20 m. The Brunt–Väisälä frequencies are $N_3 = 7$ cph, $N_2 = 10$ cph and $N_1 = 4$ cph on July 5, 2002, and $N_3 = 3.5$ cph, $N_2 = 8.4$ cph and $N_1 = 4$ cph on August 24, 2003. Reasonably taking the averages to estimate the case at SEASAT SAR imaging time, we obtain $N_3 = 5$ cph, $N_2 = 9$ cph and $N_1 = 4$ cph. In above section, we have obtained that the average wavelength of the seamount wakes is 4.9 km for the typical case of wave packet of seamount 4. The phase speed can be estimated by $c = \sqrt{g'd}$, in which $g'(= \delta\rho/\rho\, g)$ is the reduced gravitational acceleration. $c$ is estimated as $1.4\,\mathrm{m\,s^{-1}}$ for $\delta\rho/\rho = 2 \times 10^{-3}$, and $d = 100$ m. Therefore, the seamount wakes frequency $\sigma_m(= c/\lambda)$ is about 6 cph, which satisfies the relation $N_3$ (5 cph) $< \sigma$ (6 cph) $< N_2$ (9 cph). This implies that the seamount wakes may penetrate through the stratified ocean vertically to the upper layer without any dynamic constraint.

## 10.4. Complementary Observations

In order to confirm the phenomena shown on the SEASAT SAR image, we collect new generation C-band RADARSAT-1 ScanSAR images of the same Gulf Stream region. We find that the phenomena show up frequently on high resolution SAR wind field images, although the images have been processed for many times. Among ScanSAR wind field images from July to October 2006, there were at least nine images containing the same or similar wake patterns in the same area of SEASAT SAR coverage. Figure 10.5 shows four examples, which were acquired on July 21, August 21,

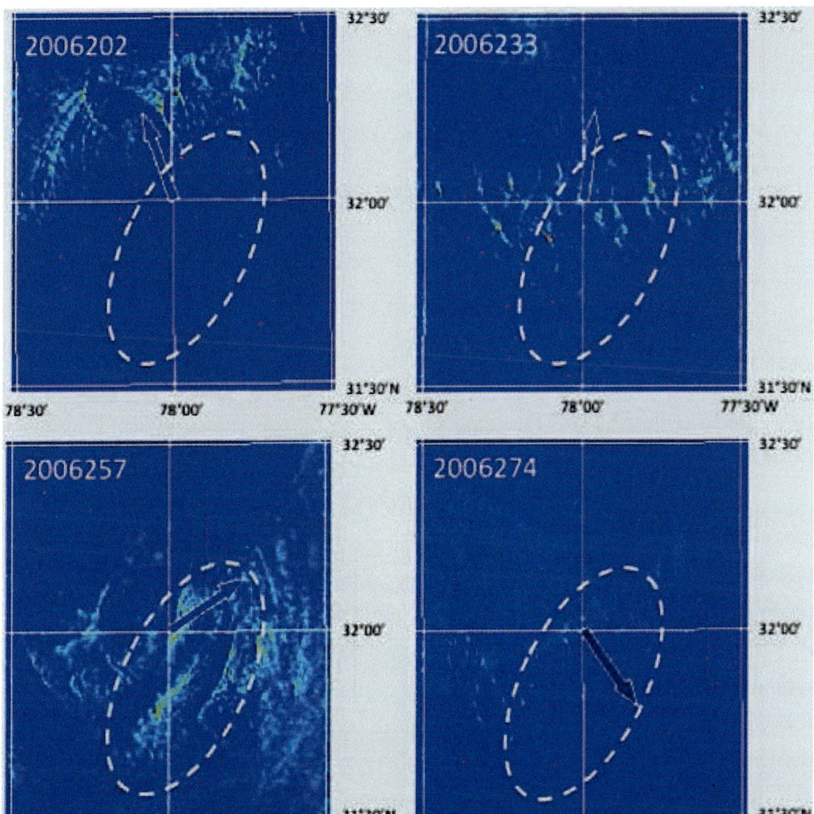


Fig. 10.5 High-resolution SAR wind field images of the Gulf Stream region derived from C band RADARSAT-1 ScanSAR images acquired on July 21 (2006202), August 21 (2006233), September 14 (2006257) and October 1, 2006 (2006274). The white-circled area represents the coverage of seamount wakes on SEASAT SAR image in Figure 10.3(a). Big arrows represent wind direction (Credit: NOAA NESDIS)

September 14 and October 1, 2006, respectively. One can see the grouped, streak-like patterns similar to the SEASAT SAR imagery. The difference is that SEASAT SAR image contains more details and fine structures. The reasons for resulting in the difference may include different spatial resolutions (25 m for SEASAT SAR and 100 m for RADARSAT-1 ScanSAR), different radiation sensitivities, and different radar wavelengths of two instruments.

## 10.5. Summary

We examine connection of the SEASAT SAR imagery to seamounts, analyze the boundary and dynamic conditions, compare the imagery to new observations, and come to the following findings and conclusions. (1) In the area covered by streak-like patterns, there exist isolated seamounts of the Hoyt Hills with the horizontal size of 2–15 km and the top height of 20–140 m above the background sea floor (~600 m). The area is located within the main stem of the Gulf Stream. The current velocity is as high as $2.5\,\mathrm{m\,s^{-1}}$ at the surface and $1.0\,\mathrm{m\,s^{-1}}$ at the depth of 500 m. This is a necessary dynamic condition for the occurrence of seamount wakes. The SAR image shows that all imagery streaks originate from seamount chains, extend downstream, and are enhanced after passing over the large seamounts. The fine structure of streaks indicates that the seamount wakes essentially comprise the SOVTs, but modified when they propagate from the ocean bottom to the surface. (2) The boundary and dynamic conditions are favorable for generation of the seamount wakes. The internal structure of seamount wakes generally agrees with the behavior of wakes generated by the hydrostatic flow passing over a solid obstacle as predicted by classic fluid dynamics. These results confirm the seamount origin of the wakes as shown on the SEASAT SAR image. (3) The ocean stratification structure completely supports the seamount wakes to propagate vertically to the upper layer. Thus, it is physically sound to interpret the grouped, streak-like patterns on the SEASAT SAR image taken over the Gulf Stream region on July 25, 1978 are imprints of seamount wakes.

Previous investigators have revealed the detailed behavior of the flow past isolated topography using laboratory experiments. For example, Richards *et al.* [1992] found that a wave packet was generated over and behind a Gaussian hill when the Rossby number, $R$ ($= U/fL$, here, $U$ is the flow speed, $f$ the Coriolis parameter and $L$ a horizontal length scale of the topography), was equal to 1.3. This is just the same as the case we observed over and behind seamount 4 ($R = 1.3$ for $U = 1\,\mathrm{m\,s^{-1}}$, $f = 0.76 \times 10^{-4}\,\mathrm{s^{-1}}$ and $L = 10\,\mathrm{km}$).

It is worth noting the following two facts. (1) The SAR may only image the sea surface manifestations of ocean phenomena, thus, the SAR imagery represents the projection of ocean phenomena on the sea surface. (2) The stratified ocean plays a role of band-pass-filter in the vertical propagation process of ocean bottom disturbance signals [Zheng *et al.*, 2006b], thus, the SAR imagery only contains signal components within a frequency band of $N_3 < \sigma < N_2$. These imply that the SAR cannot always image the seamount wakes, and the imagery patterns may vary with the seasonal variability of wind conditions, ocean stratification and the Gulf Stream current field as shown in Figure 10.5.

# Chapter 11

# Marine Atmospheric Boundary Layer Processes

## 11.1. Introduction

The marine atmospheric boundary layer (MABL) is defined as the lowest layer of the atmosphere, extending upward from the ocean surface to a height that ranges anywhere from 100 to 3000 m. The atmospheric processes in the MABL are directly influenced by the presence of the ocean surface. Meanwhile, they also directly act on the ocean surface, so-called air–sea interaction. Through interaction, the MABL processes produce their footprints on the ocean surface, which may be detected by satellite sensors. For instance, spatially varying wind field may modify the distribution patterns of small waves on the ocean surface, and satellite SAR is capable of imaging the patterns (see Chapter 1 for details). Thus, by interpreting the ocean SAR images, we may find the occurrence of MABL processes, which may never be observed, extract dynamic parameters and analyze occurrence conditions. On the SAR images, however, the imagery of MABL process sometimes may have patterns similar to that of ocean processes. In this case, we must first identify them through scale analysis methods, pattern recognition techniques and comparison to images of other sensors, such as infrared and visible imaging radiometers. In this chapter, we will give brief introductions to SAR detection of three interesting

MABL processes, island-induced atmospheric solitary wave packets, atmospheric frontal waves and rainfall on the ocean surface.

## 11.2. SAR Imaging Principles of Atmospheric Process

The sea surface is connected with the bottom of the MABL, thus, the fluctuating wind stress at the sea surface associated with atmospheric processes modulates the small wave field on the sea surface. As documented in Chapter 1, the SAR is sensitive to spatial modulation of small wave field through the resonant Bragg scattering mechanism.

Figure 11.1 shows a schematic diagram for explaining how an atmospheric wave motion modulates the small wave field on the sea surface. One can see that the vertical wave motion in the atmosphere induces upwelling and downwelling of air particles, meanwhile induces convergent and divergent zones at the sea surface. In the convergent zone, the convergent flow pushes the small waves to concentrate, so that the wave density per unit area increases. In the divergent zone, the situation is opposite. Based on SAR imaging principles, the convergent zones appear as bright streaks, and the divergent zones dark on the SAR images. Further interpretation of SAR imagery patterns allows us to extract the data and information associated with the atmospheric wave motion.

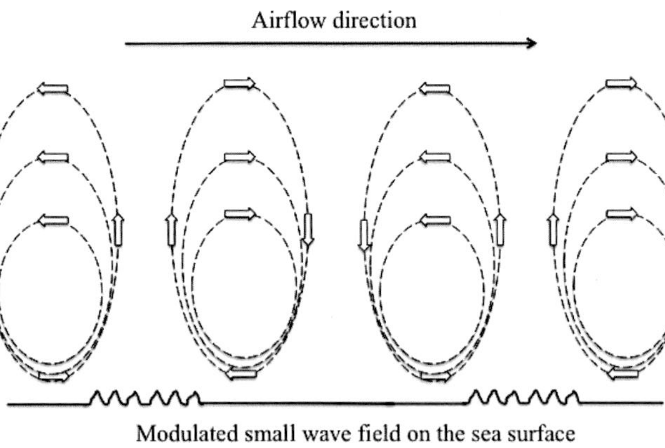


Fig. 11.1 Schematic diagram of small wave field on the sea surface modulated by atmospheric waves

## 11.3. Island-Induced Atmospheric Solitary Wave Packets

### 11.3.1. *SeaWiFS observation*

Atmospheric gravity waves can be generated by airflow over an island. These waves can propagate either downstream or upstream, depending upon the atmospheric Froude number, stratification in the MABL and topographic forcing [Baines, 1995]. When the general characteristics of these waves can be described by the classic Korteweg–de Vries (KdV), extended KdV (eKdV) or forced extended KdV (feKdV) equations [Grimshaw and Smyth, 1986; Melville and Helfrich, 1987; Miles, 1979; Shen, 1993; Wu, 1987], they are also referred to as solitary wave packets. These waves play an important role in modifying the vertical structure of the wind, moisture and temperature near the island, sometimes causing safety concerns on aviation operations [Christie, 1983]. At top of the MABL, the wind field associated with the waves causes the cloud structure to change around the island, allowing them to be observed on satellite infrared and visible images under conditions of high humidity such as sometimes found in the ocean areas [Gjevik and Marthinsen, 1978; Li *et al.*, 2001a; Mitchell *et al.*, 1990; Zheng *et al.*, 2004b].

Figure 11.2 shows a true color image of the Sea-viewing Wide Field-of-view Sensor (SeaWiFS) onboard the Orbview-2 satellite taken on March 19, 1999. Previous studies suggest that all undular cloud patterns represent signatures of atmospheric waves [Eckermann and Preusse 1999; Li *et al.*, 2001a; Zheng *et al.* 1998b]. Thus, we interpret two undular cloud patterns distributed on two sides of Hainan Island, China, represent two atmospheric wave packets. The packet stretching southward from the island P1 contains 23 waves. The concave orientation of the wave cloud lines implies that the wave packet propagates southward. The average separation distance (or wavelength) between two successive waves is 6.4 km. The maximum length of the crest lines is 330 km. The packet located northeastward 200 km off the northeast coast of the Island P2 contains more than 20 waves. The concave orientation of the wave cloud lines implies that the wave packet propagates northward.

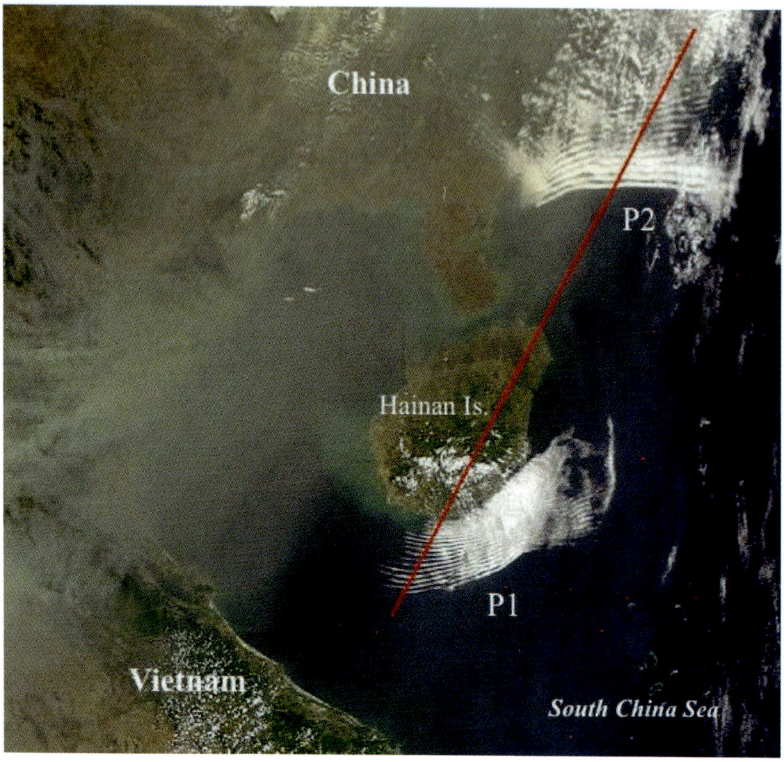


Fig. 11.2 True color SeaWiFS image taken on March 19, 1999 (Credit: US NASA). The waters of the South China Sea are shown in dark blue. Aerosol contamination is evident to the west of the Hainan Island. Two groups of wave clouds in white arrayed on the two sides of the island are interpreted as signatures of upstream (P1) and downstream (P2) solitary wave packets generated in the atmosphere by topographic disturbances. The red line represents the wind direction at 850 mb

The average wavelength of the first 10 waves is 9.8 km. The maximum length of crest lines is 370 km.

Figure 11.3 shows simultaneous weather charts of sea surface pressure and wind velocity (upper) and geo-potential height and velocity at 850 mb (lower), which show a low-pressure center moving southeastward from South China toward the imaged area. The wind was southerly 5 m $s^{-1}$ at the sea surface, and southwesterly 10 m $s^{-1}$ at 850 mb. These weather conditions further evidence that wave packet P1 propagates upstream and P2 downstream, respectively.

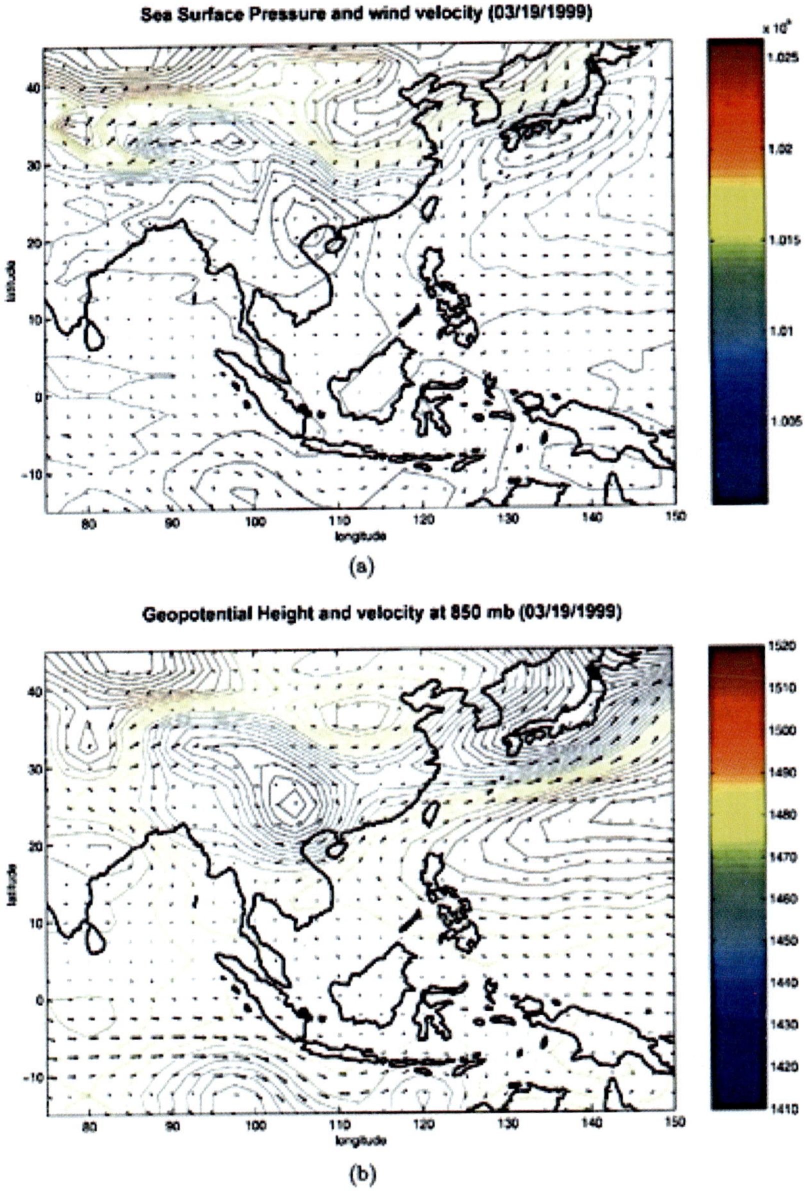


Fig. 11.3 Weather charts of Southeast Asia on March 19, 1999 (Credit: US NOAA. NCEP/NCAR). (a) Sea surface pressure and wind velocity. Units of color code are Pa. Arrows represent the wind field. (b) Geopotential height and wind velocity at 850 mb. Units of color code are in m

### 11.3.2. *SAR observation*

As discussed in the above section, satellite SARs should be the powerful sensor for detection of MABL phenomena. In fact, previous investigators have proved this point by their research results on atmospheric lee waves [Li *et al.*, 1998b; Vachon *et al.*, 1994; Zheng *et al.*, 1998a], atmospheric disturbances [Alpers and Stilke, 1996; Mourad, 1996], gravity waves [Chunchuzov *et al.*, 2000]; atmospheric boundary layer rolls and atmospheric fronts [Alpers and Brümmer, 1994; Johannessen *et al.*, 1991; Mourad, 1999], mesoscale phenomena such as polar mesoscale cyclones, storms and hurricanes [Atlas, 1994a; Du and Vachon, 2003; Friedman and Li, 2000; Li *et al.*, 2002; Sikora *et al.*, 2000], island wakes [Pan and Smith, 1999], cold air outbreak [Mourad and Walter, 1996; Thompson *et al.*, 1983], and atmospheric vortex streets [Li *et al.*, 2000b]. In particular, Li *et al.* [2004] observed upstream and downstream propagating solitary wave packets, like solitary wave packets P1 and P2 in Figure 11.2, but from SAR images.

Figure 11.4 shows a map of the satellite observation area by Li *et al.* [2004]. One can see that St. Lawrence Island is facing northward to the Bering Strait and located within the hallway of the strong cold airflow coming from the Antarctic (see Figure 11.7. There are four major mountains on the island. The highest one, Mt. Kookooligit with a height of 673 m above sea level, is located in the middle of the island. The geographic location and isolated peaky topography provide favorable conditions for generation of the atmospheric solitary wave packets.

Figure 11.5 shows a full resolution RADARSAT-1 ScanSAR wide B mode SAR image with a spatial resolution of 100 m and a pixel spacing of 50 m. One can see that St. Lawrence Island is in the middle of the image, and there are four alternate bright–dark lines over the ocean surface distributed on both north and south sides of the island near Mt. Kookooligit. The concave orientations of the wave patterns show that the north wave packet is propagating northward, while the south wave packet propagating southward. The dark lines represent the low backscatter from a smoother sea surface associated with lower wind speed, and the bright lines correspond to a higher sea

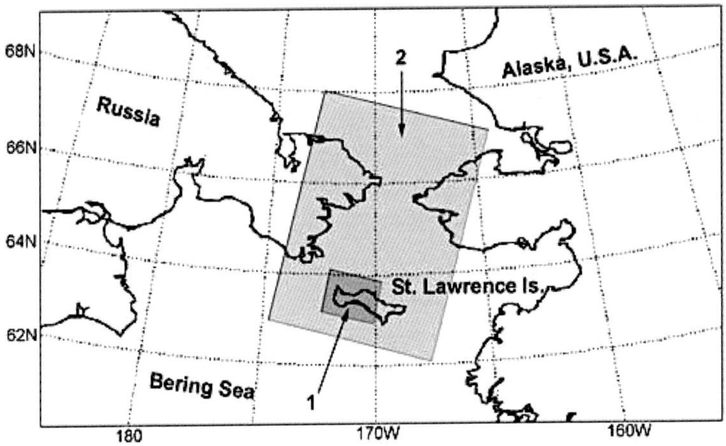


Fig. 11.4 Map of the satellite observation area. Boxes 1 and 2 show the SAR (Figure 11.5) and MODIS image (Figure 11.7) coverage, respectively. St. Lawrence Island in the Bering Sea serves as a disturbance source for atmospheric solitary wave packets [Li *et al.*, 2004]

surface wind field. These features were interpreted as a sea surface wind modulation pattern associated with the atmospheric solitary wave packets [Li *et al.*, 2004]. The soliton width, and peak-to-peak distance of the solitons measured from the SAR image are about 3 and 4.5 km, respectively.

Figure 11.6 shows simultaneous sea surface wind field retrieved from the SAR image taken on June 6, 2001 using a procedure developed by the Johns Hopkins University Applied Physics Lab (JHU/APL) [Monaldo *et al.*, 2001]. One can see that the northerly wind speed over St. Lawrence Island is as high as $20\,\mathrm{m\,s^{-1}}$. The strong upstream wave packet and downstream wave packets as long as 250 km show up even clearer than that on the SAR image in Figure 11.5.

### 11.3.3. *MODIS observation*

Figure 11.7 shows a MODIS band 1 (620–670 nm) solar reflectivity image with a spatial resolution of 250 m, taken 4 h 38 min later than the SAR image. One can see a jet-like airflow coming from the Bering Strait and a wave-like cloud pattern of 11 bright lines with

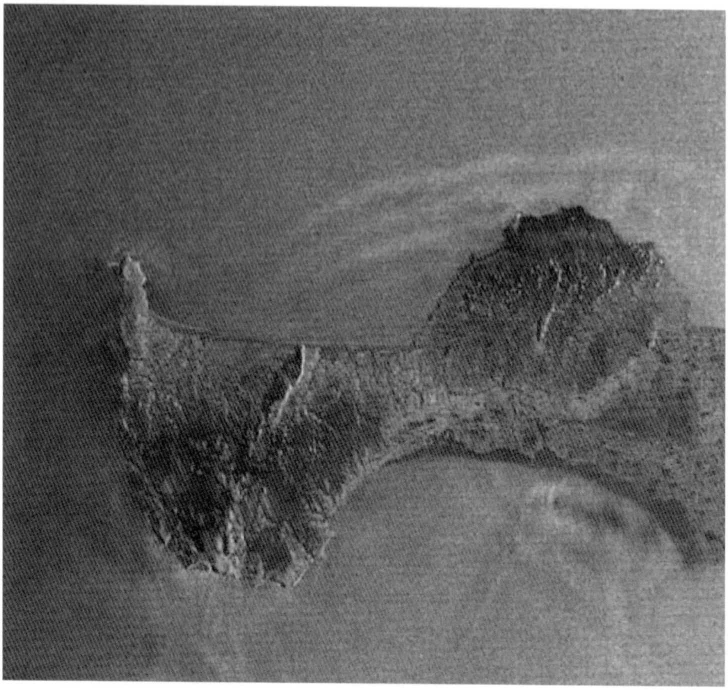

Fig. 11.5 A RADARSAT ScanSAR wide B mode SAR image of St. Lawrence Island acquired at 18:07:58 UTC on June 6, 2001. Wave-like sea surface imprints induced by two atmospheric solitary wave packets are distributed on two sides of the island. The image is centered at roughly 63.5°N and 172.0°W (Credit: Canadian Space Agency, http://www.space.gc.ca)

leading crest length as long as 125 km generated in immediate north of St. Lawrence Island.

Unlike the SAR image, the MODIS image cannot observe anything below the cloud level; instead the MODIS image shows the cloud pattern modulation by the wind variation within the MABL. The location and peak-to-peak distance of the cloud patterns match that of the atmospheric solitary wave packet showing on the SAR image in Figure 11.5, implying that they represent the same process. It is of interest to note, however, that the number (11) and leading crest length (125 km) of wave-like bright–dark cloud patterns on the MODIS image are much more and longer than that on the SAR image (4 and 60 km), indicating that the atmospheric solitary wave

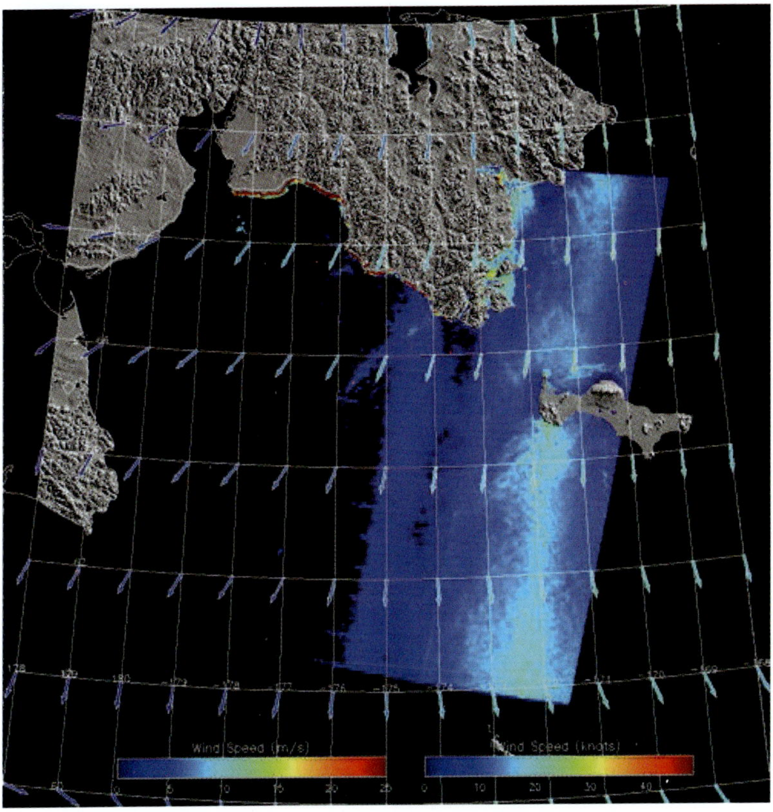


Fig. 11.6 Simultaneous sea surface wind field derived from SAR images [Li *et al.*, 2004]

packets were in the developing process during 4.5 h interval between the imaging times for two sensors. From Figure 11.7, one can see that there is a broad 125 km wide cloud band from the Bering Strait to St. Lawrence Island. Such cloud patterns are generated by strong cold jet-like airflow over the warmer seawater, which induces strong air–sea interaction and evaporation, and resembles cold air outbreak events [Mourad and Walter, 1996]. This cold airflow passed through the Bering Strait and was mostly blocked by St. Lawrence Island, implying that the airflow was in the very low level of the atmosphere. The solitary wave packet were generated and propagated against the airflow in the upstream direction [Li *et al.*, 2004].

Fig. 11.7 MODIS image of St. Lawrence Island acquired at 22:45:00 UTC on June 6, 2001, about 4.5 h later than that of the SAR image in Figure 11.5. The image is centered at roughly 63.5° N and 172.0° W (box 2 in Figure 11.4). Wave-like cloud patterns immediately north of the island represent an upstream atmospheric solitary wave packet [Li *et al.*, 2004]

### 11.3.4. *Dynamic analysis*

#### 11.3.4.1. *Two-layer forced KdV equation*

In this section, the feKdV equation is used to explain the characteristics of upstream and downstream atmospheric solitary wave packets observed in Sections 11.3.2 and 11.3.3 [Melville and Helfrich, 1987]. This phenomenon has been described by a one-layer system (fKdV)

[Akylas, 1984; Wu, 1987] and a two-layer system (feKdV) by Melville and Helfrich [1987]. Using the case of Sections 11.3.2 and 11.3.3 as examples, the local atmosphere is defined as a two-layer system with the interface at the inversion layer height based on the radiosonde data. Thus, the non-dimensional two-layer forced Korteweg–de Vries equations (fKdV) is in the form of

$$\frac{\partial \eta}{\partial t} + (F-1)\frac{\partial \eta}{\partial x} - \frac{3}{2}\delta d_{-2}\eta\frac{\partial \eta}{\partial x} - \frac{1}{6}\beta d_1\frac{\partial^3 \eta}{\partial x^3} = \frac{\gamma F}{2\delta d_{-1}}\frac{dh}{dx}, \quad (11.3.1)$$

where $\eta$ is the non-dimensional interface displacement, $h$ is the non-dimensional height of the topography, $F = U/c_0$, the Froude number, $c_0 = \sqrt{g' h_0}$, the phase speed of linear non-dispersion long wave, $g'$ is the reduced gravitational acceleration, $h_0 = (d_m d_p)/(d_m + d_p)$, the characteristic depth of the atmosphere, $d_m$ and $d_p$ are the thickness of the lower and upper layers, respectively, $\delta = a/h_0$, where $a$ is the characteristic displacement of the interface, $\delta$ is the measure of nonlinearity, $\beta = (h_0/l)^2$, where $l$ is the characteristic length of the topography, $\gamma = H_0/h_0$, where $H_0$ is the characteristic height of the topography, $d_n = d_-^n + (-1)^{n-1}d_+^n$, where

$$d_+ = \frac{d_p}{h_0}, \quad d_- = \frac{d_m}{h_0}.$$

11.3.4.2. *Solution to the fKdV equation*

Solutions of Eq. (11.3.1) are found numerically using the explicit finite difference scheme by Vliegenthart [1971]. In numerical experiments, we specify the values of parameters in Eq. (11.3.1) based on radiosonde measurements at the time the SAR image in Figure 11.5 was taken and the dimension of St. Lawrence Island. The height of the forcing term is taken as the height of the local mountain peak, *i.e.*, 673 m. The horizontal scale was taken as 5 km. A vertical profile of potential temperature shows a two-layer system with the interface at about 800 m [Li *et al.*, 2004]. Thus, it is reasonable to set $d_m = 1000$ m, $d_p = 4000$ m. The calculated characteristic depth $h_0 = 800$ m, $\beta = 0.026$ and $\gamma = 0.75$ ($H_0 = 673$ m). The reduced gravitational acceleration is $g' = 0.1\,\mathrm{m\,s^{-2}}$, so then the phase speed is $c_0 = 8.9\,\mathrm{m\,s^{-1}}$. According to the topography profile of Mt. Kookooligit, the non-dimensional island topography function, $h(x) = 1 - 4x^2$ for $|x| < 0.25$, and $h(x) = 0$ elsewhere, is used to represent the forcing term.

Li *et al.* [2004] run the model with $F = 0.95$ and $d = 0.375$. Figure 11.8 shows the non-dimensional interface displacement evolves with time $T$. In the upstream direction ($x < 0$), at $T = 12$, two solitons appear; at $T = 24$, the fourth crest starts; by $T = 60$, eight crests have developed. Meanwhile, the downstream solitary wave packet also develops and evolves with time. This wave generation pattern is consistent with the simulation by Melville and Helfrich [1987] and Baines [1995]. One can see that the solitons in general have the same amplitude and are in the elevation form. Their spacing

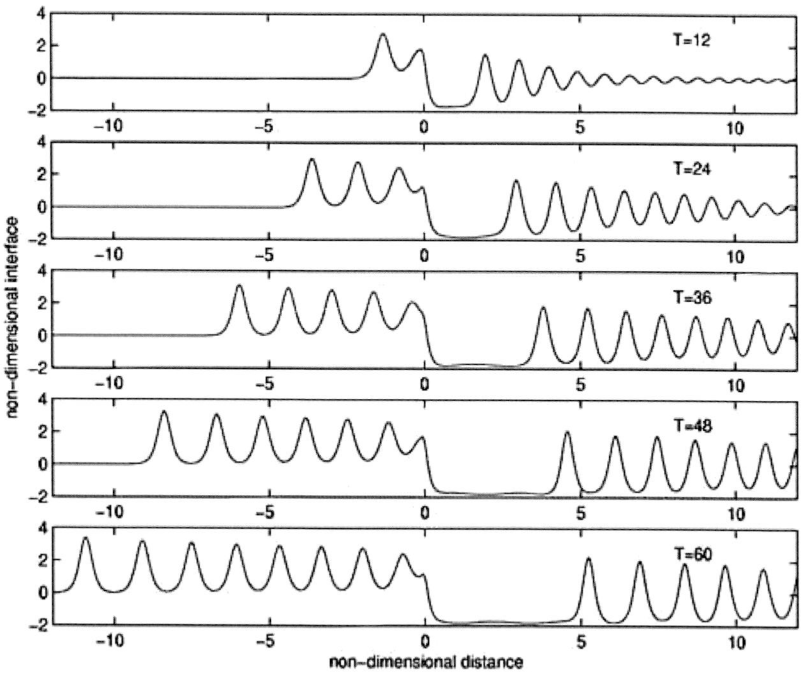


Fig. 11.8 Numerical simulation of evolution with time of upstream and downstream solitary wave packets on two sides of topographic obstacle for $f = 0.95$ and $\delta = 0.375$. Interface displacement and horizontal distance are non-dimensional. $T$ is the non-dimensional time [Li *et al.*, 2004]

is also about the same, which suggests the soliton generation rate and propagation velocity are nearly constant. The evolution of the upstream solitary wave packet in the dimensional time-space $(t - x)$ (see Li *et al.*, 2004) indicates that at $t = 3.0\,h$, the third soliton crest appears, and at $t = 7.8$ h, seven solitons have developed. It takes 4.8 h for the solitary wave packet to develop from 3 to 7 solitons, which is comparable to the satellite observations. The simulated peak-to-peak distance of the solitons is about 6.5 km and soliton width is about 2.5 km. The simulated soliton phase speed is $1.6\,\mathrm{m\,s^{-1}}$. From the SAR images in Figure 11.5, the soliton peak-to-peak distance is measured to be about 4.5 km and crest width is about 3 km. The soliton phase speed measured by successive SAR and MODIS images is $1.2\,\mathrm{m\,s^{-1}}$. One can see that the simulated values of dynamic parameters a little overestimate the actual observations, but still are in a reasonable range, implying that the fKdV equation may serve as a very good dynamical frame for the studied phenomenon.

## 11.4. Atmospheric Frontal Waves

### 11.4.1. *Coastal lee waves*

Lee waves are a kind of atmospheric internal gravity waves occurring in the lower atmospheric boundary layer. An obstacle to the airflow is a common reason for generation of lee waves. Mountain lee waves are a typical example of lee waves, which are stationary waves with wavelengths of $O$ (10 km) [Gossard and Hooke, 1975; Panchev, 1985]. On ERS-1 SAR image, Vachon *et al.* [1994] observed lee waves induced by isolated linear island Hopen in the Barents Sea. Zheng *et al.* [1998a] observed a new type of lee waves on the SAR images, which are characterized by their occurrence being associated with the land–breeze circulation front, a unique feature in the coastal environment, and defined the wave as the coastal lee waves. The observed coastal lee waves occurred over the offshore waters off the western coast of the Taiwan Strait as shown in Figure 11.9. The coastal lee waves were recognized on the SAR image taken by ERS-1 satellite at 06:26:43 UTC on December 8, 1994 as shown in Figure 11.10. One can see three groups of wave-like, alternative

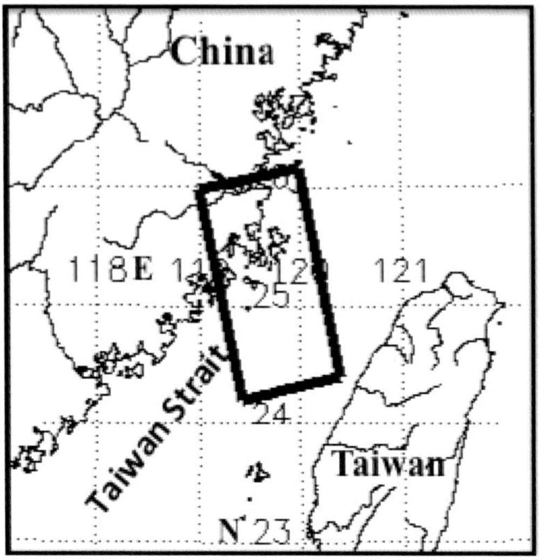


Fig. 11.9 Map of Taiwan Strait Box represents the coverage of SAR image containing the coastal lee waves (Credit: ESA)

dark–bright patterns distributed in a band with an offshore width of 20–40 km, which were interpreted as three packets of the coastal lee waves. Packet A/B contains six waves with the average wavelength of 1.7 km along line A and 2.7 km along line B. Packet C contains 12 waves with the average wavelength of 4.2 km. Packet D contains 17 waves with the average wavelength of 2.0 km. The wave crest length in packets A/B and C is about 20–40 km, and reaches 80 km in packet D. The total distribution extent is as long as 150 km along the western side of the Taiwan Strait.

The local weather chart at SAR imaging time showed that the land–sea temperature difference was 4°C and the average wind speed was 2.5 m s$^{-1}$. This was a favorable condition for establishment of the land breeze circulation. In fact, an offshore wind of 1.5–3.5 m s$^{-1}$ confirmed that the land breeze had occurred. Meanwhile, in the offshore open water, the wind speed was 3.6 m s$^{-1}$ and the wind direction was southeasterly. This implies that the onshore wind and the offshore wind had a close speed but nearly opposite directions. The two wind systems might collide and generate a front.

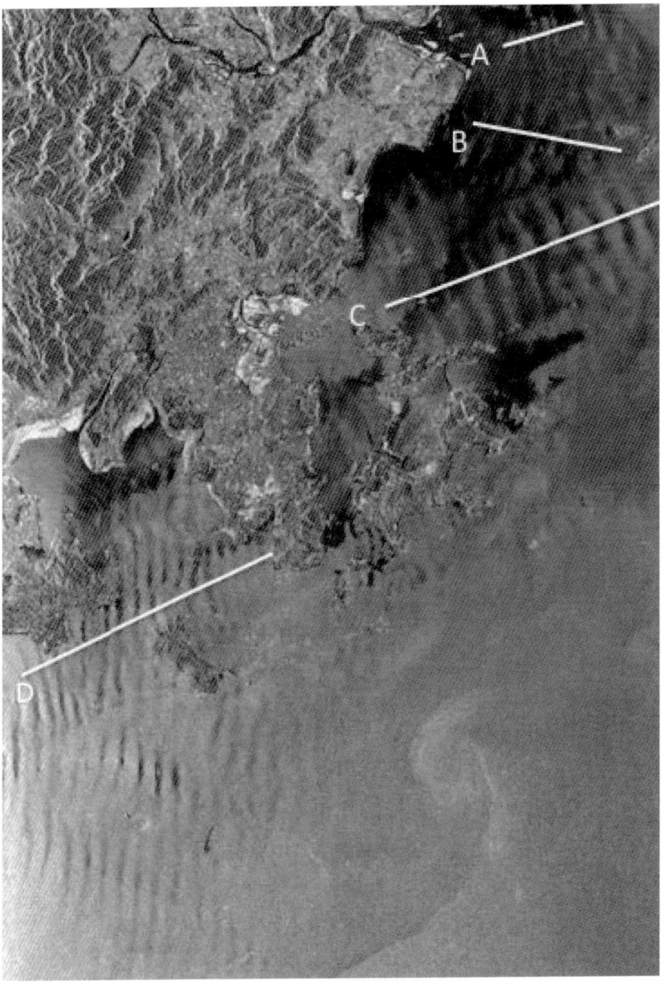


Fig. 11.10 ERS-1 SAR image of the western Taiwan Strait acquired at 06:26:43 UTC on December 8, 1994, which covers an area of about 100 by 150 km centered at 25°38′27″N 119°38′41″E with a scale of 1:560,000. Alternative dark–bright patterns along the coast represent coastal lee wave packets (Credit: ESA)

Assume that the horizontal variation of wind speed has the form of the hyperbolic tangent function

$$u(x, 0) = -u_0 \tanh \frac{x}{L}, \quad x \geq 0, \qquad (11.4.1)$$

where $L$ is defined as a characteristic length of the front, the wave motion equation has the following solution [Zheng *et al.*, 1998a]

$$\omega = \frac{u_0 H \sinh kz}{L \sinh kH} \operatorname{sech}^2\left(\frac{x}{L}\right) \exp[i(kx - \sigma t)], \qquad (11.4.2)$$

where $H$ is the thickness of the land breeze circulation and $k$ the wavenumber. The modeled waves derived from this solution are comparable to the SAR-observed coastal lee waves shown in Figure 11.9.

### 11.4.2. *Solitary wave packets in the atmospheric front*

#### 11.4.2.1. *Space shuttle observation*

A solitary wave packet is a special wave packet, of which the behavior of the waves can be described dynamically by solutions to the KdV equation [Korteweg and de Vries, 1895]. Abdullah [1949] first suggested the possibility of existence of the solitary waves in the atmosphere. Christie *et al.* [1978] believed that the morning glory is a kind of atmospheric solitary waves. Smith [1988] observed similar solitary wave packets along the west coast of Cape York Peninsula, Australia. Doviak *et al.* [1991] observed atmospheric solitary waves characterized by very large amplitude and high phase speed. Zheng *et al.* [1998b] recognized a beautiful solitary wave packet in the atmosphere over the northern Arabian Sea. Dynamically, the genesis of these atmospheric solitary wave packets was attributed to hydraulic jumps or fronts in the atmosphere. Thus, we categorize them to the atmospheric front waves in this section.

Atmospheric solitary wave packets showed up on space shuttle photographs, which were taken by the US astronaut on-board space shuttle with hand-held cameras. The photographs were taken at altitude of 264 km during mission STS55 of space shuttle Columbia in April and May 1993. Figure 11.11 shows photograph coded STS55-78-71, one of seven photographs containing atmospheric solitary wave packets. The photograph covers an area of 100 by 100 km centered at 23.3°N 63.7°E, which is located on the continental shelf of the northern Arabian Sea, about 200 km south off the Pakistani shoreline near Pasni or about 320 km southwest of Karachi. The imaging time

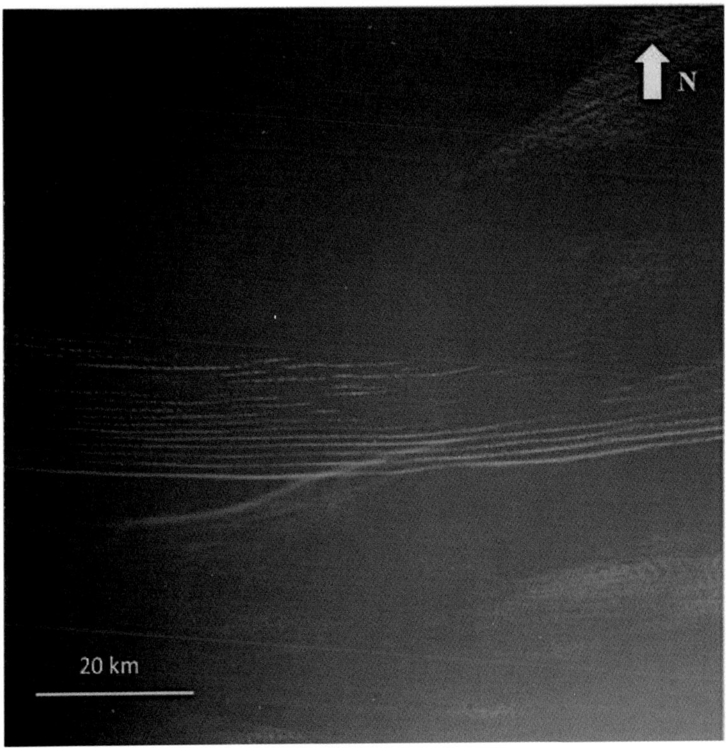


Fig. 11.11 Packet of cloud lines in the northern Arabian Sea imaged on space shuttle photograph STS55-78-71, taken by space shuttle Columbia at around 01:30 UTC on April 29, 1993 (Credit: US NASA). The imagery was interpreted as manifestation of atmospheric solitary wave packet [Zheng *et al.*, 1998b]

was around 01:30 UTC (local time 5:30 AM) on April 29, 1993. The spatial resolution of the photograph is estimated to be about 2 m [Zheng *et al.*, 1998b].

From the image in Figure 11.11, one can see an isolated packet of undular lines of low altitude cumuli clouds crossing the middle of the photograph. The imagery was interpreted as manifestation of atmospheric solitary wave packet [Zheng *et al.*, 1998b]. The packet consists of 19 parallel bright lines, which line up within a band about 20 km wide. An average wavelength (separation distance between two successive lines) is 1 km. The total length of the leading line observed by seven successive photographs is as long as 250 km.

The leading line in the front is on the south side of the packet and all lines are remarkably curved northward, implying that the waves were propagating southward or in the offshore direction at the imaging time. Further examining the cloud patterns, one can see that the wavelength and the brightness that is associated with the wave amplitude decrease from the front to the rear of the packet, implying its nature of solitary wave packet to be analyzed in Section 11.4.2.3.

#### 11.4.2.2. *SAR observation*

According to the SAR imaging principles of the atmospheric wave motions as analyzed in Section 11.2, it is expectable that the atmospheric solitary wave packets may appear on the SAR images. Liu *et al.* [2010] collected SAR images of nine cases of atmospheric solitary wave packets in the Bohai Sea and the Yellow Sea from 2002 to 2008. In the Bohai Sea, three packets were distributed within the unique zone from 119° to 120.5°E and from 38° to 39.5°N. There were six packets distributed in the Yellow Sea. All SAR images were taken in the local winter and spring season from January to May. During this period, atmospheric fronts associated with northerly winds are frequently formed. The SAR images show that the number of waves in a packet ranges from 2 to 8 with an average of 5 per packet. The average wavelength is 3.82 km with a maximum of 6.5 km and a minimum of 1.1 km. The average length of the wave crest line is $80 \pm 50$ km. Figure 11.12 shows a RADRSASAT ScanSAR wide mode image taken at 22:00:03 UTC on March 22, 2003, one of the nine cases. The image is centered at 119.9°E 39.5°N.

From the SAR image, one can see a packet of sea surface imprints of eight atmospheric waves. The average wavelength is 3.7 km, much larger than that in the case observed by space shuttle Columbia in the northern Arabian Sea. The leading wave crest length is 66 km, much shorter than that in the Arabian Sea case. But, the two cases show the same feature of internal structure, *i.e.*, the wavelength and imagery brightness decrease from the front to the rear of the packet monotonically, implying the same nature of atmospheric solitary wave packet. Figure 11.13 shows the wind profile across the straight

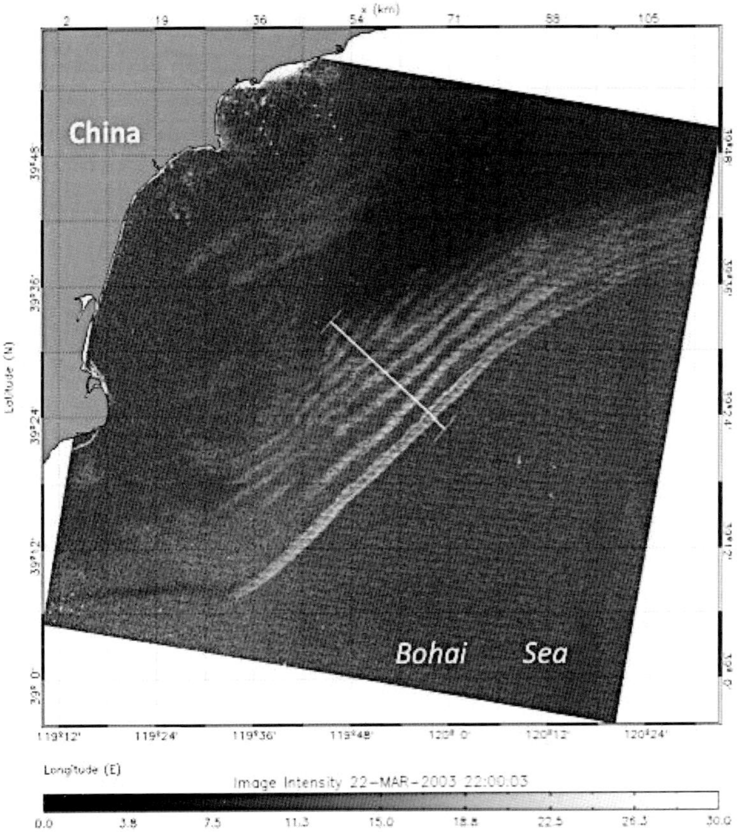


Fig. 11.12 RADRSASAT ScanSAR wide mode image showing atmospheric solitary wave packet in an atmospheric front in the Bohai Sea taken at 22:00:03 UTC on March 22, 2003. The image size is 250 by 400 km. Straight line shows the propagation direction of waves [Liu, *et al.*, 2010]

line in Figure 11.12 calculated with sea surface wind retrieval model from SAR image. One can see that the solitary wave packet occurs within a weak atmospheric frontal zone with a wind velocity jump from 2.2 to 3.9 $\mathrm{m\,s^{-1}}$.

### 11.4.2.3. *Comparison to solutions to KdV equation*

In order to confirm the nature of solitary wave packets in above two observed cases, we compare them to solitary wave packet solutions to

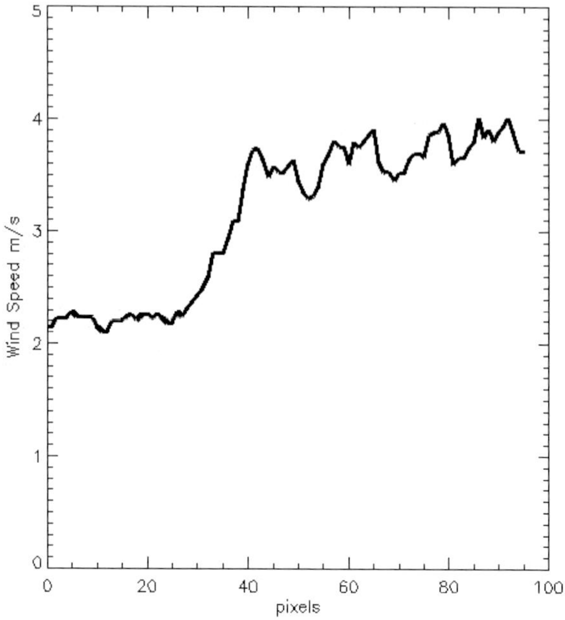


Fig. 11.13 Wind profile across the atmospheric frontal zone along the straight line in Figure 11.12 [Liu *et al.*, 2010]

KdV equation in this section. The time evolution of small but finite dispersive shallow-water waves can be described, in the first order approximation, by the KdV equation in the form of

$$\frac{\partial A}{\partial t} + c_0 \left( \frac{\partial A}{\partial x} + \alpha A \frac{\partial A}{\partial x} + \beta \frac{\partial^3 A}{\partial x^3} \right) = 0, \tag{11.4.3}$$

where $A(x,t)$ is the wave amplitude, $c_0$ the phase speed of the linear waves, $\alpha$ and $\beta$ are dependent on fluid stratification [Korteweg and de Vries; 1895; Ostrovsky and Stepanyants, 1989]. This equation has a dnoidal solution

$$A(x,t) = A_0 \left\{ 2 \frac{dn^2}{s} \left[ \frac{1}{2} k_0 (x - Ct) \right] - (1 - s^2) \right\}, \tag{11.4.4}$$

where $k_0/2$ and $C$ are the wavenumber and the nonlinear phase speed, respectively, and $dn_s(X)$ represents the Jacobi elliptic function of argument $X$ and nonlinear parameter $s$, which is determined by

$$\frac{1}{3}(1+s^2) - \frac{2s^2(1-s^2)K(s)}{3E(s)-(1-s^2)K(s)} = \frac{x}{c_0 t}, \quad (11.4.5)$$

where $K(s)$ and $E(s)$ are complete elliptic integrals. Other important relations follow: the relation of the short-length wavenumber to parameters $A_0$, $\alpha$ and $\beta$

$$\frac{1}{2}k_0 = \left(\frac{\alpha A_0}{12\beta}\right)^2, \quad (11.4.6)$$

the stretched wavelength, which is defined as the distance between two successive solitons

$$\lambda = \frac{4K(s)}{k_0}, \quad (11.4.7)$$

and the nonlinear phase speed

$$C = c_0\left(1 + \frac{1+s^2}{6}\alpha A_0\right). \quad (11.4.8)$$

Zheng *et al.* [1998b] have compared the spatial variations of the wave amplitude and the wavelength taken from the wave packet observed by space shuttle as shown in Figure 11.11 to the solutions (11.4.4) and (11.4.7). Their results show very good agreement between the two, implying that observed cloud patterns on the space shuttle photograph are manifestation of atmospheric solitary wave packet.

In the case of SAR-observed atmospheric wave packet in the Bohai Sea shown in Figure 11.12, the image intensity and the wavelength along the straight line marked in Figure 11.12 were measured. Comparisons of the measured data to the solitary wave packet solutions (11.4.4) and (11.4.7) are shown in Figures 11.14 and 11.15, respectively. One can see that the solitary wave packet solutions fit the SAR-observed data very well, implying that observed atmospheric wave imprints on the sea surface are manifestation of atmospheric solitary wave packet.

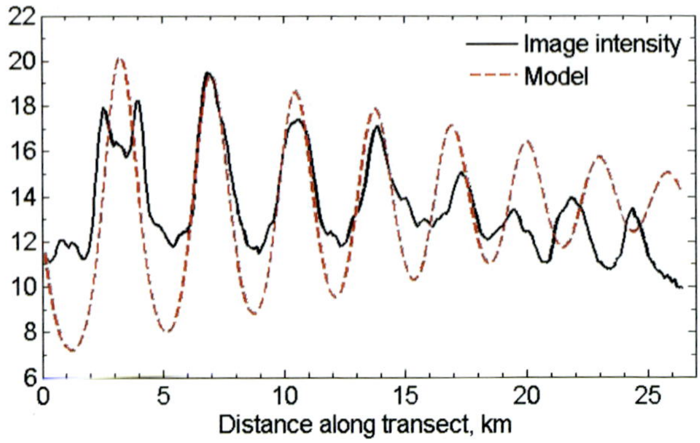


Fig. 11.14 Comparison of SAR image intensity profile taken across the atmospheric wave packet imprints on the sea surface shown in Figure 11.12 (solid line) with the dnoidal solution to the KdV equation (dashed line)

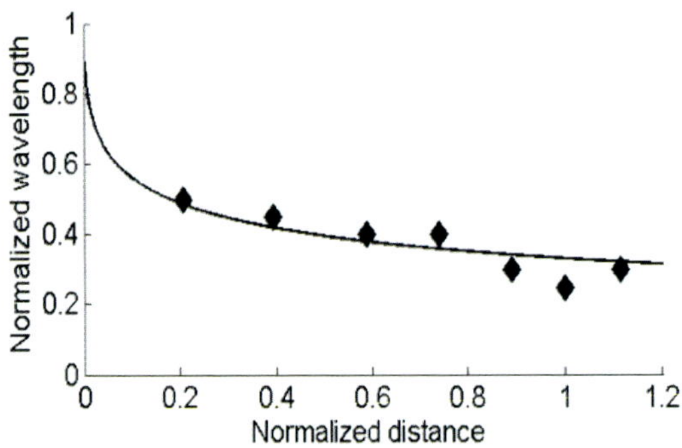


Fig. 11.15 Comparison of the wavelengths taken across the atmospheric wave packet imprints on the sea surface shown in Figure 11.11 (data points) and the stretched wavelength solution (11.4.7) (curve)

## 11.5. Ocean Surface Rainfall

### 11.5.1. *Introduction*

Rainfall on the ocean surface markedly affects the transfer of heat, mass and momentum between the atmosphere and the ocean. It has been argued that about 70 to 85% of the heat energy that the global atmosphere received is latent heat released from precipitation [Alpers and Melsheimer, 2004; Salby, 1996; Simpson *et al.*, 1996] and that approximately 79% of the global total precipitation occurs over the ocean [Barry and Chorley, 2003]. Rainfall over the ocean greatly enhances the production of sea salt aerosols, the gas exchange between the air–water interface [Harrison *et al.*, 2012; Zappa *et al.*, 2009], the turbulence in the upper ocean layer [Green and Hou, 1979; Lange *et al.*, 2000] and the wind stress acting on the sea surface [Caldwell and Elliot, 1971, 1972]. Freshwater generated by oceanic rainfall directly changes the salinity distribution that affects the horizontal pressure gradients and the equatorial thermocline that maintain the Pacific climate and its low-frequency variability [Delcroix and Picaut, 1998; Delcroix *et al.*, 2007; Fedorov *et al.*, 2004; Hackert *et al.*, 2011; Maes *et al.*, 2002; Zhang and Busalacchi, 2009]. The freshwater caused by rainfall over the ocean may play a potential role in modulating ENSO and global hydrological cycles [Reason, 1992].

Knowledge about oceanic rainfall is poor due to numerous technical difficulties in the measurement of this fast changing phenomenon. The intensity of rainfall and frequency of rainstorms on the ocean may be estimated with the traditional accumulation-type rain gauges installed along the coast and on islands, underwater acoustic rain gauges [Nystuen, 2001] and ground-based weather radars. But, it is difficult to estimate the areal extent of rainstorms on the ocean by these techniques. Since the time of the SEASAT mission in 1978 [Atlas, 1994a, 1994b; Fu and Holt, 1982], a number of satellite sensors have been launched into orbits to obtain rainfall information. A typical example of these sensors is the precipitation radar aboard the US–Japanese Tropical Rainfall Measuring Mission satellite [Simpson *et al.*, 1996].

The radar signatures of a rainfall event over the ocean are mainly generated from the scattering and attenuation of the microwaves by raindrops and ice particles in the atmosphere and from the scattering by the sea surface features that are associated with raindrop impacts. The effects of raindrops and ice particles on the scattering and attenuation of the microwaves in the atmosphere have been extensively studied by radar meteorologists [Oguchi, 1983]. The scattering is described by a volume backscattering coefficient that is a function of radar wavelength, dielectric constants of water and the reflectivity of raindrops, while the attenuation is depicted by a volume attenuation coefficient that depends on the rain rate with a power law [Ulaby *et al.*, 1982]. The previous investigators have analyzed effects of rainfall on radar scattering of different microwave frequencies from the water surface [Braun *et al.*, 2002; Hansen, 1986; Moore *et al.*, 1979; Sobieski *et al.*, 1999].

When raindrops impinge on the ocean surface, they disrupt the ocean surface by generating ring (gravity–capillary) waves and producing splashes including crowns, stalks, cavities and secondary drops [Bliven *et al.*, 1997; Fedorchenko and Wang, 2004; Le Méhauté, 1988; Lemaire *et al.*, 2002; Sobieski *et al.*, 1999; Wetzel, 1990]. However, roles of these secondary products in generation of Radar scattering signals are debated. Sobieski *et al.* [1999] compared the radar backscattering powers from crowns, craters, stalks and ring waves and concluded that ring waves are the dominant scattering contributors, though the scatterings from crowns, craters and stalks are not negligible. By developing a scattering model from the splash products of a single raindrop impact, Wetzel [1990] showed that the stalk has significant effects on the major scattering features. Contreras and Plant [2006] developed a numerical model of the ocean surface wind waves modified by rainfall and the backscattering of microwaves from the modified surface wave field.

On SAR images rain cells on the ocean surface usually show up as circular or oval bright patterns with sharp edges [Alpers and Melsheimer, 2004; Fu and Holt, 1982]. In most cases, there is a dark patch centered at the SAR signature of a rain cell. Altas [1994a, 1994b] interpreted these bright patterns as the results of the cool downdraft accompanying the rain reaching the ocean surface and diverging to become winds. These winds generate surface waves that have Bragg wavelengths and are detectable by the radar. The formation of dark patches at the centers of rain cells is due to the damping of surface waves by rain-induced turbulence. It is difficult, however, to use the downdraft-Bragg scattering hypotheses to explain the strong dependence of radar return on the radar frequency. Melsheimer *et al.* [1998, 2001] and Alpers and Melsheimer [2004] interpreted a number of multi-band, multi-polarization rain cell SAR images in the Gulf of Mexico and the Strait of Malacca taken by SIR-C/X-SAR on April 17, 1994. SIR-C/X-SAR flew aboard the space shuttle Endeavour in April and August 1994 [Jameson *et al.*, 1997]. The images show that the L band radar signature from rain cell is the weakest, and a clear dark spot shows up on the image. The C band radar signature is the strongest, and there is no clearly distinguishable dark spot on the image. The X band radar signature is in-between the two. If the downdraft-Bragg scattering mechanism still works, the impinging raindrops would be capable of selectively damping the L and X band Bragg waves with wavelengths of 21.5 and 1.9 cm, and meanwhile enhancing the C band Bragg waves with wavelength of 5.3 cm. This seems not physically sound. Thus, to explain this strong frequency dependence of radar return from ocean rain cell, Liu *et al.* [2016] further considered a new kind of radar scatterers, analyzed the physical mechanism and developed innovative radar backscattering model.

### 11.5.2. *Radar backscattering from ocean surface rainfall*

Laboratory simulations by Liu *et al.* [2015], as shown in Figure 11.16, indicate that the water surface under rainfall shows the following key points.

(1) Raindrops generate different surface features including ring waves, stalks, crowns and secondary drops. Among these features, stalks are dominant in terms of the magnitude, lifetime, water mass and energy they carry.
(2) Stalks are randomly distributed on the rainfall ocean surface.
(3) The height of individual stalk ($h$) follows a Rayleigh distribution and the average stalk height exceeds the Rayleigh criterion, *i.e.*,

$$h > \frac{\lambda}{8\cos\theta'}, \qquad (11.5.1)$$

where $\lambda$ is the radar wavelength and $\theta$ the incident angle of radar waves.

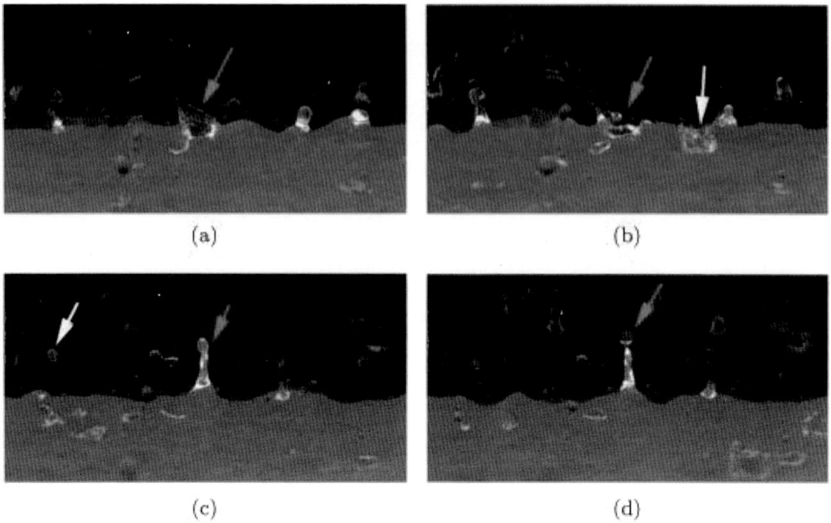


Fig. 11.16 Example images of a drop impact on the water surface within the light sheet. The size of each image is 16.2 cm wide and 8.1 cm high and a raindrop impinges on the water surface at the center of the image ($t = 0$). (a) A crown with an irregular upper boundary is formed, $t = 16$ ms. (b) The crown decays and leaves a central depression at its center of the site, $t = 32$ ms. (c) The stalk reaches a maximum height of 23 mm with a diameter of 3.8 mm, $t = 76$ ms. (d) A water ball is formed just before the separation from the water column, $t = 92$ ms [Liu *et al.*, 2016]

(4) The distance between two successive stalks decreases with increasing rain rate.

Based on the above simulation results, Liu *et al.* [2016] proposed a physics model for radar backscattering considering multiple raindrop splashes as shown in Figure 11.17. Small cylinders (compared to the radar wavelength) are used to represent stalks and crowns generated by raindrops because of similar radar scattering features in the microwave band. Ideally, distances among these small cylinders in the model would be random; however, the sum of radar backscattering from these small cylinders is extremely difficult to calculate due to the randomness of distances. Therefore, a uniform distance $L$ between two successive stalks is considered.

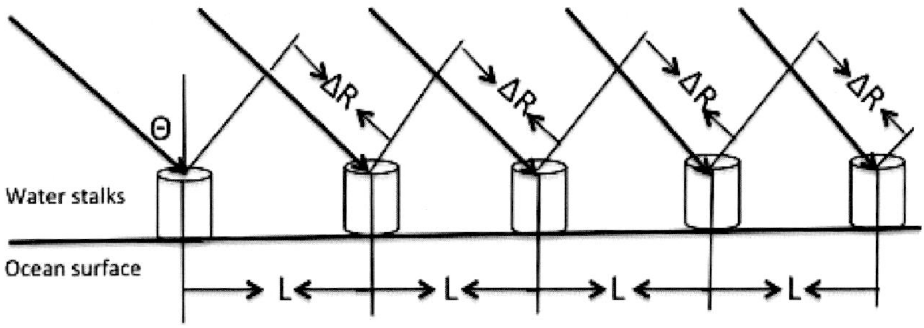


Fig. 11.17 Schematic showing the radar backscattering from stalks generated by rainfall on the ocean surface

Since the cylinder height $h$ and diameter $d$ satisfy a criterion, $d < h \ll \lambda$, the volume scattering of the radar backscattering is ignored. Thus, the received voltage by SAR from the rain field is the phase-coherent sum of the voltages from the individual cylinders [Ulaby *et al.*, 1982],

$$V_r = \sum_{n=1}^{N} V_n e^{-i2k(R_0+n\Delta R)}, \tag{11.5.2}$$

where $N$ is the total number of stalks in the resolution cell, $V_n$ is the received voltage by SAR from the $n$th stalk, $k$ $(=2\pi/\lambda)$ is the wavenumber of radar waves, $R_0$ is the radiant distance to the radar antenna, $\Delta R = L\sin\theta$ is the single-trip difference of two successive ray paths and $\theta$ is the incident angle of radar waves.

It should be mentioned that the received voltage of radar backscattering from the ocean rain field is attenuated by many factors including raindrops and secondary drops in the air, the water cylinders themselves, waves, bubbles and foams on the water surface. Because of the variations of these factors between the radar antenna and each individual stalk, the received voltage $(V_n)$ varies from one stalk to another. Thus, a damping coefficient $\beta$ is introduced into Eq. (11.5.2) and the received voltage $(V_n)$ is given as follows

$$V_n = V_0\beta^n, \tag{11.5.3}$$

where $V_0$ is the incident voltage. Thus, Eq. (11.5.2) becomes

$$V_r = V_0 e^{-i2kR_0} \beta^n \sum_{n=1}^{N} e^{-i2kn\Delta R}. \tag{11.5.4}$$

Using the following mathematic identity for a large $N$

$$\lim_{N\to\infty} \left( \sum_{n=1}^{N} \beta^n e^{-in\phi} \right) = \frac{(\beta\cos\phi - \beta^2) - i\beta\sin\phi}{1 - 2\beta\cos\phi + \beta^2} \tag{11.5.5}$$

yields

$$V_r = V_0 \beta \frac{\cos(2k\Delta R) - \beta - i\beta\sin(2k\Delta R)}{1 - 2\beta\cos 2k\Delta R + \beta^2} e^{-i2kR_0}, \tag{11.5.6}$$

where $\phi = 2k\Delta R = 2kL\sin\theta = 4\pi L/\lambda \sin\theta$.

Furthermore, the radar backscattering coefficient is defined as

$$\sigma_0(\theta, L/\lambda) = \frac{|V_r|^2}{|V_0|^2} = \frac{V_r V_r^*}{|V_0|^2} = \frac{\beta^2}{1 - 2\beta\cos(4\pi\frac{L}{\lambda}\sin\theta) + \lambda^2}. \tag{11.5.7}$$

Liu *et al.* [2016] derived an empirical relation of $L$ to the rain rate $R$ in the form of

$$L = \left( \frac{5.17\pi \times 10^5}{1 + 7.79R^{-0.21}} \right)^{1/3} R^{-0.12}. \tag{11.5.8}$$

Substituting Eq. (11.5.8) into Eq. (11.5.7) yields the radar backscattering coefficient varying with the rain rate. Figure 11.18 shows the radar backscattering coefficient $\sigma_0$ versus the rain rate for L, C and X bands at an incident angle of 38°. One can see that for all three frequencies the radar backscattering coefficients increase with increasing rain rate for light/moderate rain (at low rain rates), but decrease with increasing rain rate for heavy rain (at high rain rates). The maximum radar backscattering coefficients occur at certain rain rates that vary with the radar wave frequency. The rain rates corresponding to the maximum radar backscattering coefficients of L, C and X band radar waves are 12, 16 and 24 mm h$^{-1}$, respectively.

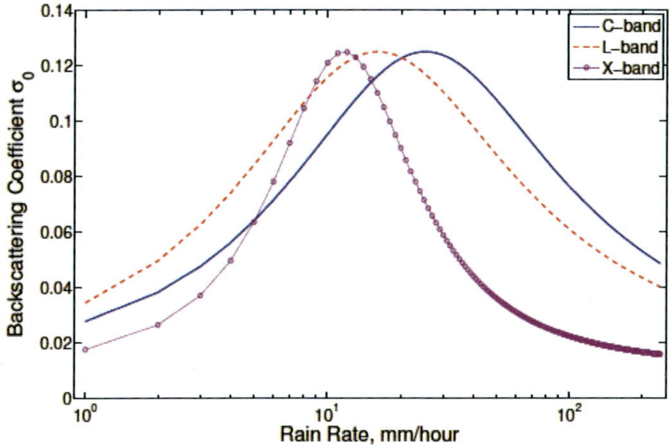


Fig. 11.18 Radar backscattering coefficient versus the rain rate for L, C and X bands for incident angle of 38° [Liu *et al.*, 2016]

### 11.5.3. *Interpretation of SAR images of ocean rainfall*

In this section, the radar backscattering model proposed in the previous section is used to interpret the radar signatures of a rainfall event simultaneously observed by the ENVISAT ASAR and ground weather radar in the Northwestern Pacific. Figure 11.19 shows a ground weather radar image taken at 13:53:02 UTC on November 19, 2006. The image covers a circular area with a radius of 460 km from the origin at Station RCKT46779 of Taiwan. The spatial resolution of the image is about 4 by 1.1 km and the elevation angle of radar antenna beam is 0.5° above the ground. One can see a rain cell on the ocean surface southeast of Taiwan, where is the open water of the Northwestern Pacific. The rain cell has a width of about 100 km, and extends northeastward from the Luzon Strait to the open ocean of 350 km. The image shows the radar reflectivity of the rain cell varying from 9.0–13.7 dBZ along the outer edges to 55.7–60.3 dBZ in a high rain rate core located at the center of the rain cell, *i.e.*, the rain rate from about 0.1 mm $h^{-1}$ along the outer edges to 167 mm $h^{-1}$ at the center of the rain cell.

Figure 11.20 shows a mosaic of ENVISAT ASAR images of the Northwestern Pacific near the Luzon Strait. The image was acquired

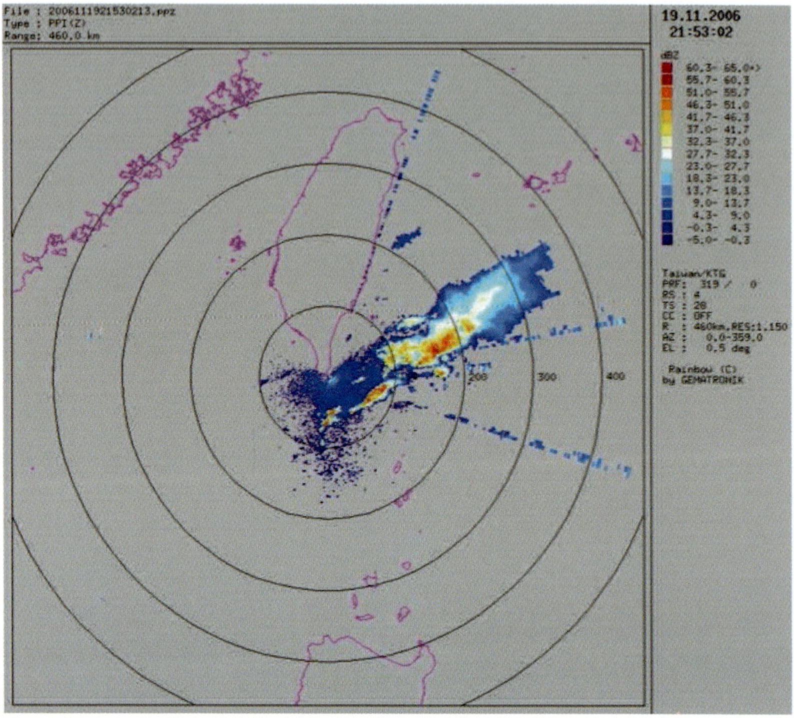


Fig. 11.19 Weather radar image taken at Station RCKT46779 of Taiwan at 13:53:02 UTC on November 19, 2006. Numerals on the radiant radius are distance in km from the weather radar station (Credit: Central Weather Bureau, Taiwan)

at 13:53:45 UTC on November 19, 2006, only 43 s later than the time the weather radar image was taken. Thus, it can be treated as a simultaneous observation with the weather radar image shown in Figure 11.19.

In Figure 11.20, the southern tip of Taiwan Island and the northern tip of Luzon Island show up and are labeled as landmarks. Compared to Figure 11.19, one can see that the bright area southeast of Taiwan on the central top of the image has analogous imagery patterns to the rain cell and occupies the same area in Figure 11.19. The imagery features are also analogous to the SAR images of rain cells interpreted by the previous investigators [Alpers and Melsheimer, 2004; Jameson *et al.*, 1997; Melsheimer *et al.*, 1998; 2001]. Thus, we recognize that the bright area on the ASAR images

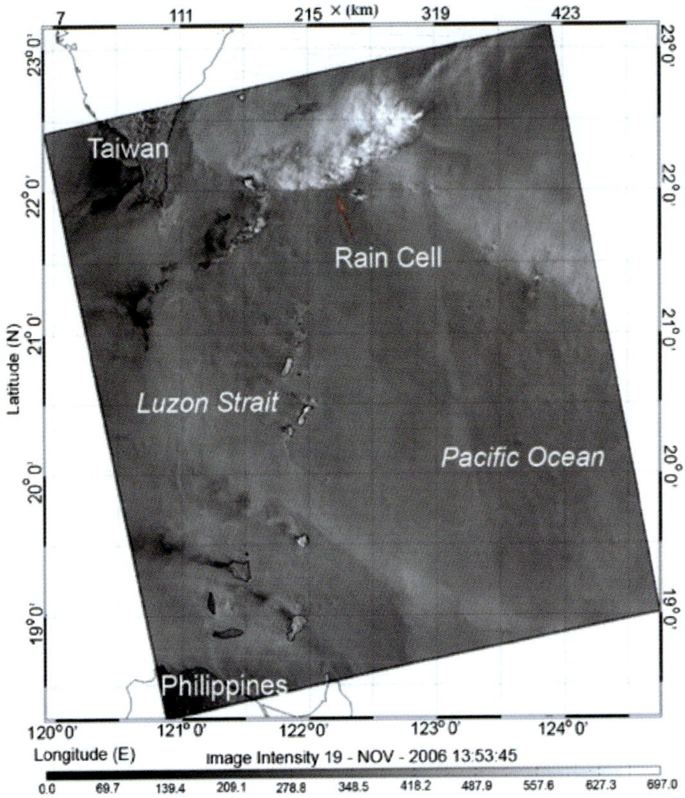


Fig. 11.20 Mosaic of ASAR images of the Northwestern Pacific near the Luzon Strait acquired at 13:53:45 UTC on November 19, 2006. (Credit: the Hong Kong Satellite Remote Sensing Ground Receiving Station, the Chinese University of Hong Kong)

is the rain cell signature. The SAR signature gives the west–east length of the rain cell is about 250 km and north–south width about 100 km. Its central location is 22°30′N 122°30′E. Within the rain cell, the SAR signature shows a number of fine structures. There is a dark spot with a diameter of about 10 km centered at 22°22′N 122°21′E. The south boundary of the rain cell appears as arch-shaped imagery with a sharp contrast to the surrounding water area, implying that it was an intense squall line. Embedded with this line were squall thunderstorms that brought heavy rain.

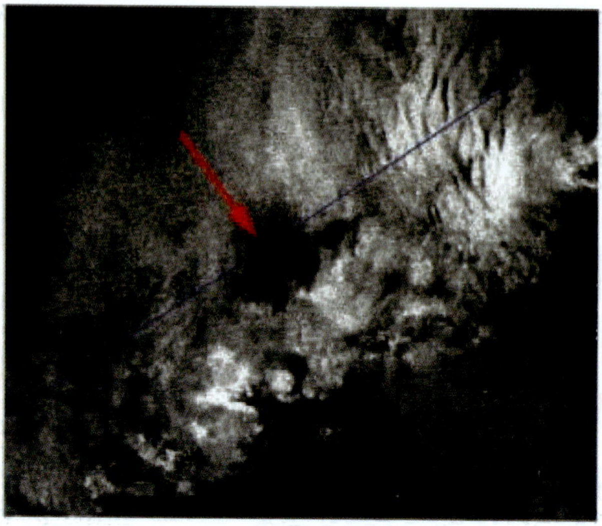

(a)

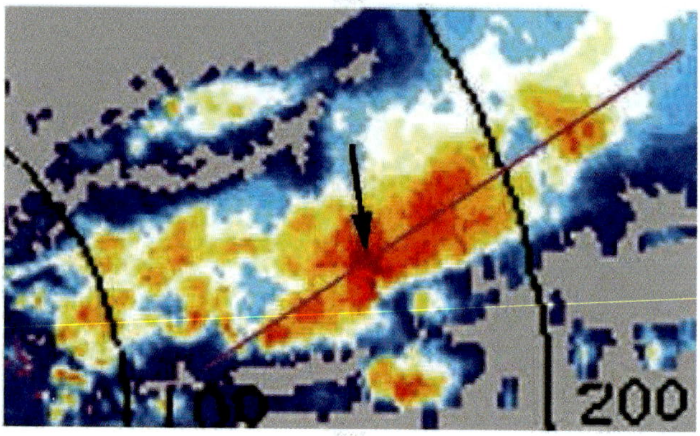


(b)

Fig. 11.21 Zoomed images from Figures 11.20(a) and (b) showing detailed features of the rain cell. Two straight lines have the same geographic coordinates. Arrows show the rain cell centers [Liu *et al.*, 2016]

Detailed features around the center of this rain cell (pointed by arrow) in both zoomed ASAR and ground weather radar images are shown in Figures 11.21(a) and (b), respectively. One can see that the rain intensity patterns in these two images are quite similar.

To quantitatively compare the data on the two images, the radar backscattering coefficient from the ASAR image and corresponding

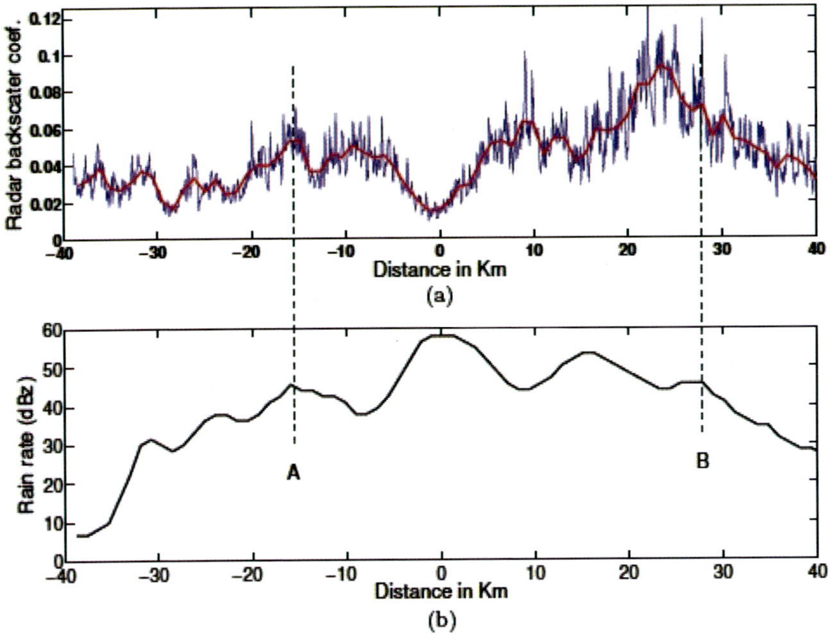


Fig. 11.22 (a) Variations of radar backscattering coefficient taken along the central line of the rain cell marked on zoomed ASAR image shown in Figures 11.21(a) and (b) variation of the rain rate along the line marked on zoomed ground weather radar image shown in Figure 11.21(b). The origin of the horizontal axis corresponds to the center of the rain cell. Vertical lines A and B show the rain cell center boundaries [Liu *et al.*, 2016]

the radar reflectivity from the ground-based weather radar along a straight line are shown in Figures 11.22(a) and (b). The horizontal axis is the distance with respect to the center of the rain cell along the lines, which are shown in Figures 11.21(a) and (b). One can see that the radar backscattering coefficients are nearly constant in the range of −40 km to −32 km while the radar reflectivity measured from the ground weather radar increases from 7 to 30 dBZ. The radar backscattering coefficient varies with the radar reflectivity in the ranges of −32 to −15 km and 28 to 40 km, *i.e.*, the radar backscattering coefficient increases as the radar reflectivity increases, and vice versa.

Around the rain cell center in the range of −15–28 km between two vertical lines A and B in Figure 11.22, the radar backscattering coefficient measured from the ASAR image decreases in the ranges of −10–0 km, 10–15 km and 25–28 km, while the radar reflectivity measured from the ground weather radar image increases. In contrast, the radar backscattering coefficient increases in the ranges of −12–10 km, 0–10 km and 15–25 km, while the radar reflectivity decreases. This anti-phase phenomenon indicates that near the center of the rain cell, the radar backscattering coefficient is negatively correlated with the radar reflectivity from ground weather radar when the radar reflectivity is above 45 dBZ. Figure 11.23 shows the average value and the standard deviation of the radar backscattering coefficients measured from Figure 11.22(a) versus the corresponding radar reflectivity from Figure 11.22(b). One can see that the radar backscattering intensity from SAR images does not increase monotonically with the radar reflectivity from ground weather radar, but its response can be divided into three regions. (a) From 10 to

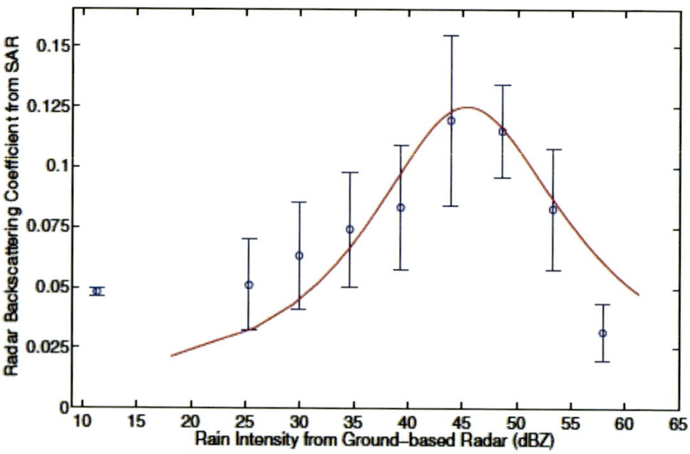


Fig. 11.23 Observed dependence of C band ASAR radar backscattering coefficient on the rain rate on the ocean surface data points represent statistical means and standard deviations of radar backscattering coefficient. Red solid line represents the calculated radar backscattering coefficient from the model. To compare with the satellite data, the calculated values are calibrated by a factor of 1/8 [Liu *et al.*, 2016]

25 dBZ (0.1–1.0 mm $h^{-1}$), SAR return intensity has little change. (b) From 25 to 45 dBZ (1.0–24 mm $h^{-1}$), SAR return intensity increases gently with the weather radar reflectivity and (c) From 45 to 58 dBZ (24–167 mm $h^{-1}$), radar return intensity decreases quickly with the weather radar reflectivity. For heavy rain at the center of the rain cell (167 mm $h^{-1}$), SAR return is even 23% lower than that of light rain (0.1 mm $h^{-1}$).

Figure 11.23 also shows a curve of the radar backscattering coefficient versus the radar reflectivity calculated from the model developed in section 11.5.2 with $\beta = 0.5$, $\lambda = 5.7$ cm and $\theta = 38°$. One can see that the theoretical curve fits the data quite well in the range of 35–53 dBZ. For the weather radar reflectivity less than 35 dBZ the theoretical prediction is a little lower than the SAR data, while for heavy rain at the center of the rain cell, the calculation is a little higher than the SAR data. In general, the model explains the radar response to the rain rate simultaneously observed by C band ASAR and ground weather radar well, implying that the physical analysis and radar backscattering model developed here are sound.

# Bibliography


Abdullah, A. J. (1949). Cyclogenesis by a purely mechanical process, *J. Meteor.*, 6, 86–97.

Akylas, T. R. (1984). On the excitation of long nonlinear water waves by a moving pressure distribution, *J. Fluid Mech.*, 141, 455–466.

Alford, M. H., Lien, R.-C., Simmons, H., Klymak, J., Ramp, S., Yang, Y. J., Tang, D. and Chang, M.-H. (2010). Speed and evolution of nonlinear internal waves transiting the South China Sea, *J. Phys. Oceanogr.*, 40, 1338–1355.

Alford, M. H., MacKinnon, J. A., Nash, J. D., Simmons, H., Pickering, A., Klymak, J. M., Pinkel, R., Sun, O., Rainville, L., Musgrave, R., Beitzel, T., Fu, K.-H. and Lu, C.-W. (2011). Energy flux and dissipation in Luzon Strait: Two tales of two ridges, *J. Phys. Oceanogr.*, 41, 2211–2222.

Alpers, W. and Hennings, I. (1984). A theory of imaging mechanism of underwater bottom topography by real and synthetic aperture radar, *J. Geophys. Res.*, 89, 10,529–10,546.

Alpers, W. and Brümmer, B. (1994). Atmospheric boundary layer rolls observed by the synthetic aperture radar aboard the ERS 1 satellite, *J. Geophys. Res.*, 99, 12,613–12,621.

Alpers, W. and Stilke, G. (1996). Observations of a nonlinear wave disturbance in the marine atmosphere by synthetic aperture radar aboard ERS-1 satellite, *J. Geophys. Res.*, 101, 6513–6525.

Alpers, W., Campbell, G., Wensink, H. and Zheng, Q. (2004). Underwater topography. *Synthetic Aperture Radar Marine User's Manual*, C. R. Jackson and J. R. Apel, eds., (US Department of Commerce, NOAA NESDIS ORA), pp. 245–262.

Alpers, W. and Melsheimer, C. (2004). *Synthetic Aperture Radar Marine User's Manual*, Jackson C. R. and J. R. Apel, eds., Chapter 17 "Rainfall", (US Department of Commerce, Washington, D.C.) pp. 355–371.

Apel, J. R. and Gonzalez, F. I. (1983). Nonlinear features of internal waves off Baja California as observed from the SEASAT imaging radar, *J. Geophys. Res.*, 88, 4459–4466.

Apel, J. R. (1987). *Principles of Ocean Physics* (Academic Press, London, UK).

Apel, J. R. (1995). *Principles of Ocean Physics* (Academic Press, USA).

Apel, J. R., Gasparovic, R. F., Thompson, D. R. and Gotwolds, B. L. (1998). Signature of surface wave/internal wave interactions: Experiment and theory, *Dynam. Atmos. Oceans,* 12, 89–106.

Atlas, D. A. (1994a). Origin of storm footprints on the sea by synthetic aperture radar, *Science*, 266, 1363–1366.

Atlas, D. A. (1994b). Footprints of storms on the sea: A view from spaceborne synthetic aperture radar, *J. Geophys. Res.*, 99, 7961–7969.

Arbic, B., Richman, J., Shriver, J., Timko, P., Metzger, E. and Wallcraft, A. (2012). Global modeling of internal tides within an eddying ocean general circulation model, *Oceanography,* 25, 20–29, doi:10.5670/oceanog.2012.38.

Bai, S., Liu, Q. D., Li, H. and Wu, H. (1999). Sea ice in the Bohai Sea of China, *Mar. Forecasts*, 16, 1–9.

Baines, P. G. (1995). *Topographic Effects in the Stratified Flows* (Cambridge University Press, New York).

Barry, R. G. and Chorley, R. J. (2003). *Atmosphere, Weather and Climate*, 8th edn. (Routledge, New York).

Bidokhti, A. A. and Tritton, D. J. (1992). The structure of a turbulent free shear layer in a rotating fluid, *J. Fluid Mech.*, 241, 469–502.

Bliven, L., Sobieski, P. W. and Craeye, C. (1997). Rain generated ring-waves: Measurements and modeling for remote sensing, *Int. J. Remote Sens.*, 18, 221–228.

Bole, J. B., Ebbesmeyer, C. C. and Romea, R. D. (1994). Soliton currents in the South China Sea: Measurements and theoretical modeling, *Offshore Technology Conference*, Texas, 367–376.

Boyer, T. and Levites, S. (1994). Quality control and processing of historical oceanographic temperature, salinity, and oxygen data, *NOAA Technical Report NESDIS*, 81, Washington, DC, 1–64.

Braun, N., Gade, M. and Lange, P. A. (2002). The effect of artificial rain on wave spectra and multi-polarization X band radar backscatter, *Int. J. Remote Sens.*, 23, 4305–4322.

Bresciano, S., Espedal, H. A., Jenkins, A. D. and Johannessen, O. M. (1998). On the occurrence of surface slicks in the oceans, *NERSC Technical Report*, 152.

Brown, G. L. and Roshko, A. (1974). On density effects and larger structure in turbulent mixing layer, *J. Fluid Mech.*, 64, 775–816.

Buijsman, M. C., Kanarska, Y. and McWilliams, J. C. (2010a). On the generation and evolution of nonlinear internal waves in the South China Sea, *J. Geophys. Res.*, 115, C02012, doi:10.1029/2009JC005275.

Buijsman, M. C., McWilliams, J. C. and Jackson, C. R. (2010b). East–west asymmetry in nonlinear internal waves from Luzon Strait, *J. Geophys. Res.*, 115, C10057, doi:10.1029/2009JC006004.

Cai, A., Zhu, X., Li, Y. and Cai, Y. (1992). Sedimentary environment in Taiwan Shoal, *Chinese J. Oceanol. Limnol.*, 10, 331–339.

Cai, S., Long, X. and Gan, Z. (2002). A numerical study of the generation and propagation of internal solitary waves in the Luzon Strait, *Oceanol. Acta*, 25, 51–60.

Caldwell, D. R. and Elliott, W. P. (1971). Surface stresses produced by rainfall, *J. Phys. Oceanogr.*, 1, 145–148.

Caldwell, D. R. and Elliott, W. P. (1972). The effect of rainfall on the wind in the surface layer, *Bound.-Lay Meteoro.*, 3, 146–151.

Centurioni, L. R., Niiler, P. P. and Lee, D. K. (2004). Observations of inflow of Philippine Sea surface water into the South China Sea through the Luzon Strait, *J. Phys. Oceanogr.*, 34, 113–121.

Chaigneau, A., Pizarro, O. and Rojas, W. (2008). Global climatology of near-inertial current characteristics from Lagrangian observations, *Geophys. Res. Lett.*, 35, L13603, doi:10.1029/2008GL034060.

Chang, M.-H., Tang, T. Y., Ho, C. R. and Chao, S. Y. (2013). Kuroshio-induced wake in the lee of Green Island off Taiwan, *J. Geophys. Res.*, 118, doi:10.1002/jgrc.20151.

Chen, Z., Hu, J., Zhang, C., Zhang, X., Lin, F., Liang, H. and Hong, J. (1999). The characteristics of tidal current and residual current in the southern Taiwan Strait in August, 1997, *J. Xiamen Univ.* (*Natural Science*), 268–272.

Chen, G., Xue, H., Wang, D. and Xie, Q. (2013). Observed near-inertial kinetic energy in the northwestern South China Sea, *J. Geophys. Res.*, 118(10), 4965–4977.

Cheney, R. E. and Richardson, P. L. (1976). Observed decay of a cyclonic Gulf Stream ring, *Deep-Sea Res.*, 23, 143–145.

Chopra, K. P. and Hubert, L. F. (1964). Kármán vortex streets in earth's atmosphere, *Nature*, 203, 1341–1343.

Christie, D. R., Muirhead, K. J. and Hales, A. L. (1978). On solitary waves in the atmosphere, *J. Atmos. Sci.*, 35, 805–825.

Christie, D. R. (1983). Solitary waves as aviation hazard, *Eos Trans. AGU*, 64, 67.

Chu, P., Veneziano, J. M., Fan, C., Carron, M. J. and Liu, W. T. (2000). Response of the South China Sea to tropical cyclone Ernie, *J. Geophys. Res.*, 105, 13991–14009.

Chuang, W. S. (1986). A note on the driving mechanism of the current in the Taiwan Strait, *J. Oceanogr. Soc. Jpn.*, 42, 355–361.

Chunchuzov, I., Vachon, P. W. and Li, X. (2000). Analysis and modeling of atmospheric gravity waves observed in RADARSAT SAR images, *Remote Sens. Environ.*, 74, 343–361.

Clemente-Colón, P. (2001). *Coastal Oceanography Applications of Spaceborne Synthetic Aperture Radar* (*SAR*) *in the Middle Atlantic Bight* (*MAB*), Ph.D. Dissertation, University of Delaware, 1–231.

Colwell, R. N. (1983) *Manual of Remote Sensing I,* 2nd edn., Simonett, D. S. and F. T. Ulaby, eds., Chapter 9 "Radar Fundamentals and Scatterometers" (American Society of Photogrammetry, Falls Church), pp. 369–426.

Contreras R. F. and Plant, W. J. (2006). Surface effect of rain on microwave backscatter from the ocean: Measurements and modeling, *J. Geophys. Res.*, 111, C08019, doi:10.1029/2005JC003356.

Corcos, G. M. and Sherman, F. S. (1976). Vorticity concentration and the dynamics of unstable free shear layers, *J. Fluid Mech.*, 73, 241–264.

Corcos, G. M. and Sherman, F. S. (1984). The mixing layer: Deterministic models of a turbulent flow, Part I, Introduction and the two-dimensional flow, *J. Fluid Mech.*, 139, 29–65.

Cox, C. and Munk, W. (1954). Measurement of the roughness of the sea surface from photographs of the Sun's glitter, *J. Opt. Soc. Am.*, 44, 838–850.

da Silva, J. C. B., Ermakov, S. A., Robinson, I. S., Jeans, D. R. G. and Kijashko, S. V. (1998). Role of surface films in ERS SAR signatures of internal waves on the shelf 1. Short-period internal waves, *J. Geophys. Res.*, 103, 8009–8031.

Delcroix, T. and J. Picaut (1998). Zonal displacement of the western equatorial Pacific fresh pool, *J. Geophys. Res.*, 103, 1087–1098.

Delcroix T., Cravatte, S. and McPhaden, M. J. (2007). Decadal variations and trends in tropical Pacific sea surface salinity since 1970, *J. Geophys. Res.*, 112, C03012, doi:10.1029/2006JC003801.

Djordjevic, V. D. and Redekopp, L. G. (1978). The fission and disintegration of internal solitary waves moving over two-dimensional topography, *J. Phys. Oceanogr.*, 8, 1016–1024.

Donato, T. F., Askari, F., Marmorino, G. O., Trump, C. L. and Lyzenga, D. R. (1997). Radar imaging of sand waves on the continental shelf east of Cape Hatteras, NC, U.S.A., *Cont. Shelf Res.*, 17, 989–1004.

Doviak, R. J., Chen, S. S. and Christie, D. R. (1991). A thunderstorm-generated solitary wave observation compared with theory for nonlinear waves in a sheared atmosphere, *J. Atmos. Sci.*, 48, 87–111.

Du, Y. and Vachon, P. W. (2003). Characterization of hurricane eyes in RADARSAT-1 images with wavelet analysis, *Can. J. Remote Sens.*, 29, 491–498.

Duda, T. F., Lynch, J. F., Irish, J. D., Beardsley, R. C., Ramp, S. R., Chiu, C.-S., Tang, T. Y. and Yang, Y. J. (2004). Internal tide and nonlinear internal wave behavior at the continental slope in the northern South China Sea, *IEEE J. Oceanic Eng.*, 29, 1105–1130.

Eckermann, S. D. and Preusse, P. (1999). Global measurements of stratospheric mountain waves from space, *Science*, 286, 1534–1537.

Editorial Group of Marine Climate Atlas of Institute of Geography, Chinese Academy of Sciences and Publishing & Compilation Group of Institute of Marine Scientific and Technological Information, National Bureau of Oceanography, China (1982). *Climatic Atlas of Chinese Offshore Areas and Northwest Pacific*, Vol. 1, *Wind, Wave and Storm* (Ocean Press, Beijing), pp. 65–115.

Egbert, G. D. and Ray, R. D. (2000). Significant dissipation of tidal energy in the deep ocean inferred from satellite altimeter data, *Nature*, 405, 775–778.

Egbert, G. D., and Erofeeva, S. Y. (2002). Efficient inverse modeling of barotropic ocean tides, *J. Atmos. Oceanic Tech.*, 19, 183–204.

Egbert, G. D. and Ray, R. D. (2003). Semi-diurnal and diurnal tidal dissipation from TOPEX/Poseidon altimetry, *Geophys. Res. Lett.*, 30, 1907, doi:10.1029/2003GL017676.

Elachi, C. and Apel, J. R. (1976). Internal wave observations made with an airborne synthetic aperture imaging radar, *Geophys. Res. Lett.*, 3, 647–650.

ESA (1991), ERS-1 Special Issue, *ESA Bulletin* 65, (ESA Publications Division, Paris).

Espedal, H. A., Johannesson, O. M., Johannessen, J. A., Dano, E., Lyzenga, D. R. and Knust, J. C. (1998). COASTWATCH '95: ERS 1/2 SAR detection of natural film on the ocean surface, *J. Geophys. Res.*, 103, 24969–24982.

Fang, G. and Yang, J. (1985). A two-dimensional numerical model of the tidal motions in the Bohai Sea, *Chinese J. Oceanol. Limnol.*, 3, 135–152.

Farmer, D., Li, Q. and Park, J.-H. (2009). Internal wave observations in the South China Sea: The role of rotation and non-linearity, *Atmos. Ocean*, 47, 267–280, doi:10.3137/OC313.2009.

Farmer, D. M., Alford, M. H., Lien, R.-C., Yang, Y. J., Chang, M.-H. and Li, Q. (2011). From Luzon Strait to Dongsha Plateau: Stages in the life of an internal wave, *Oceanography*, 24, 64–77, doi:10.5670/oceanog.2011.95.

Fedorchenko, A. I. and Wang, A. (2004). On some common features of drop impact on liquid surfaces, *Phys. Fluids*, 16, 1349–1365.

Fedorov, A. V., Pacanowski, R. C., Philander, S. G. H. and Boccaletti, G. (2004). The effect of salinity on the wind-driven circulation and the thermal structure of the upper ocean, *J. Phys. Oceanogr.*, 34, 1949–1966.

Flater, D., Poskanzer, J., Greenbaum, J., Miracle, R., Kuenning, G., DePriest, D., Uno, S., Pentcheff, D., Dairiki, J. and Rosen, E. (1997). *Xtide: Harmonic tide clock and tide predictor* Version 1.6.2. Available at: http://www.universe.digex.net/~dave/files/.

Friedman, K. and Li, X. (2000). Storm patterns over the ocean with wide swath SAR, *J. Hopkins Apl. Tech. D.*, 21, 80–85.

Fu, L.-L. and Holt, B. (1982). *Seasat Views Oceans and Sea Ice with Synthetic Aperture Radar* (National Aeronautics and Space Administration, Jet Propulsion Laboratory, JPL Publication), pp. 81–120.

Fu, L.-L. and Holt, B. (1984). Internal waves in the Gulf of California: Observations from a spaceborne radar, *J. Geophys. Res.*, 89, 2053–2060.

Garcia-Pineda, O., Zimmer, B., Howard, M., Pichel, W., Li, X. and MacDonald, I. (2009). Using SAR images to delineate ocean oil slicks with a Texture Classifying Neural Network Algorithm (TCNNA), *Can. J. Rem. Sens.*, 35, 411–421.

Garvine, R. W. (1984). Radial spreading of buoyant, surface plumes in coastal waters, *J. Geophys. Res.*, 89, 1989–1996.

Garvine, R. W. (1987). Estuary plumes and fronts in shelf waters: A layer model, *J. Phys. Oceanogr.*, 17, 1877–1896.

Garvine, R. W. (1991). Subtidal frequency estuary-shelf interaction: Observations near Delaware Bay, *J. Geophys. Res.*, 96, 7049–7064.

Gasparovic, R. R., Apel, J. R. and Kasichke, E. S. (1988). An overview of the SAR internal wave signature experiment, *J. Geophys. Res.*, 93, 12304–12316.

Gill, A. D. (1982). *Atmosphere-Ocean Dynamics* (Academic Press, San Diego, CA).

Gjevik, B. and Marthinsen, T. (1978). Three-dimensional lee-waves pattern, *Q. J. R. Meteorol. Soc.*, 104, 947–957.

Gossard, E. E. and Munk, W. H. (1954). On gravity waves in the atmosphere, *J. Meteorol.*, 11, 259–269.

Gossard, E. E. and Hooke, W. H. (1975) *Waves in the Atmosphere* (Elsevier Scientific Publishing Co., New York, USA).

Green, T. and Hou, D. F. (1979). The mixing of rain with near-surface water, *J. Fluid Mech.*, 90, 569–588.

Grimshaw, R. H. J. and Smyth, N. (1986). Resonant flow of a stratified fluid over topography, *J. Fluid Mech.*, 169, 429–464.

Guan, B. (1985). The main characteristics of hydrology in the shallow seas of the Yellow and East China Seas, *J. Oceangr. Huanghai and Bohai Seas*, 3, 1–10.

Guan, B. and Fang, G. (2006). Winter counter-wind currents off the southeastern China coast: A review, *J. Oceanogr.*, 62, 1–24.

Guan, S., Zhao, W., Huthnance, J., Tian, J. and Wang, J. (2014). Observed upper ocean response to typhoon Megi (2010) in the Northern South China Sea, *J. Geophys. Res.*, doi:10.1002/2013JC009661.

Guo, X. and Yanagi, T. (1998). Three-dimensional structure of tidal current in the East China Sea and the Yellow Sea, *J. Oceanogr.*, 54, 651–668.

Guo, B., Hu, X., Xiong, X. and Ge, R. (2003). Study on interaction between the coastal water, shelf water and Kuroshio water in the Huanghai Sea and East China Sea, *Acta Oceanol. Sin.*, 22, 351–367.

Guo, B., Huang, Z., Li, P., Ji, W., Liu, G. and Xu, J. (2004). *Marine Environment of Offshore and Adjacent Ocean Areas of China* (Ocean Press, Beijing), pp. 189–223.

Gurevich, A. V. and Pitaevskii, L. P. (1973). Decay of initial discontinuity in the Korteweg–de Vries equation, *JETP Lett.*, 17, 193–195.

Hackert, E., Ballabrera-Poy, J., Busalacchi, A. J., Zhang, R.-H. and Murtugudde, R. G. (2011). Impact of sea surface salinity assimilation on 315 coupled forecasts in the tropical Pacific, *J. Geophys. Res.*, 116, C05009, doi:10.1029/2010JC006708.

Hansen, J. P. (1986). Rain backscatter tests dispel old theorics, *Microwaves RF*, 6, 97–102.

Hansen, D. V. and Poulain, P. M. (1996). Quality control and interpolation of WOCE/TOGA drifter data, *J. Atmos. Ocean. Tech.*, 13, 900–909.

Harrison, E. L., Veron, F., Ho, D. T., Reid, M. C., Orton, P. and McGillis, W. R. (2012). Nonlinear interaction between rain- and wind-induced air-water gas exchange, *J. Geophys. Res.*, 117, C03034, doi:10.1029/2011JC007693.

Hayes, R. M. (1981). Detection of the Gulf Stream, *Spaceborne Synthetic Aperture Radar for Oceanography*, The Johns Hopkins Oceanography Study, No. 7 (The Johns Hopkins University Press, Baltimore), pp. 146–160.

Hayne, G. S. (1980). Radar altimeter mean return waveforms from near-normal incidence ocean surface scattering, *IEEE T. Antennas Propag.*, AP 28, 687–692.

Helfrich, K. R. and Grimshaw, R. H. J. (2008). Nonlinear disintegration of the internal tide, *J. Phys. Oceanogr.*, 38, 686–701, doi:10.1175/2007JPO3826.1.

Hennings, I. (1990). Radar imaging of submarine sand waves in tidal channels, *J. Geophys. Res.*, 95, 9713–9721.

Ho, C.-R., Lee, L.-S., Kuo, N.-J., Li, H.-W. and Chen, C.-T. (2001). Intercomparison of spaceborne ocean color measurements between OCI and SeaWiFS, *Geophys. Res. Lett.*, 28, 1255–1258, doi:10.1029/2000GL012363.

Ho, C.-R., Zheng, Q., Kuo, N.-J., Tsai, C.-H. and Huang, N. E. (2004). Observation of the Kuroshio Intrusion region in the South China Sea from AVHRR data, *Int. J. Remote Sen.*, 25, 4583–4591.

Holliday, D., St-Cyr, G. and Woods, N. E. (1986). A radar ocean imaging model for small to moderate incidence angles, *Int. J. Remote Sen.*, 7, 1809–1834.

Hoskins, B. J. and Bretherton, F. P. (1972). Atmosphere frontogenesis models: Mathematical formulation and solution, *J. Atmos. Sci.*, 29, 11–37.

Hsu, M.-K., Mitnik, L. M. and Liu, C.-T. (1995). Upwelling area northeast of Taiwan on ERS-1 SAR images, *Acta Oceanogr. Taiwan.*, 34, 27–38.

Hsu, M.-K., Mitnik, L. M. and Shih, S.-M. (1997). Mapping of sand waves and channels in the Taiwan Tan area with ERS SAR, *Proc. 3rd ERS Symp. On Space at the Service of Our Environment, Florence, Italy, 17–21 March 1997*, ESA SP-414, 3, pp. 453–456.

Hsu, M.-K. and Liu, A. K. (2000). Nonlinear internal waves in the South China Sea, *Can. J. Remote Sen.*, 26, 72–81.

Hsueh, Y., Chern, C.-S. and Wang, J. (1993). Blocking of the Kuroshio by the continental shelf northeast of Taiwan, *J. Geophys. Res.*, 98, 12351–12360.

Hu, J., Kawamura, H., Hong, H. and Qi, Y. (2000). A review on the currents in the South China Sea: Seasonal circulation, South China Sea Warm Current and Kuroshio intrusion, *J. Oceanogr.*, 56, 607–624.

Hu, C., Li, X., Pichel, W. G. and Muller-Karger, F. E. (2009). Detection of natural oil slicks in the NW Gulf of Mexico using MODIS imagery, *Geophys. Res. Lett.*, 36, L01604.

Huang, R. X. (1988). On boundary value problems of the ideal-fluid thermocline, *J. Phys. Oceanogr.*, 18, 619–641.

Huang, N. E., Shen, Z., Long, S. R., Wu, M. C., Shih, H. H., Zheng, Q., Yen, N.-C., Tung, C. C. and Liu, H. H. (1998). The empirical mode decomposition and the Hilbert spectrum for nonlinear and non-stationary time series analysis. *P. Roy. Soc. Lond. A Mat.*, 454, 903–995.

Hughes, B. A. and Dawson, T. W. (1988). Joint Canada–U.S. Ocean Investigation Project: An overview of the Georgia Strait Experiment, *J. Geophys. Res.*, 93, 12219–12234.

Jameson, A. R., Li, F. K., Durden, S. L., Haddad, Z. S., Holt, B., Fogarty, T., Im, E. and Moore, R. K. (1997). SIR-C/X-SAR observations of rain storms, *Remote Sens. Environ.*, 59, 267–279.

Jan, S., Chern, C.-S., Wang, J. and Chao, S.-Y. (2004). The anomalous amplification of M2 tide in the Taiwan Strait, *Geophys. Res. Lett.*, 31, L07308, doi:10.1029/2003GL019373.

Jan, S., Lien, R.-C. and Ting, C.-H. (2008). Numerical study of baroclinic tides in Luzon Strait, *J. Oceanogr.*, 64, 789–802.

Johannessen, J. A., Shuchman, R. A., Johannessen, O. M., Davidson, K. L. and Lyzenga, D. R. (1991). Synthetic aperture imaging of upper ocean circulation features and wind fronts, *J. Geophys. Res.*, 96, 10411–10422.

Johnston, T. M. S., Rudnick, D. L., Alford, M. H., Pickering, A. and Simmons, H. L. (2013). Internal tidal energy fluxes in the South China Sea from density and velocity measurements by gliders, *J. Geophys. Res.*, 118, 3939–3949.

Jordan, R. L., Huneycutt, B. L. and Werner, M. (1991). The SIR-C/X-SAR synthetic aperture radar system, *Proc. IEEE*, 79, 827–838.

Klein, L. A. and Swift, C. T. (1977). An improved model for the dielectric constant of sea water at microwave frequencies, *IEEE T. Antenn. Propag.*, AP-25, 104–111.

Kloosterziel, R. C. (1990). *Barotropic Vortices in a Rotating Fluid*, Ph.D. Thesis, University of Utrecht, The Netherlands.

Klymak, J. M., Pinkel, R. Liu, C.-T., Liu, A. K. and David, L. (2006). Prototypical solitons in the South China Sea, *Geophys. Res. Lett.*, 33, L11607, doi:10.1029/-2006GL025932.

Korteweg, D. J. and de Vries, H. (1895). On the change of form of long waves advancing in a rectangular canal, and on a new type of long stationary waves, *Philos. Mag.*, 39, 422–443.

Kundu, P. K. (1990) *Fluid Dynamics*, (Academic Press, San Diego, CA) pp. 321–338.

Kunze, E. (1985). Near-inertial wave propagation in geostrophic shear, *J. Phys. Oceanogr.*, 15, 544–565.

Kuo, N.-J. and Ho, C.-R. (2004). ENSO effect on the sea surface wind and sea surface temperature in the Taiwan Strait, *Geophys. Res. Lett.*, 31, L13309, doi:10.1029/2004GL020303.

Kuo, Y., Chern, C., Wang, J. and Tsai, L. (2011). Numerical study of upper ocean response to a typhoon moving zonally across the Luzon Strait, *Ocean Dynam.*, 61, 1783–1795.

Lame, D. B. and Born, G. H. (1982). SEASAT measurement system evaluation: Achievements and limitations, *J. Geophys. Res.*, 87, 3175–3178.

Lange, P. A., Graaf, G. V. D. and Gade, M. (2000). Rain-induced subsurface turbulence measured using image processing methods, *Proc. Int. Geosci. Remote Sens. Sympos. (IGARSS)'00*, 7, IEEE (Piscataway, NJ) pp. 3175–3177.

Le Méhauté, B. (1988). Gravity-capillary rings generated by water drops, *J. Fluid Mech.*, 197, 415–427.

Lemaire, D., Bliven, L., Craeye, C. and Sobieski, P. W. (2002). Drop size effects on rain- generated ring-waves with a view to remote sensing applications, *Int. J. Remote Sens.*, 23, 2345–2357.

Lesieur, M., Yanase, S. and Métais, O. (1991). Stabilizing and destabilizing effects of a solid-body rotation upon quasi-two-dimensional shear layers, *Phys. Fluids A*, 3, 403–407.

Li, X., Yang, X., Zheng, Q., Pietrafesa, L. J., Pichel, W. G., Li, Z. and Li, X. (2010). Deep-water bathymetry feature imaged by spaceborne SAR in the Gulf Stream region, *Geophys. Res. Lett.*, 37, L19603, doi:10.1029/2010GL044406.

Li, L. and Li, D. (1989). Summer hydrographic features of channel west of Taiwan Shoals and the coastal upwelling, *J. Oceanogr. Taiwan Strait*, 8, 353–359.

Li, F. L. and Raney, R. K. (1991). Prolog to special section on spaceborne radars for earth and planetary observations, *Proc. IEEE*, 79, 773–776.

Li, L., Nowlin, Jr., W. D. and Su, J. (1998a). Anticyclonic rings from Kuroshio in the South China Sea, *Deep-Sea Res. I*, 45, 1469–1482.

Li, X., Pichel, W. G., Friedman, K. S. and Clemente-Colón, P. (1998b). The sea surface imprint of island lee waves as observed by RADARSAT synthetic aperture radar, *Proc. Int. Geosci. Remote Sens. Sympos.*, Seattle, Washington, pp. 763–766.

Li, L., Guo, X. and Wu, R. (2000a). Oceanic fronts in southern Taiwan Strait, *J. Oceanogr. Taiwan Strait*, 19, 147–156.

Li, X., Clemente-Coln, P., Pichel, W. G. and Wachon P. W. (2000b). Atmospheric vortex streets on a RADARSAT SAR image, *Geophys. Res. Lett.*, 27, 1655–1658.

Li, C. (2001). 3D analytic model for testing numerical tidal models, *J. Hydrol. Eng.*, 127, 709–717.

Li, X., Zheng Q., Pichel, W. G., Yan, X.-H., Liu, W. T. and Clemente-Colón, P. (2001a). Analysis of coastal lee waves along the coast of Texas observed in advanced very high resolution radiometer images, *J. Geophys. Res.*, 106, 7017–7025.

Li, Y., Ma, L., Yang, J. and Shi, A. (2001b). Study on stability of sand waves by satellite sensing, *Proceedings of the First Asian and Pacific Coastal Engeering Conference*, 850–856.

Li, X., Pichel, W. G., He, M., Wu, S., Friedman, K. S., Clemente-Colón, P. and Zhao, C. (2002). Observation of hurricane-generated ocean swell refraction at the Gulf Stream North Wall with the RADARSAT-1 synthetic aperture radar, *IEEE T. Geosci. Remote*, 40, pp. 2131–2142.

Li, Y., Hu, J., Li, J., Fu, B. and Ma, L. (2003). Optical image modulation above the submarine bottom topography: A case study on the Taiwan Banks, China, *Proc. SPIE*, 382–390.

Li, X., Dong, C., Clemente-Colón, P., Pichel, W. G. and Friedman, K. S. (2004). Synthetic aperture radar observation of the sea surface imprints of upstream atmospheric solitons generated by flow impeded by an island, *J. Geophys. Res.*, 109, C02016, doi:10.1029/2003JC002168.

Li, X., Li, C., Xu, Q. and Pichel, W. G. (2009). Sea surface manifestation of along-tidal-channel underwater ridges imaged by SAR, *IEEE T. Geosci. Remote Sens.*, 47, 2467–2477, doi:10.1109/TGRS.2009.2014154.

Li, Q. and Farmer, D. M. (2011). The generation and evolution of nonlinear internal waves in the deep basin of the South China Sea, *J. Phys. Oceanogr.*, 41, 1345–1363, doi:10.1175/2011JPO4587.1.

Li, X., Li, C., Yang, Z. and Pichel, W. G. (2013). SAR imaging of ocean surface oil seep trajectories induced by near inertial oscillation, *Remote Sens. Environ.*, 130, 182–187.

Liang, S.-J., Lin, C. Y., Hsu, T. W., Ho, C.-R. and Chang, M. H. (2013). Numerical study of vortex characteristics near Green Island, Taiwan, *J. Coast. Res.*, 29, 1436–1444.

Lien, R.-C., Tang, T. Y., Chang, M. H. and D'Asaro, E. A. (2005). Energy of nonlinear internal waves in the South China Sea, *Geophys. Res. Lett.*, 32, L05615, doi:10.1029/2004GL022012.

Liu, A. K., Chang, Y. S., Hsu, M.-K. and Liang, N. K. (1998). Evolution of nonlinear internal waves in the East and South China Seas, *J. Geophys. Res.*, 103, 7995–8008.

Liu, Z., Xia, D., Berne, S., Wang, K., Marsset, T., Tang, Y. and Bourillet, J. F. (1998). Tidal deposition systems of China's continental shelf, with special reference to the eastern Bohai Sea, *Mar. Geol.*, 145, 225–253.

Liu, A. K., Ramp, S. R., Zhao, Y. and Tang T. Y. (2004). A case study of internal solitary wave propagation during ASIAEX 2001, *IEEE J. Ocean. Eng.*, 29, 1144–1156.

Liu, Z., Berne, S., Saito, Y., Yu, H., Trentesaux, A., Uehara, K., Yin, P., Liu, J. P., Li, C., Hu, G. and Wang, X. (2007). Internal architecture and mobility of tidal sand ridges in the East China Sea, *Cont. Shelf Res.*, 27, 1820–1834.

Liu, S., Li, Z., Yang, X., Pichel, W. G., Yu, Y., Zheng, Q. and Li, X. (2010). Atmospheric frontal gravity waves observed in satellite SAR images of the Bohai Sea and Huanghai Sea, *Acta Oceanol. Sin.*, 29, 35–43 doi: 10.1007/s13131-010-0061-8.

Liu, X., Zheng, Q., Liu, R., Wang, D., Duncan, J. H. and Huang, S.-J. (2016). A study of radar backscattering from water surface in response to rainfall, *J. Geophys. Res. Oceans*, 121, doi: 10.1002/2015JC010975.

Lodge, D. W. (1983). Surface expressions of bathymetry on SEASAT synthetic aperture radar images, *Int. J. Remote Sens.*, 4, 639–653.

Long, S. R. and Huang, N. E. (1976). On the variation and growth of wave-slope spectra in the capillary-gravity range with increasing wind, *J. Fluid Mech.*, 67, 209–228.

Lyzenga, D. R. and Marmorino, G. O. (1998). Measurement of surface currents using sequential synthetic aperture radar images of slick patterns near the edge of the Gulf Stream, *J. Geophys. Res.*, 103, 18769–18777.

MacDonald, I. R., Guinasso, Jr., N. L., Ackleson, S. G., Amos, J. F., Duckworth, R., Sassen, R. and Brooks, J. M. (1993). Natural oil slicks in the Gulf of Mexico, *J. Geophys. Res.*, 98, 16351–16364.

MacDonald, I., Reilly, J., Best, S., Venkataramaiah, R., Guinasso, N. and Amos, J. (1996). Remote sensing inventory of active oil seeps and chemosynthetic communities in the Northern Gulf of Mexico, *AAPG Memoir*, 66, 27–37.

MacVean, M. K. and Woods, J. D. (1980). Redistribution of scalars during upper ocean frontogenesis: A numerical model, *Q. J. Roy. Meteoroy. Soc.*, 106, 293–311.

Maes C., Picaut, J. and Belamari, S. (2002). Salinity barrier layer and onset of El Niño in a Pacific coupled model, *Geophys. Res. Lett.*, 29, doi:10.1029/2002GL016029.

McCandless, S. W. and Jackson, C. R. (2004). *Synthetic Aperture Radar Marine User's Manual*, C. R. Jackson and J. R. Apel, eds., Chapter 1 "Principles of Synthetic Aperture Radar", (U.S. Department of Commerce National Oceanic and Atmospheric Administration, Washington, DC), pp. 1–23.

McGoldrick, L. F. (1965). Resonant interaction among capillary-gravity wave, *J. Fluid Mech.*, 21, 305–331.

Melsheimer, C., Alpers, W. and Gade, M. (1998). Investigation of multifrequency/multipolarization radar signatures of rain cells over the ocean using SIR-C/X- SAR data, *J. Geophys. Res.*, 103, 18867–18884, doi:10.1029/98JC00779.

Melsheimer, C., Alpers, W. and Gade, M. (2001). Simultaneous observations of rain cells over the ocean by the synthetic aperture radar aboard the ERS satellites and by surface-based weather radars, *J. Geophys. Res.*, 106, 4665–4677.

Melville, W. K. and Helfrich, K. R. (1987). Transcritical two-layer flow over topography, *J. Fluid Mech.*, 178, 31–52.

Meng, J. (2002). *A Study of Information Extraction Technology of Ocean Internal Waves from SAR Images*, Ph.D. Dissertation, Ocean University of Qingdao, Qingdao, China, pp. 1–114.

Metzger, E. J. and Hurlburt, H. E. (1996). Coupled dynamics of the South Chine Sea, the Sulu Sea, and the Pacific Ocean, *J. Geophys. Res.*, 101, 12331–12352.

Métais, O., Flores, C., Yanase, S., Riley, J. J. and Lesieur, J. (1995). Rotating free-shear flows, Part 2, Numerical simulations, *J. Fluid Mech.*, 293, 47–80.

Miles, J. W. (1979). On internal solitary waves, *Tulles*, 31, 456–462.

Milne-Thomson, L. M. (1960). *Theoretical Hydrodynamics*, 4th edn. (Macmillan, London).

Minerals Management Service (1999). *DeSoto Canyon Eddy Intrusion Study Annual Report: Year 3, MMS*, Pub. No. 99-56.

Mitchell, R. M., Cechet, R. P., Turner, P. J. and Elsum, C. C. (1990). Observation and interpretation of wave clouds over Macquarie Island, *Q. J. Roy Meteor. Soc.*, 116, 741–752.

Monaldo, F. M., Thompson, D. R., Beal, R. C., Pichel, W. G. and Clemente-Colón P. (2001). Comparison of SAR-derived wind speed with model predictions and ocean buoy measurements, *IEEE T. Geosci. Remote*, 39, 2587–2600.

Mooers, C. N. K. (1975). Several effects of a baroclinic current on the cross-stream propagation of inertial-internal waves, *Geophys. Fluid Dyn.*, 6, 245–275.

Moore, R. K., Yu, Y. S., Fung, A. K., Kaneko, D., Dome, G. J. and Werp, R. E. (1979). Preliminary study of rain effects on radar scattering from water surface, *IEEE J. Oceanic Eng.*, *OE-4*, 31–32.

Mourad, P. D. (1996). Inferring multiscale structure in atmospheric turbulence using satellite-based synthetic aperture radar imagery, *J. Geophys. Res.*, 101, 18433–18449.

Mourad, P. D. and Walter, B. A. (1996). Viewing a cold air outbreak using satellite-based synthetic aperture radar and advanced very high resolution radiometer imagery, *J. Geophys. Res.*, 101, 16391–16400.

Munk, W., Armi, L., Fischer, K. and Zachariasen F. (2000). Spirals on the sea, *Proc. Roy Soc. Lond. A Mat.*, 456, 1217–1280.

Münchow, A., Garvine, R. W. and Pfeiffer, T. F. (1992). Subtidal currents from a shipboard acoustic Doppler current profiler in tidally dominated waters, *Cont. Shelf Res.*, 12, 499–515.

Münchow, A. and Garvine, R. W. (1993a). Dynamical properties of a buoyancy driven coastal current, *J. Geophys. Res.*, 98, 20063–20077.

Münchow, A. and Garvine, R. W. (1993b). Buoyancy and wind forcing of a coastal current, *J. Mar. Res.*, 51, 293–322.

Natural Resources Canada (2013). *Polarization in Radar Systems*, Available at: www.nrcan.gc.ca/earthsciences/geomatics/satellite.

Nemoto, Y., Nishino, H., Ono, M., Mizutamari, H., Nishikawa, K. and Tanaka, K. (1991). Japanese Earth Resources Satellite-1 Synthetic Aperture Radar, *Proc. IEEE*, 79, 800–809.

Newell, A. C. (1985). *Soliton in Mathematics and Physics*, (Society for Industrial and Applied Mathematics, Philadelphia), pp. 1–246.

Niwa, Y. and Hibiya, T. (2004). Three-dimensional numerical simulation of M2 internal tides in the East China Sea, *J. Geophys. Res.*, 109, C04027, doi:10.1029/2003JC001923.

Niiler, P. P., Sybrandy, A. S., Bi, K., Poulain, P. M. and Bitterman, D. (1995). Measurements of the water-following capability of holey-sock and TRISTAR drifters, *Deep-Sea Res.*, 42, 1951–1964.

Nystuen, J. (2001). Listening to raindrops from underwater: an acoustic disdrometer, *J. Atmos. Oceanic Tech.*, 18, 1640–1657.

O'Brien, J. J. and Pillsbury, R. D. (1974). Rotary wind spectra in a sea breeze regime, *J. Appl. Meteorol.*, 13, 820–825.

Oguchi, T. (1983). Electromagnetic wave propagation and scattering in rain and other hydrometeors, *Proc. IEEE.*, 71, 1029–1078.

Orr, M. H. and Mignerey, P. C. (2003). Nonlinear internal waves in the South China Sea: observation of the conversion of depression internal waves to elevation internal waves, *J. Geophys. Res.*, 108, 3064, doi:10.1029/2001JC 001163,C3.

Orvik, K. A. and Mork. M. (1995). A case study of Doppler-shifted inertial oscillations in the Norwegian Coastal Current, *Cont. Shelf Res.*, 15, 1369–1379.

Osborne, A. R. and Burch, T. L. (1980). Internal solitons in the Andaman Sea, *Science*, 208, 451–460.

Ostrovsky, L. A. and Stepanyants, Y. A. (1989). Do internal solitons exist in the ocean, *Rev. Geophys.*, 27, 293–310.

Pan, F. and Smith, R. B. (1999). Gap winds and wakes: SAR observations and numerical simulations, *J. Atmos. Sci.*, 56, 905–923.

Panchev, S. (1985) *Dynamic Meteorology*, (D. Reidel, Norwell, MA).

Park, J. J., Kim, K. and King, B. A. (2005). Global statistics of inertial motions, *Geophys. Res. Lett.*, 32, L14612, doi: 10.1029/2005GL023258.

Park, J.-H. and Farmer, D. (2013). Effects of Kuroshio intrusions on nonlinear internal waves in the South China Sea during winter, *J. Geophys. Res.*, 118, 7081–7094, doi:10.1002/2013JC008983.

Parker, C. E. (1971). Gulf Stream rings in the Sargasso Sea, *Deep-Sea Res.*, 18, 981–993.

Pedley, T. J. (1969). On the stability of viscous flow in a rapidly rotating pipe, *J. Fluid Mech.*, 36, 177–222.

Pedlosky, J. (1979). *Geophysical Fluid Dynamics* (Springer, New York) pp. 2–3.

Pedlosky, J. (1987). *Geophysical Fluid Dynamics*, 2nd edn. (Springer, Berlin, Germany), pp. 185–194.

Peltzer, R. D., Griffin, O. M., Barger, W. R. and Kaiser, J. A. C. (1992). High-resolution measurements of surface-active film redistribution in ship wakes, *J. Geophys. Res.*, 97, 5231–5252.

Phillips, O. (1977). *The Dynamics of the Upper Ocean*, 2nd edn. (Cambridge University Press, London) pp. 1–336.

Pinkel, R., Buijsman, M. and Klymak, J. M. (2012). Breaking topographic lee waves in a tidal channel in Luzon Strait, *Oceanography*, 25, 160–165, doi:10.5670/oceanog.2012.51.

Plant, W. J. and Wright, J. W. (1977). Growth and equilibrium of short gravity waves in a wind-wave tank, *J. Fluid Mech.*, 82, 767–793.

Plant, W. J. (1980). On the steady-state energy balance of short gravity wave system, *J. Phys. Oceanogr.*, 10, 1340–1352.

Plant, W. J. (1982). A relationship between wind stress and wave slope, *J. Geophys. Res.*, 87, 1961–1967.

Plant, W. J., Keller, W. C. and Cross, A. (1983). Parametric dependence of the ocean wave-radar modulation transfer function, *J. Geophys. Res.*, 88, 9747–9756.

Plant, W. J. (1990). Bragg scattering of electromagnetic waves from the air/sea interface, *Surface Waves and Fluxes, Vol. II — Remote Sensing*, G. L. Geernaert and W. J. Plant, eds. (Kluwer Academic Publishers, London), pp. 41–108.

Pohlmann, T. (1987). A three dimensional circulation model of the South China Sea, *Three dimensional Models of Marine and Estuarine Dynamics*, J. J. Nihoul and B. M. Jamart eds. (Elsevier, New York), pp. 245–268.

Polis, D. F. and Kupferman, S. L. (1973). *Physical Oceanography, Delaware Bay Report Series*, Vol. 4, (University of Delaware, Newark), pp. 1–143.

Potylitsin, P. G. and Peltier, W. R. (1998). Stratification effects on the stability of columnar vortices on the f-plane, *J. Fluid Mech.*, 355, 45–79.

Price, J. F. (1981). Upper ocean response to a hurricane, *J. Phys. Oceanogr.*, 11, 153–175.

Rainville, L., Lee, C. M., Rudnick, D. L. and Yang, K.-C. (2013). Propagation of internal tides generated near Luzon Strait: Observations from autonomous gliders, *J. Geophys. Res.*, 118, 4125–4138, doi:10.1002/jgrc.20293.

Ramp, S. R., Tang, T. Y., Duda, T., Lynch, J. F., Liu, A. K., Chiu, C. S., Bahr, F., Kim, H. R. and Yang, Y. J. (2004). Internal solitons in the northeastern South China Sea Part I: sources and deep water propagation, *IEEE J. Oceanic Eng.*, 29, 1157–1181.

Raney, R. K., Luscombe, A. P., Langham, E. J. and Ahmed, S. (1991). RADARSAT, *Proc. IEEE*, 79, 839–849.

Reason, C. J. C. (1992). On the effect of ENSO precipitation anomalies in a global ocean GCM, *Clim. Dynam.*, 8, 39–47.

Richards, K. J., Smeed, D. A., Hopfinger, E. J. and D'Hières, G. C. (1992). Boundary-layer separation of rotating flows past surface-mounted obstacles, *J. Fluid Mech.*, 237, 343–371.

Roberts, H. (2006). Seafloor reflectivity. An important seismic property for interpreting fluid/gas expulsion geology and the presence of gas hydrate, *The Leading Edge*, 25620–25628.

Romeiser, R. and Alpers, W. (1997). An improved composite surface model for the radar backscattering cross section of the ocean surface, 2, Model response to surface roughness variations and the radar imaging of underwater bottom topography, *J. Geophys. Res.*, 102, 25251–25267.

Romeiser, R., Schmidt, A. and Alpers, W. (1994). A three-scale composite surface model for the ocean wave-radar modulation transfer function, *J. Geophys. Res.-Atmos.*, 99, 9785–9801.

Rudnick, D. L., Johnston, T. M. S. and Sherman, J. T. (2013). High-frequency internal waves near the Luzon Strait observed by underwater gliders, *J. Geophys. Res.*, 118, 774–784, doi:10.1002/jgrc.20083.

Qiao, F., Zheng, Q. Ge, R., Yu, F. and Fang, G. (2005). Cruise observations of a cold core ring and $\beta$-spiral on the East China Sea continental shelf, *Geophys. Res. Lett.*, 32, L02601, doi:10.1029/2004GL021591.

Salby, M. L., (1996). *Fundamentals of Atmospheric Physics*, (Academic Press, San Diego).

Sanders, T. M. and Garvine, R. W. (1996). Frontal observations of the Delaware Coastal current source region, *Cont. Shelf Res.*, 16, 1009–1021.

Saxton, J. A. and Lane, J. A. (1952). Electrical properties of sea water, *Wireless Eng.*, 291, 269–275.

Schott, F. and Stommel, H. (1978). Beta spirals and absolute velocities in different oceans, *Deep-Sea Res.*, 25, 961–1010.

Shay, L. K. and Elsberry, R. L. (1987). Near inertial ocean current response to hurricane Frederic, *J. Phys. Oceanogr.*, 17, 1249–1269.

Shen, S. S. P. (1993). *A Course on Nonlinear Waves*, (Kluwer Academic Publishers, London).

Shuchman, R. A., Lyzenga, D. R. and Meadows, G. A. (1985). Synthetic aperture radar imaging of ocean-bottom topography via tidal-current interactions, theory and observations, *Int. J. Remote Sens.*, 6, 1179–1200.

Sikora, T. D., Friedman, K. S., Pichel, W. G. and Clemente-Colón, P. (2000). Synthetic aperture radar as a tool for investigating polar mesoscale cyclones, *Wea. Forcast.*, 15, 745–758.

Simmons, H. L. (2008). Spectral modification and geographic redistribution of the semi-diurnal internal tide, *Ocean Model.*, 21, 126–138.

Simpson, J., Kummerow, C., Tao, W.-K. and Adler, R. F. (1996). On the Tropical Rainfall Measuring Mission (TRMM), *Meterogol. Atmos. Phys.*, 60, 19–36.

Skolnik, M. I. (1970). *Radar Handbook*, (McGraw-Hill, New York).

Smith, R. K. (1988). Traveling waves and bores in the low atmosphere: the 'morning glory' and related phenomena, *Earth-Sci. Rev.*, 25, 267–290.

Sobieski, P. W., Craeye, C. and Bliven, L. F. (1999). Scatterometric signatures of multivariate drop impacts on fresh and salt water surfaces, *Int. J. Remote Sens.*, 20, 2149–2166.

St. Laurent, L. C., Stringer, S., Garrett, C. and Perrault-Joncas, D. (2003). The generation of internal tides at abrupt topography, *Deep-Sea Res.*, Part I, 50, 987–1003.

Stevenson, R. E. (1989). *Oceanography from the Space Shuttle*, (Office of Naval Research, and the University Corporation for Atmospheric Research, Washington, DC).

Stevenson, R. E. (1998). Spiral eddies: The discovery that changed the face of the oceans, *21st Century Sci. Technol.*, 11, 58–71.

Stevenson, R. E. (1999). A view from space: the discovery of nonlinear waves in the ocean's near surface layer, *21st Century Sci. Technol.*, 12, 34–47.

Stommel, H. (1976). *The Gulf Stream*, 2nd edn. (University of California Press, Berkeley, CA), pp. 44–74.

Stommel, H. and F. Schott (1977). The beta spiral and the determination of the absolute velocity field from hydrographic station data, *Deep-Sea Res.*, 24, 325–329.

Stuart, J. T. (1967). On finite amplitude oscillations in laminar mixing layers, *J. Fluid Mech.*, 29, 417–440.

Sun, Z., Hu J., Zheng, Q., Li, C. (2011). Strong near-inertial oscillations in geostrophic shear in the northern South China Sea, *J. Oceanogr.*, doi:10.1007/s10872-011-0038-z.

Sun, L., Zheng, Q., Wang, D., Hu, J., Tai, C.-K. and Sun, Z. (2011). A case study of near-inertial oscillation in the South China Sea using mooring observations and satellite altimeter data, *J. Oceanogr.*, 67, 677–687.

Sun, L, Zheng, Q., Tang, T. Y., Chuang, W. S., Li, L., Hu, J. and Wang, D. (2012). Upper ocean near-inertial response to 1998 Typhoon Faith in the South China Sea, *Acta Oceanol. Sin.*, 31, 25–32.

Sun, Z., Hu, J., Zheng, Q. and Gan, J. (2015). Comparison of typhoon-induced near-inertial oscillations in shear flow in the northern South China Sea, *Acta Oceanol. Sin.*, 34, 38–45, doi: 10.1007/s13131-015-0746-0.

Susanto, R. D., Mitnik, L. and Zheng Q. (2005). Ocean internal waves observed in the Lombok Strait, *Oceanography*, 18, 80–89.

Taneda, S. (1965). Experimental investigation of vortex street, *J. Phys. Soc. Jpn*, 20, 1714–1721.

Teague, W. J., Jarosz, E., Wang, D. W. and Mitchell, D. A. (2007). Observed oceanic response over the upper continental slope and outer shelf during Hurricane Ivan, *J. Phys. Oceanogr.*, 37, 2181–2206.

The Ohio State University (2014). *Radar Antenna*, Available at: http://esl.eng.ohio-state.edu/~cbaker/cb~files/classes/Antennas_2014.pdf.

Thompson, T. W., Liu, W. T. and Weissman, D. E. (1983). Synthetic aperture radar observation of ocean roughness from rolls in an unstable marine boundary layer, *Geophys. Res. Lett.*, 10, 1172–1175.

Thompson, D. R. (2004) *Synthetic Aperture Radar Marine User's Manual*, C. R. Jackson and J. R. Apel, eds., Chapter 4 "Microwave Scattering from the Sea", (U.S. Department of Commerce National Oceanic and Atmospheric Administration, Washington, DC), pp. 117–138.

Thomson, R. E., Gower, J. F. R. and Bowker, N. W. (1977). Vortex street in the wake of the Aleutian Islands, *Mon. Weather Rev.*, 105, 873–884.

Tsuchiya, K. (1969). The clouds with the shape of Kármán vortex street in the wake of Cheju Island, Korea, *J. Meteorol. Soc. Japan*, 47, 457–464.

Turner, J. and Warren, D. E. (1988). A multi-spectral satellite image of a vortex street in the lee of Jan Mayen, *Weather*, 43, 363–369.

Ulaby, F. T., Moore, R. K. and Fung, A. K. (1982). *Microwave Remote Sensing Active and Passive*, Vol. II, (Addison-Wesley Publishing Company, London).

Vachon, P. W., Johannessen, O. M. and Johannessen, J. A. (1994). An ERS-1 synthetic aperture radar image of atmospheric lee waves, *J. Geophys. Res.*, 99, 22483–22490.

Valenzuela, G. R., Chen, D. T., Garrett, W. D. and Kaiser, A. A. C. (1983). Shallow water bottom topography from radar imagery, *Nature*, 303, 687–689.

Valenzuela, G. R., Plant, W. J., Schuler, D. L., Chen, D. T. and Keller, W. C. (1985). Microwave probing of shallow water bottom topography in the Nantucket Shoals, *J. Geophys. Res.*, 90, 4931–4942.

van der Kooij, M. W. A., Vogelzang, J. and Calkoen, C. J. (1995). A simple analytical model for brightness modulations caused by submarine sand waves in radar imagery, *J. Geophys. Res.*, 100, 7069–7082.

Vesecky, J. F. and Stewart, R. H. (1982). The observation of ocean surface phenomena using imagery from the SEASAT synthetic aperture radar: An assessment, *J. Geophys. Res.*, 87, 3397–3430.

Vliegenthart, A. C. (1971). On finite-difference methods for the Korteweg–de Vries equation, *J. Eng. Math.*, 5, 137–155.

Vogelzang, J. (1989). The mapping of bottom topography with imaging radar: A comparison of the hydrodynamic modulation in some existing models, *Int. J. Remote Sens.*, 10, 1053–1518.

Vogelzang, J, Wensink, G. J., De Loor, G. P., Peters, H. C. and Pouwells, H. (1992). Sea bottom topography with X-band SLAR: The relation between radar imagery and bathymetry, *Int. J. Remote Sens.*, 13, 1943–1958.

Vogelzang, J. (1997). Mapping submarine sand waves with multiband imaging radar 1. Model development and sensitivity analysis, *J. Geophys. Res.*, 102, 1163–1181.

von Kármán, T. (1954). *Aerodynamics*, (McGraw-Hill, New York).

Wang, Y., Fang, G., Wei Z., Qiao, F., and Chen, H. (2006). Interannual variation of the south china Sea circulation and its relation to El Niño, as seen from a variable grid global ocean model, *J. Geophys. Res.*, 111, C11S14, doi:10.1029/2005JC003269.

Wang, J., Huang, W., Yang J., Zhang, H., and Zheng, G. (2012). Study of the propagation direction of the internal waves in the South China Sea using satellite images, *Acta Oceanol. Sin.*, 32, 42–50, doi:10.1007/s13131-013-0312-6.

Wetzel, L. B. (1990). On the theory of electromagnetic scattering from a raindrop splash, *Radio Sci.*, 25, 1183–1197.

Winant, C. D. and Browand, F. K. (1974). Vortex pairing: The mechanism of turbulent mixing layer growth at moderate Reynolds numbers, *J. Fluid Mech.*, 63, 237–255.

Wong, K.-C. and Münchow, A. (1995). Buoyant forced interaction between estuary and inner shelf: Observation, *Cont. Shelf Res.*, 15, 59–88.

Woods, J. D., Wiley, R. I. and Briscoe, M. G. (1977). Vertical circulation at fronts in the upper ocean, *Deep-Sea Res.*, 24, 253–275.

Wright, J. W., Plant, W. J., Keller, W. C. and Jones, W. L. (1980). Ocean wave-radar modulation transfer function from the West Coast Experiment, *J. Geophys. Res.*, 85, 4957–4966.

Wu, T. Y.-T. (1987). Generation of upstream advancing solitons by moving disturbances, *J. Fluid Mech.*, 184, 75–99.

Xie, X., Shang, X., Chen, G. and Sun, L. (2009). Variations of diurnal and inertial spectral peaks near the bi-diurnal critical latitude, *Geophys. Res. Lett.*, 36, L02606, doi:10.1029/2008GL036383.

Xu, Z., Yin, B., Hou, Y. and Liu, A. K. (2014). Seasonal variability and north–south asymmetry of internal tides in the deep basin west of the Luzon Strait, *J. Marine Syst.*, 134, 101–112, doi:10.1016/j.jmarsys.2014.03.002.

Yin, F. (1984). Tides Around Taiwan, *Ocean Hydrodynamics of the Japan and East China Seas*, T. Ichiye, ed., (Elsevier, New York USA), pp. 301–315.

Yuan, Y. (1997). Representation of high frequency spectra of ocean waves and the basis for analyzing SAR images, *Chinese J. Oceanol. Limnol.*, 28, 1–5.

Yuan, Y., Liu, Y. and Su, J. (2001). Variability of the Kuroshio in the East China Sea during El Niño to La Niña phenomenon of 1997 and 1998, *Chinese J. Geophys.*, 44, 196–207.

Yuan, Y., Zheng, Q., Dai, D., Hu, X., Qiao, F. and Meng, J. (2006). The mechanism of internal waves in the Luzon Strait, *J. Geophys. Res.*, 111, C11S17, doi:10.1029/2005JC003198.

Zappa, C. J., Ho, D. T., McGillis, W. R., Banner, M. L., Dacey, J. W. H., Bliven, L. F., Ma, B. and Nystuen, J. (2009). Rain-induced turbulence and air–sea gas transfer, *J. Geophys. Res.*, 114, C07009, doi:10.1029/2008JC005008.

Zebker, H. A. and Van Zyl, J. J. (1991). Imaging radar polarimetry: A review, *Proc. IEEE*, 79, 1583–1606.

Zhang, R.-H. and Busalacchi, A. J. (2009). Freshwater flux (FWF)-induced oceanic feedback in a hybrid coupled model of the tropical Pacific, *J. Climate*, 22, 853–879.

Zhang, X. and Cox, C. S. (1994). Measuring the two-dimensional structure of a wavy water surface optically: A surface gradient detector, *Exp. Fluids*, 17, 225–237.

Zhang, X., Liang, X. and Tian, J. (2005). Study on internal tides and near-inertial motions in the upper 450 m ocean in the northern South China Sea, *Chinese Sci. Bull.*, 50, 2027–2031.

Zhang, Z., Fringer, O. B. and Ramp, S. R. (2011). Three-dimensional, non-hydrostatic numerical simulation of nonlinear internal wave generation and propagation in the South China Sea, *J. Geophys. Res.*, 116, C05022, doi:10.1029/2010JC006424.

Zhang, S., Xie, L, Cao, R. and Zhao, H. (2012). Observation of upper ocean mixing in the region west of the Luzon Strait in spring,*J. Coastal Res.*, 28, 1208–1213.

Zhang, S., Xie, L., Hou, Y., Zhao, H, Qi, Y. and Yi, X. (2014). Tropical storm-induced turbulent mixing and chl.a enhancement in the continental shelf southeast of Hainan Island, *J. Marine Sys.*, 129, 405–414.

Zhao, Z., Klemas, V., Zheng, Q. and Yan, X.-H. (2003). Satellite observation of internal solitary waves converting polarity, *Geophys. Res. Lett.*, 30, doi: 10.1029/2003GL018286.

Zhao, Z., Klemas, V., Zheng, Q. and Yan, X.-H. (2004). Remote sensing evidence for baroclinic tide origin of internal solitary waves in the northeastern South China Sea, *Geophys. Res. Lett.*, 31, doi: 10.1029/2003GL019077.

Zhao Z. (2005). *A study of nonlinear internal waves in the northeastern South China Sea.* [Dissertation], (University of Delaware, Newark, USA).

Zhao, Z. (2014). Internal tide radiation from the Luzon Strait, *J. Geophys. Res.*, 119, doi: 10.1002/2014JC010014.

Zheng, Q. and Klemas, V. (1982). Determination of winter temperature patterns, fronts, surface currents in the Yellow Sea and East China Sea from satellite imagery, *Remote Sens. Environ.*, 12, 201–218.

Zheng, Q., Klemas, V. and Huang, N. E. (1984). Dynamics of the slope water off New England and its influence on the Gulf Stream as inferred from satellite IR data, *Remote Sen. Environ.*, 15, 135–153.

Zheng, Q. and Yuan, Y. (1989). A study on analytical model of decay of mesoscale eddies on the continental shelf, *Sci. China* (*Ser. B*), 32, 1135–1143.

Zheng. Q., Yuan, Y. and Zheng, X. (1990). Remote sensing and dynamical studies of typical meso-scale phenomena in the Kuroshio current system, *Selected Papers of the Kuroshio Investigation I*, Su, J. ed., (Ocean Press, Beijing), pp. 313–323.

Zheng, Q., and Xiu, S. (1992). Studies on satellite remote sensing data of the Kuroshio current system I. Dynamics of meso-scale processes of the Kuroshio in the East China Sea in spring, *Selected Papers of the Kuroshio Investigation IV*, J. Su, ed. (Ocean Press, Beijing), pp. 79–85.

Zheng, Q., Yan, X.-H. and Klemas V. (1993a). Derivation of Delaware Bay tidal parameters from space shuttle photographs, *Remote Sens. Environ.*, 45, 51–59.

Zheng, Q., Yan, X.-H. and Klemas, V. (1993b). Statistical and dynamic analysis of internal waves on the continental shelf of the Middle Atlantic Bight from space shuttle photographs, *J. Geophys. Res.*, 98, 8495–8504.

Zheng, Q., Yan, X.-H., Klemas, V., Ho, C.-R., Kuo, N.-J. and Wang, Z. (1998a). Coastal lee waves on ERS-1 SAR images, *J. Geophys. Res.*, 103, 7979–7993.

Zheng, Q., Yan, X.-H., Liu, W. T., Klemas, V., Greger, D. and Wang, Z. (1998b). A solitary wave packet in the atmosphere observed from space, *Geophys. Res. Lett.*, 25, 3559–3562.

Zheng, Q., Klemas, V., Yan, X.-H. and Pan, J. (2001a). Nonlinear evolution of ocean internal solitons as propagating along an inhomogeneous thermocline, *J. Geophys. Res.*, 106, 14083–14094.

Zheng, Q., Yuan, Y., Klemas, V. and Yan, X.-H. (2001b). Theoretical expression for an ocean internal soliton SAR image and determination of the soliton characteristic half width, *J. Geophys. Res.*, 106, 31415–31423.

Zheng, Q., Clemente-Colón, P., Yan, X.-H., Liu, W. T. and Huang N. E. (2004a). Satellite SAR detection of Delaware Bay plumes: Jet-like feature analysis, *J. Geophys. Res.*, 109, C03031, doi: 10.1029/2003JC002100.

Zheng, Q., Shen, S. S. P., Yuan, Y., Huang, N. E., Klemas, V., Yan, X.-H., Shi, F., Zhang, X., Zhao, Z., Li, X. and Clemente-Colón, P. (2004b). Evidence of the coexistence of upstream and downstream solitary wave trains in the real atmosphere, *Int. J. Remote Sens.*, 25, 4433–4440.

Zheng, Q., Lai, R. J., Huang, N. E., Pan, J. and Liu, T. W. (2006a). Observation of ocean current response to 1998 Hurricane Georges in the Gulf of Mexico, *Acta Oceanol. Sin.*, 25, 1–14.

Zheng, Q., Li, L., Guo, X., Ge, Y., Zhu, D. and Li, C. (2006b). SAR Imaging and hydrodynamic analysis of ocean bottom topographic waves, *J. Geophys. Res.*, 111, C09028, doi:10.1029/2006JC003586.

Zheng, Q., Susanto, R. D., Ho, C.-R., Song, Y. T. and Xu, Q. (2007). Statistical and dynamical analyses of generation mechanisms of solitary internal waves in the northern South China Sea, *J. Geophys. Res.*, 112, doi:10.1029/2006JC003551.

Zheng, Q., Lin, H., Meng, J., Hu, X., Song, Y. T., Zhang, Y. and Li, C. (2008a). Sub-mesoscale ocean vortex trains in the Luzon Strait, *J. Geophys. Res.*, 113, C04032, doi:10.1029/2007JC004362.

Zheng, Q., Song, Y. T., Lin, H., Hu, X., Meng, J. and Wang, D. (2008b). On generation source sites of internal waves in the Luzon Strait, *Atca Oceanol. Sin.*, 27, 38–50.

Zheng, Q., Holt, B., Li, X., Liu, X., Zhao, Q., Yuan, Y. and Yang, X. (2012a). Deep-water seamount wakes on SEASAT SAR image in the Gulf Stream region, *Geophys. Res. Lett.*, 39, L16604, doi:10.1029/2012GL052661.

Zheng, Q., Zhao, Q., Yuan, Y. Liu, X., Hu, J., Liu, X., Yin, L. and Ye, X. (2012b). Shear-flow induced secondary circulation in parallel underwater topographic corrugation and its application to satellite image interpretation, *J. Ocean Univ. China*, 11, 427–435, doi: 10.1007/s11802-012-2093-5, 2012.

Zheng, Z.-W. and Zheng, Q. (2014). Variability of island-induced ocean vortex trains, in the Kuroshio region southeast of Taiwan Island, *Cont. Shelf Res.*, 81, 1–6.

# Index